# Environmental and
# Biological Control of Photosynthesis

# Environmental and Biological Control of Photosynthesis

Proceedings of a conference held at the 'Limburgs Universitair Centrum', Diepenbeek, Belgium, 26-30 August 1974.

Edited by R. Marcelle
Laboratory of Plant Physiology, Research Station of Gorsem, Sint-Truiden, Belgium.

Dr. W. Junk b.v., Publishers, The Hague, 1975.

*Sponsored by:*

Het Ministerie van Nationale Opvoeding
Het Nationaal Fonds voor Wetenschappelijk Onderzoek
Het Bestuur van de Provincie Limburg
De Vlaamse Leergangen te Leuven
Het Limburgs Universitair Centrum
Het Opzoekingsstation van Gorsem (I.W.O.N.L.)

ISBN 90 6193 179 7

© Dr. W. Junk b.v., Publishers, The Hague
Cover design: Max Velthuijs

# CONTENTS

* The manuscripts of these two papers were received so late that it was only possible to
include them at the end of the book in order to avoid a delay in the publication of the
proceedings.

# PREFACE

This book reports the proceedings of a meeting held in the 'Limburgs Universitair Centrum', Diepenbeek, Belgium, August 26 to 30, 1974. In convening this meeting, my aim was to bring together a small number of specialists working on photosynthesis of course but also always keeping in mind that plants are influenced by their environment (temperature, light quality and intensity, air composition, daylength.....) and can differently react according to their stage of development. In general, all these specialists work on whole plants cultivated in well-known conditions (they are not 'market spinach specialists') but, when necessary, they don't give up the idea of measuring photochemical activities in isolated chloroplasts, enzyme kinetics... etc. It is noticeable that about 50% of them are working in laboratories directly involved with applied research in agriculture or forestry.

The format of the meeting was intentionally kept small but it allowed generous time for discussion; thanks are due to Drs. O. BJÖRKMAN, J.W. BRADBEER, M.M. LUDLOW and C.B. OSMOND for taking the chairs during these discussions. In such a small meeting, the choice of invited scientists was really a personnal one and thus reflected my own fields of interest. When planning the conference, I was continually divided between the wish for inviting other interesting people and the necessity of keeping time free for discussions.

The meeting would never took place without the active collaboration of the 'Limburgs Universitair Centrum' which not only gave us the hospitality in its new building but also undertook much of the responsibility for raising funds. This support and the sponsorships of all organisms listed before are gratefully acknowledged. Particular thanks are due to Dr. H. CLIJSTERS and Mr. G. OBEN for their help in the planning and arrangements of the meeting.

R. MARCELLE
Sint-Truiden
March 15, 1975

# ENVIRONMENTAL AND BIOLOGICAL CONTROL OF PHOTOSYNTHESIS: INAUGURAL ADDRESS

O. BJÖRKMAN

*Carnegie Institution of Washington, Stanford, California, 94305 U.S.A.*

It is certainly a pleasure, and a privilege to be invited to this meeting on Environmental Control of Photosynthesis, a topic that lies very close to my heart. Regardless of what our individual motivations for working in this particular field of research may be I am sure that all of the members of the audience agree that this is indeed a very important field for many reasons.

As is well known, nearly all chemical energy and organic carbon and every bit of food that enter into any ecosystem on earth is provided by photosynthesis. As a result, all biological activity from the simplest virus to man is ultimately limited by how well photosynthesis is able to operate in all of the very diverse environments that exist on this earth. It is therefore very important to workers in many branches of biology to know what the environmental restraints and the adaptive limits of photosynthesis are as well as to understand what the various mechanisms that underlie photosynthetic adaptation are and how they work.

Such information is of course also of utmost importance to applied research concerned with improvement of the productivity and efficiency of agricultural crops. Increasing population pressures have resulted in a world-wide demand for an increased production of food. Unfortunately, with currently available crop plants and modern agricultural practices an increased production would require a greatly increased input of oil and other limited and costly sources of fossil fuel energy and would also require that land, now considered marginal or even unsuitable for cultivation, be utilized. Estimates show that in today's technologically highly advanced agriculture in the United States the energy content of fossil fuels (products of past photosynthetic activity) used in the production of food crops may exceed the energy content of this food. An increase in food production, or even maintaining it at the present level, will therefore require an increase in the efficiency of primary productivity in using limited resources. It is a sobering thought that when our fossil fuel reserves are depleted current photosynthesis may have to provide not only the food, but also much of the fuel, the clothing and many essential raw materials. Knowledge of the environmental limitations and adaptive mechanisms of photosynthesis would provide a basis for the selection and breeding of plants with improved productivity and improved efficiency in utilizing water and fertilizers in different climatic regions. Spending money on research with the objective of providing information on environmental and biological control of photosynthesis is certainly not extravagant.

During the past 25 years tremendous progress has been made in uncovering the basic mechanisms of the photosynthetic process. However, whereas in the early days environmental and adaptive aspects of photosynthesis were in the mainstream of photosynthesis research, in the past-World War II period they have been treated somewhat like a step-child. Fortunately, I believe this trend has now been

1

reversed and the fact that we are gathered here to exchange results and views in a conference entirely devoted to Environmental and Biological Control of Photosynthesis is, indeed, a sure sign that this is so.

One of the main reasons for the relatively slow development of our research area in the past-World War II period was the necessity of specialization in order to do research on the basic mechanisms of photosynthesis. This led not only to a compartmentation of research efforts but it also necessitated a strict limitation in the choice of experimental materials and the bulk of information on the mechanism of photosynthesis has been obtained from work on spinach chloroplasts and laboratory strains of green algae, notably *Chorella.* This limitation of experimental materials was probably one of the factors that contributed to the rapid progress in research on the basic mechanisms of the process at the subcellular level. However, in part because of this limitation this research did not shed much light on environment-process interactions, nor did it provide much information on the extent and mechanisms of genetically based adaptive differentiations that exist among plants occupying contrasting environments. Unquestionably, this limited choice of plants was the primary reason why $C_4$ photosynthesis which characterizes very large and important groups of plants, including many valuable crops, remained undiscovered for so long.

On the other side of the fence are the whole-plant physiologists, plant breeders and plant ecologists concerned with environmental aspects of photosynthesis and productivity. These workers are faced with a multitude of problems requiring the applications of a wide range of methods for measurements of micrometeorological factors, energy, water, carbon and nutrient fluxes in the field, and must take into account a host of interacting environmental variables, and the effects of plant structure and developmental stage, stomatal behavior and plant water relations, to mention only a few. It is easy to understand that these workers are often unable, poorly equipped or at least reluctant to carry their studies to lower levels than the whole plant or leaf.

It would seem that the ideal solution to this problem would be a close collaboration between the environmentally oriented plant scientists, working at higher levels of organization, and the mechanism-oriented biochemists and biophysists working at the cellular and molecular levels. Unfortunately, such collaboration has been rare, probably mainly because these specialists have come too far apart in concepts, terminology and knowledge of one anothers fields even to be able to communicate in a meaningful way. Building bridges between the workers in these branches of photosynthesis research would probably be very profitable to our understanding of environmental and adaptive aspects of photosynthesis. As in bridge construction in general, this bridge building should be undertaken from both sides of the abyss. As one of those who have tried to put in a brick here and there I am very glad to see that such a bridge construction is no longer a rare event. Undoubtedly, the discovery of the $C_4$ pathway of photosynthesis was an important factor in stimulating a resurgence of interest in studies of environmental, evolutionary and adaptive aspects of photosynthesis in general, even among specialists who are primarily concerned with the basic photochemical and biochemical mechanisms of the process.

Our knowledge of environmental responses and adaptive differentiation of photosynthesis has been derived largely from studies on economically important, cultivated plants and until only a few years ago these studies were mostly restricted to plants of the temperate regions of northern Europe and North America. The reasons for this are not difficult to understand and need not be enumerated. As far as short-term objectives are concerned it certainly seems reasonable that priority is given to research that yields information on those economically important plants which constitute the backbone of the agriculture and forestry of the countries supporting the research.

However, if the objective is to gain an understanding of the environmental and evolutionary limits of adaptation of the photosynthetic process and of the physical, structural and molecular mechanisms involved, then we should obviously choose plants which are native to environments which are extreme in one respect or another as far as photosynthesis is concerned. Such information would provide a good foundation for the development of new crop plants, particularly for the world's marginal lands and in evaluating the extent to which the photosynthetic efficiency of already existing ones can be improved.

The use of wild plants for these studies also has an advantage additional to providing a greater choice of environmental extremes and therefore a higher probability of discovering adaptive mechanisms. It also avoids a problem that may exist with cultivated plants, namely that man in his breeding for varieties that possess certain desirable features may have altered adaptive characteristics that are important in a stress environment.

When comparing photosynthetic efficiencies of different species or genotypes it is important to remember that the criteria for efficiency depend on the particular environment under consideration, and one must also take into account the biosynthetic cost of producing and maintaining the photosynthetic machinery itself. Photosynthetic rates, determined under some arbitrary environmental conditions, or expressed on some arbitrary basis such as leaf area or fresh weight alone, may be misleading when used as a criterion for comparing photosynthetic efficiencies. For example, in a densely shaded environment it is the efficiency with which the plant is able to absorb and utilize light of low intensities which is important. What the photosynthetic capacity at high light may be or how much water is spent per amount of $CO_2$ fixed are at most of secondary importance. Also, if two plants have the same light-harvesting capacities and quantum yields per unit leaf area, then it is of course the plant which has invested the least in proteins and other constituents of the photosynthetic apparatus that is the most efficient one under these particular conditions (cf. paper by Charles-Edwards, this volume).

The criteria for photosynthetic efficiency in another extreme of habitats such as hot arid deserts and semi-deserts are obviously quite different. Here radiant energy is abundant but water supply is limited and the high thermal load together with a low atmospheric humidity results in a very high water vapor pressure gradient between the leaves and the surrounding air. The water loss for a given stomatal conductance to gaseous diffusion thus becomes extremely high. In this environment it is the ability of the photosynthetic machinery to tolerate and to operate effectively at high temperatures and high irradiance levels, and the efficiency with which the plants is capable of fixing $CO_2$ in relation to transpirational water loss which are of overriding importance.

I believe that intensive comparative studies on carefully selected, wild plants, native to extreme natural habitats, is a most powerful and effective approach that we have only begun to make use of. Such studies should ideally include a critical analysis of the respective environments in terms that are pertinent to photosynthesis. They should also include studies of photosynthetic performance in the field as well as in a range of controlled environments. They should also be compared with the response of growth. Kinetic analyses of photosynthetic gas exchange characteristics, although valuable in themselves, become much more powerful when they are combined with studies of the component reactions of photosynthesis, the structure of the photosynthetic apparatus and composition of its constituents.

For the remainder of my time this evening I will take the opportunity to review a few examples that illustrate how strikingly different the photosynthetic characteristics of plants from contrasting environments can be. I will limit myself to certain aspects of adaptation to extremes of light and temperature. I will first discuss the well-known differentiation of sun and shade plants.

Light, the driving force of photosynthesis, shows a very wide variation among habitats occupied by higher plants. Recent studies in Queensland rain forest have shown that plants native to the densely shaded forest floor such as *Alocasia macrorrhiza*, a $C_3$ plant of the *Arum* family, are capable of net photosynthesis and sustained growth in sites where the average daily quantum flux (400-700 nm) is only 22 $\mu$einstein cm$^{-2}$day$^{-1}$, including the contribution of sunflecks (Björkman & Ludlow, 1972). This compares with a daily flux of about 4500 $\mu$einstein cm$^{-2}$day$^{-1}$ in an open habitat on the California coast, which is occupied by a greater number of species, including *Atriplex hastata* ($C_3$, Chenopodiaceae). Even higher values, about 6600 $\mu$einstein cm$^{-2}$day$^{-1}$, have been recorded on the sun-baked desert floor of Death Valley, California, during the summer months when growth of *Tidestromia oblongifolia,* a $C_4$ plant of the Amaranth family, is at the peak (Björkman, unpublished). The quantum flux received by this plant thus exceeds that received by *Alocasia* in its rain forest habitat by a factor of about 300.

Figure 1 shows the striking differences that exist in the light dependence of these three plants when grown under the light intensity regimes of their natural habitats. The arrows indicate the average maximum light intensities to which the plants were exposed during growth. In the extreme shade plant dark respiration is very low and light compensation is reached at extremely low light levels. Although light saturation occurs at very low light levels it is never reached in this extreme rainforest habitat. The sun plants, because of their higher respiratory rates would never even reach light compensation under these conditions. On the other hand the sun plants have much higher capacity of utilizing high light intensities. In *Tidestromia* light saturation is not reached even in full noon sunlight and the light saturated rate is about 15 times higher than in *Alocasia*. While it is not difficult to see that high capacity for photosynthesis is advantageous in high light intensity environments it does not directly follow that it would be disadvantageous in low intensity environments. However, as I will show in a moment, a high capacity of light-saturated photosynthesis requires a greater investment in several leaf constituents including photosynthetic enzymes and electron carriers and this cost would obviously be disadvantageous since there would be no return on this greater investment. Moreover, maintenance of this high-capacity machinery presumably requires an increased rate of oxidative phosphorylation and hence an increased

4

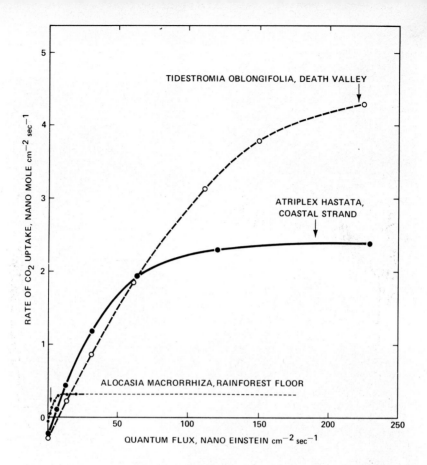

*Fig. 1.* Light dependence of net $CO_2$ uptake by single attached leaves, grown under the contrasting light intensity regimes of their natural habitats. Rates were determined at near optimum temperature for each species and at 320 $\mu$bar $CO_2$ and 21% $O_2$. The arrows indicate the average maximum light intensities to which the plants were exposed during growth (Data from Björkman et al., 1971, 1972; Mooney & Björkman, unpublished).

respiratory loss.

The chlorophyll content on a leaf area basis is similar in these three different plants and so is the overall light absorbing efficiency of the leaves. However, the distribution of chlorophyll differs strikingly between these shade and sun plants (Figure 2). As shown in this EM picture, taken by Dr. Goodchild, the development of the grana region is extreme in this shade plant and the grana stacks reach prodigous proportions. In the sun plant, *Atriplex hastata*, the grana region is much less pronounced and the stroma occupy a larger fraction of the chloroplast volume. As a result the chlorophyll concentration per chloroplast volume as well as per cell becomes very high in the shade plant, but this is compensated by fewer chloroplast-containing cell layers so that the chlorophyll content per unit leaf area remains essentially the same in the two types of plants.

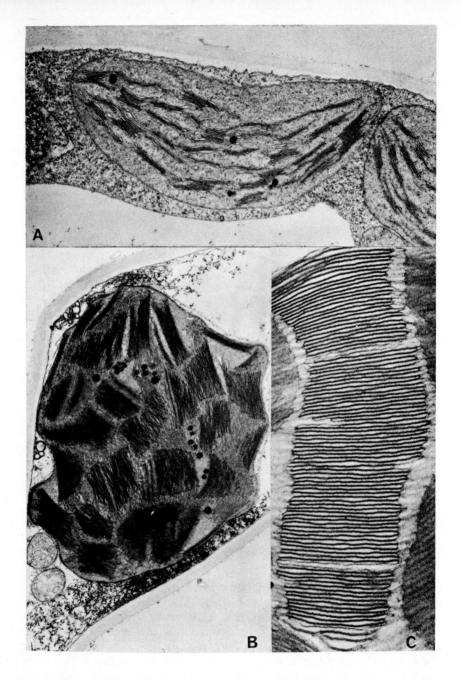

*Fig. 2.* Electron micrographs showing grana arrangements in chloroplasts of the 'sun' plant *Atriplex hastata,* grown at a high light intensity (A), the 'shade' plant *Alocasia macrorrhiza,* grown in its native rain forest habitat (B), and a large granal stack of the latter species (C). (From Björkman et al., 1972 and Goodchild et al., 1972).

6

It is well known that in plants which normally grow under high light intensities the number of palisade cell layers as well as the overall thickness of the leaves tend to decrease when the plants are grown under a low light intensity and many plants growing in the cool temperate forest of Europe and North America often have thin leaves. However, thin leaves are not necessarily characteristics of plants growing in the extreme shade of tropical and subtropical rainforest. Indeed, many of them have thick leaves and relatively high ratio of dry matter to leaf area (Goodchild, Björkman & Pyliotis, 1972).

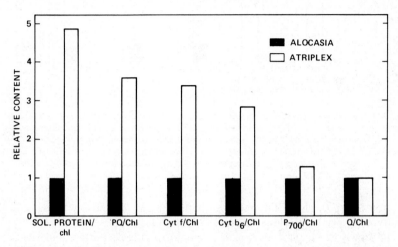

*Fig. 3.* Quantity of soluble protein, photosynthetic electron carriers (plastoquinone, cytochrome f, cytochrome b$_6$), P 700 (the reaction center of photosystem I) and Q (the quencher of fluorescence from photosystem II) in the 'sun' plant *Atriplex hastata*, grown at a high light intensity and in the 'shade' plant *Alocasia macrorrhiza*, grown in its native rainforest habitat (after Björkman et al., 1972 and Boardman et al., 1972).

As shown in Figure 3 there were very important differences between sun plants and the extreme shade plants from the rainforest floor in the quantity of a number of components all of which may potentially determine the capacity for light-saturated photosynthesis. The amount of total soluble protein and also Fraction I protein, or RuDPcarboxylase, was many times higher in the sun plant, *Atriplex hastata,* than in the extreme shade plant, *Alocasia macrorrhiza.* This is in agreement with our findings with many other sun and shade plants. But the same was also true for several of the carriers of the electron transport chain which can be expected to limit the capacity of photosynthetic electron transport. These carriers include PQ,a plastoquinone, and cytochrome f, both of which are involved in the transport of electrons from photosystem II to photosystem I. They also include cytrochrome b$_6$ which is thought to be involved in cyclic electron flow driven by photosystem I. However, contrary to our expectations there were no significant differences in the amount of P$_{700}$, the reaction center of photosystem I, nor in the amount of Q, the fluorescence quencher which is thought to be identical or closely associated with the primary reductant of photosystem II. This indicates that although there are important differences in the capacity of the

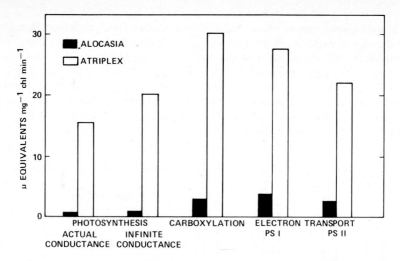

*Fig. 4.* Capacity for light-saturated $CO_2$ uptake *in vivo* and for carboxylation and light-saturated electron transport *in vitro* in *Atriplex hastata,* grown at a high light intensity and in *Alocasia macrorrhiza,* grown in its native rainforest habitat. All rates are expressed as microequivalents of electron transferrred per mg of chlorophyll per minute. It has been assumed that fixation of 1 $\mu$mole of $CO_2$ corresponds to 4 $\mu$ equivalents. (From Björkman et al., 1972 and Boardman et al., 1972).

electron transport chain, the photosynthetic unit size is probably the same in the two types of plants (Björkman et al., 1972; Boardman et al., 1972).

As shown in Figure 4 the greater content of soluble protein and fraction I protein in the sun plant is related to a higher activity of the carboxylation enzyme. Similarly, the greater content of electron carriers is associated with a higher capacity of electron transport both that driven by photosystem I and that driven by photosystem II. These activities are again closely correlated with the light-saturated rates of the intact leaves. This is true both of the rates measured under actual stomatal conductances of the leaves and of the rates that would exist in normal air if the diffusion barriers between the external air and the intercellular spaces were entirely removed.

Similar but less striking differences in all of these characteristics can also be obtained in a single genotype, grown under different light intensity regimes. For example, growth of the sun species *Atriplex hastata* under low light intensity resulted in an approximately 3.5-fold decrease in light-saturated photosynthetic rate, dark respiration rate, RuDP carboxylase activity and electron transport capacity, and an approximately 2-fold decrease in the contents of soluble protein, PQ and cytochromes f and $b_6$, all on a chlorophyll basis. $P_{700}$ and Q remained constant as did also the ratio of leaf area to chlorophyll. In this case the decrease in soluble protein and electron carriers were matched by a similar 2-fold decrease in leaf thickness and in the ratio of leaf dry weight to leaf area (Björkman et al., 1972, 1973; cf. paper by Wild & Grahl, this Volume).

The potential range of environmentally induced acclimation in sun plants such as *Atriplex hastata* is, however, limited. This and other sun plants which we have worked with are incapable of growth at light levels as low as those tolerated by extreme shade plants and in none of the characteristics analyzed did low-light-grown *A. hastata* come near those of the rainforest species. Conversely, other evidence indicates that species or ecotypes, limited in nature to densely shaded habitats, lack the genetic ability to produce a photosynthetic machinery as efficient at high light intensities as that of sun plants. Growing these shade plants at high light intensities results in a severe inhibition of photosynthesis, apparently because the photochemical reaction centers are inactivated by excess excitation energy. The greater resistance to photoinhibition in sun plants is probably a result of their higher photosynthetic capacity. This diverts a greater fraction of the excitation energy to be used in photosynthesis under high light intensities, thus minimizing the excess excitation energy that causes inactivation of the reaction centers (Björkman et al., 1972).

These studies indicate that maximum photosynthetic efficiency at one extreme of light regime precludes maximum efficiency at the other extreme. They further show that photosynthetic adaptation to contrasting light climates cannot be ascribed to any single limiting factor but that it involves several component processes all of which can be expected, on theoretical grounds, to affect the ability to photosynthesize efficiently in their respective environments. These factors need not be restricted to the subcellular level; they often also include stomatal frequency and size as well as other leaf structural characteristics.

These examples illustrate that striking differences in overall photosynthetic characteristics do not necessarily require a qualitative change in the mechanism of photosynthesis but that this can be achieved by quantitative changes in the relative capacities of its component steps. It is, however, clearly evident that $C_4$ photosynthesis and Crassulacean acid metabolism represent changes in the mechanism of photosynthesis that endow the plants with photosynthetic characteristics which could not have been achieved by quantitative changes alone. Before discussing the adaptive significance of $C_4$ photosynthesis I would like to illustrate the striking differences in the temperature dependence of photosynthesis that exist among wild plants from habitats with contrasting thermal regimes.

Figure 5 compares the temperature dependence of photosynthesis of *Atriplex glabriuscula* with that of *Tidestromia oblongifolia,* grown under the temperature regimes of their respective habitats. *A. glabriuscula* is a $C_3$ plant native to cool oceanic habitats in Northwestern Europe and Northeastern North America. It is close relative of *A. hastata* but it extends into even cooler environments. Its growth is rapid in our cool transplant garden at Bodega Head on the coast of Northern California, but it does not survive the hot summer in our Death Valley garden even when supplied with ample water. Similarly, under controlled conditions it grows rapidly at low and moderate day temperatures ($15 - 25°C$), poorly at high temperatures ($35 - 40°C$) and it dies within a day if exposed to a temperature of $42°C$ or above. As shown in Figure 5 this plant is also capable of high photosynthetic rates at low moderate temperatures but the rate does not increase with increased temperature beyond $22°C$, it declines rapidly above $35°C$ and falls to zero at $44°C$.

This contrasts strikingly with the responses of growth and photosynthesis to temperature of *Tidestromia,* a $C_4$ plant, native to the floor of Death Valley, one of the hottest places on earth. This remarkable plant, in contrast to other plants native to Death Valley floor, is dormant during the milder part of the year and

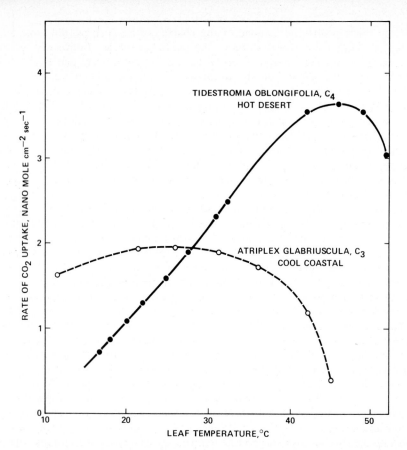

*Fig. 5.* Temperature dependence of photosynthesis in *Tidestromia oblongifolia* and *Atriplex glabriuscula* at a high light intensity of 160 nano-einstein $cm^{-2}sec^{-1}$ a $CO_2$ partial pressure of 320 μbar, and an $O_2$ concentration of 20% Stomatal conductances were almost identical in the two species. The plants were grown under the light and temperature regimes of their respective habitats. (Björkman, Mooney & Ehlringer, unpublished).

confines its growth to the hottest summer months. Peak growth in the Death Valley garden occurs when air and leaf temperatures reach 40° to 50° C. Growth experiments under controlled conditions, simulating the Death Valley temperature regime, showed that this plant is capable of a phenomenal relative growth rate of almost 30% per day at a day temperature of 45° C. This is equivalent to a doubling of its total biomass in less than 3 days. In contrast, growth in this plant is rather slow in the 20° to 30° C range and it does not grow at all below 20° C. As shown in Figure 5 photosynthesis in *Tidestromia* is also markedly inferior to that in *A. glabriuscula* at low and moderate temperatures but the rate increases very steeply with temperature to reach a very high maximum value at about 45° C. Even at 50° C does the rate exceed the maximum rate obtained with *A. glabriuscula* under any condition. These differences are in no way caused by stomatal responses or by other factors related to the water status of the plants.

They are clearly caused by intrinsic differences in photosynthetic characteristics.

At first sight it might be tempting to ascribe these differences to the presence and absence of $C_4$ photosynthesis. For reasons that I will discuss below there is little doubt that the very high photosynthetic rates in *Tidestromia* at high temperatures could not occur if $C_4$ photosynthesis were absent. However, there are several differences between the temperature dependence characteristics of these two plants which cannot be explained on the basis of presence or absence of $C_4$ photosynthesis. For one thing, the rapid decline in photosynthesis above 40°C in *Atriplex glabriuscula* but not in *Tidestromia,* is associated with a loss in the semipermeability of the cell membranes and is thus unrelated to $C_4$ photosynthesis *per se*. For another, the poor photosynthetic performance of *Tidestromia* at low temperatures cannot be attributed to the presence of $C_4$ photosynthesis. Indeed, there is nothing about the intrinsic characteristics of $C_4$ photosynthesis that would be expected to result in an inferior photosynthetic performance at low temperatures. While it is certainly true that the performance of many $C_4$ plants adapted to a hot climate may be poor at low temperatures it does not follow that $C_4$ photosynthesis is disadvantageous when present in a plant which is adapted to a cool climate.

This is illustrated in Figure 6. This time *Atriplex glabriuscula* is compared, not with *Tidestromia* but with another $C_4$ plant, *Atriplex sabulosa*. Like *Atriplex glabriuscula* this plant is native to cool oceanic environments and the two plants often grow together on coastal strands. The growth rate of this $C_4$ plant at low temperatures is at least as high as that of *Atriplex glabriuscula,* and it also dies at temperatures exceeding 42°C. Again the photosynthetic responses to temperature are compared under identical conditions and at the same stomatal conductances. They were grown together under a cool temperature regime simulating that of their natural habitats. The photosynthetic performance of this $C_4$ plant, even at low temperatures, is clearly superior rather than inferior to that of its $C_3$ counterpart. And, as shown in Figure 7, although *Atriplex sabulosa* is as full-fledged a $C_4$ plant as *Tidestromia* it lacks high-temperature stability. As, in $C_3$ *A. glabriuscula*, growing *Atriplex sabulosa* at 40°C results in an inferior photosynthetic performance at any temperature but especially at low temperature and an increase to 43°C or above rapidly results in a cessation of photosynthesis and within less than one hour also causes a loss of leaf turgor. Experiments clearly show that this loss of turgor is not caused by an increased water vapor pressure difference between the leaf and the air as a consequence of increased temperature, but is probably due to a high temperature-induced loss of semi-permeability of cell membranes. This raises the possibility that wilting which is frequently observed in many plants in the field during exceptionally hot days, may often be a direct effect of temperature on the integrity of cell membranes rather than a result of excessive water loss and reduced leaf water potential.

As is also shown in Fig. 7, the photosynthetic performance of *Tidestromia* is in sharp contrast with that of *A. sabulosa*. Total inhibition of photosynthesis does not occur until 54°C and no turgor loss or other evidence of loss of membrane semi-permeability is observed even at this extremely high temperature. On the other hand, when *Tidestromia* is kept at temperatures at 16°C for several days or longer, the activity of its photosynthetic machinery is severely inhibited. This inhibition is present both at low, intermediate and high temperatures, irradiances

11

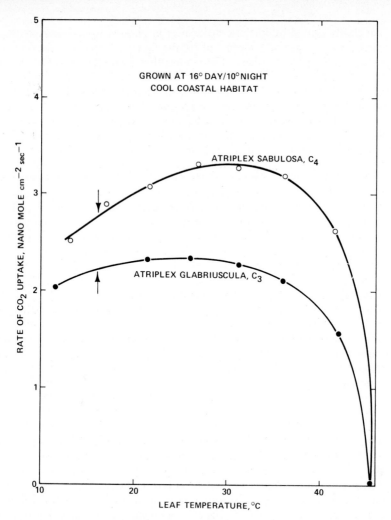

*Fig. 6.* Temperature of photosynthesis in a $C_3$ and a $C_4$ species, native to cool coastal habitats. Conditions for measurements were the same as in Figure 5. The plants were grown under a temperature regime simulating these habitats. Arrows indicate the day temperature during growth. (Björkman, Mooney & Ehlringer, unpublished).

and $CO_2$ concentrations. Studies to date indicate that neither changes in stomatal conductance or other factors influencing the capacity for $CO_2$ diffusion nor in the content of photosynthetic pigments, chloroplast protein, carboxylation enzymes or leaf anatomical characteristics are responsible for this inhibition (Berry, unpublished). Preliminary studies suggest that the apparent quantum requirement for photosynthesis is unusually high in these plants (Björkman, unpublished) and it therefore seems possible that the photochemical activity which is associated with the granal membranes is impaired. If this is so it could mean that the composition or structural organization necessary for maintaining structural in-

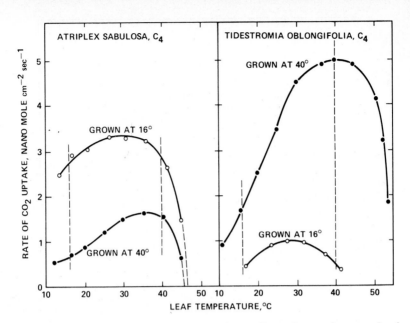

*Fig. 7* Temperature dependence of photosynthesis in two $C_4$ species, one from a cool and one from a hot habitat. Each species were grown under contrasting temperature regimes. Conditions for measurements were the same as in Figures 5 and 6. (Björkman, Mooney & Ehlringer, unpublished).

tegrity and functional efficiency of these membranes in *Tidestromia* at very high temperatures precludes their effective function at low temperatures.

It is evident that the mechanisms underlying photosynthetic adaptation to extremes of temperature regime are several and qualitative as well as quantitative in nature. As we have seen, $C_4$ photosynthesis is in itself insufficient to explain, for example, the remarkable performance of *T. oblongifolia* at very high temperatures. But this and other studies also leave little doubt that the $C_4$ pathway is an important and probably even an essential prerequisite factor. Photosynthetic gas exchange studies on many $C_3$ and $C_4$ plants show without exception that $C_4$ photosynthesis is markedly superior in utilizing low $CO_2$ concentrations in the intercellular spaces and can be considered to be the basic and universal function of $C_4$ photosynthesis. As shown in Figure 8, this characteristic is shared by both *T. oblongifolia* and *A. sabulosa* in spite of their very different photosynthetic performances in other respects.

There is now general agreement that this is accomplished by the utilization of phosphoenolpyruvate carboxylase as the primary acceptor of atmospheric $CO_2$ and the subsequent transfer of carbon to the carbon reduction cycle. The primary carboxylation step acts in effect as a $CO_2$ concentrating mechanism for the subsequent fixation of $CO_2$ by ribulose diphosphate carboxylase. As a result, $C_4$ photosynthesis provides a more efficient mechanism for $CO_2$ fixation at low $CO_2$ concentrations in the intercellular spaces that does $C_3$ photosynthesis. An important consequence is that the inhibition of net photosynthesis by atmospheric oxygen is abolished in $C_4$ plants (cf. paper by Ogren et al., this Volume).

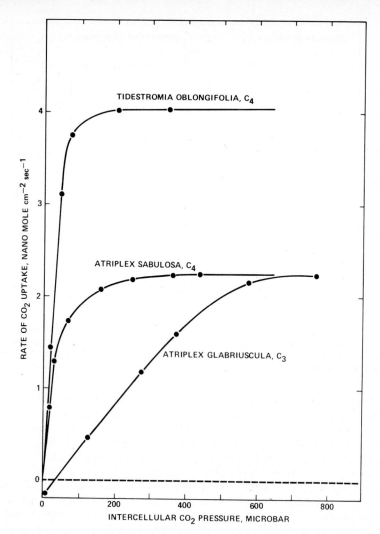

*Fig. 8.* Photosynthesis as a function of the $CO_2$ concentration in the intercellular spaces in $C_3$ and $C_4$ species, grown under a temperature regime of 40°C day/30°C night. Measurements were made at a leaf temperature of 40°C, a light intensity of 160 nano-einstein $cm^{-2}sec^{-1}$ and an $O_2$ concentration of 21%(Björkman, Mooney & Ehlringer, unpublished).

One would thus predict that $C_4$ photosynthesis would confer a adaptive advantage under any condition where photosynthesis would be markedly limited by the $CO_2$ concentration in the intercellular spaces. Where this concentration exerts no effect or only a small effect on photosynthesis, as is the case at low temperatures, the advantage would of course be as best marginal, but there is no reason to assume that $C_4$ photosynthesis *per se* would be disadvantageous. All of our experimental results to date are in full agreement with these predictions. Conversely, one would also predict that the advantages of $C_4$ photosynthesis would be maximal under conditions where photosynthesis is limited largely by the inter-cellular $CO_2$ concentration. Such conditions are present when photosynthesis is operating at high light intensities and temperatures and in particular when stomatal conductance to the exchange of $CO_2$ and water vapor is low. Thus, under a given set of environmental conditions and given rate of water loss $C_4$ plants should have a higher photosynthetic rate and consequently a higher water use efficiency. However, this can only be realized if other adaptations are also present. For example, the high thermal stability of the photosynthetic apparatus of *T. oblongifolia* permits a full expression of the adaptive advantage of $C_4$ photo-synthesis at very high temperatures. In *A. sabulosa* which lacks this thermal stabil-ity, the photosynthetic rate at high temperatures is limited by factors other than intercellular $CO_2$ concentration, and the potential advantage of $C_4$ photosynthesis cannot be realized.

These studies are in accordance with the hypothesis that $C_4$ photosynthesis has evolved in habitats where light intensities and temperatures are high and water supply is limiting. But it is also evident that in order for photosynthesis to operate effectively under these conditions, other independent adaptations, such as high thermal stability, must also be present and these adaptations may be of great disadvantage in a low temperature environment. If these adaptations are genetic-ally linked to $C_4$ photosynthesis, as they are likely to be, then they would consti-tute a considerable barrier to colonization of cool habitats. Nevertheless, it is evident that in certain instances $C_4$ plants have overcome this barrier and managed to inhabit cool oceanic habitats (e.g., *A. sabulosa*) as well as alpine environments (e.g., *Muhlenbergia montana*). But then, as we have seen, these plants apparently no longer possess a superior performance under heat stress.

In conclusion, it is evident that although a single species or genotype may have a considerable potential for phenotypic adjustment of its photosynthetic charac-teristics to a change in environmental conditions, the range of such adjustment is limited and reflect a genotypic adaptation to the particular conditions prevailing in its native habitat. It is further evident that such genotypic adaptations which enable the plant to photosynthesize with an unusual high efficiency under one environmental extreme of light or temperature require a specialization of its pho-tosynthetic machinery which precludes a high photosynthetic efficiency at the other extreme. It is my belief that this is a very important consideration in any assessment of the possibilities of increasing photosynthetic efficiency and primary productivity. In my view it is unlikely that man will be able to improve substantially upon the intrinsic photosynthetic efficiency under any given envi-ronmentally regime over that of plants which have evolved naturally in response to the stresses of such a regime. Comparative studies on plants from extreme environments should enable us to determine what the ultimate limits of photo-synthetic adaptation and what the various underlying mechanisms are. With this information we should be better equipped to design programs with objectives to select and to breed plants for food and fuel with an improved efficiency in different climatic regions.

Finally, I am absolutely certain that all of us invited to come here from different parts of the world are most grateful to the members of the Organizing Committee, and especially to its chairman, Dr. R. Marcelle, because they conceived this excellent idea in the first place and for the considerable effort involved in organizing the meeting, and I suspect, also in raising the funds for it. On behalf of all participants I would also like to point out how honored we are to have the privilege of holding this conference on the brand new campus of the Limburgs Universitaire Centrum. Our best wishes for the success of this new university.

The impressive titles of the many papers to be presented during the coming week promise that we will hear a broad and expert coverage of the many diverse aspects of Environmental and Biological Control of Photosynthesis. I am looking forward to a week that will not only be efficient and productive but also very enjoyable.

# REFERENCES

Björkman, O. & M. Ludlow (1972): Characterization of the light climate on the floor of a Queensland rainforest. *Carnegie Inst. Wash. Year Book* 71: 85-94.

Björkman, O., R.W. Pearcy, A.T. Harrison & H.A. Mooney (1971): Photosynthetic adaptation to high temperatures: a field study in Death Valley, California. *Science* 175: 786-789.

Björkman, O., N.K. Boardman, Jan M. Anderson, S.W. Thorne, D.J. Goodchild & N.A. Pyliotis (1972): Effect of light intensity during growth of *Atriplex patula* on the capacity of photosynthetic reactions, chloroplast components and structure. *Carnegie Inst. Wash. Year Book* 71: 115-135.

Björkman, O., J. Troughton & M.A. Nobs (1973): Photosynthesis in relation to leaf structure. In Basic Mechanisms in Plant Morphogenesis, Brookhaven Symposia in Biology No. 25. Pp. 206-226. Upton, New York.

Boardman, N.K., Jan M. Anderson, S.W. Thorne & O. Björkman (1972): Photochemical reactions of chloroplasts and components of the photosynthetic electron transport chain in two rainforest species. *Carnegie Inst. Wash. Year Book* 71: 107-114.

Goodchild, D.J., O. Björkman & N.A. Pyliotis (1972): Chloroplast ultrastructure, leaf anatomy, and content of chlorophyll and soluble protein in rainforest species. *Carnegie Inst. Wash. Year Book 71: 102-107.*

# RELATIONSHIPS BETWEEN CARBON DIOXIDE TRANSFER RESISTANCES AND SOME PHYSIOLOGICAL AND ANATOMICAL FEATURES

J.L. PRIOUL, A. REYSS & P. CHARTIER *

*Laboratoire Structure et Métabolisme des Plantes associé au C.N.R.S. (L.A. 40), Université de PARIS-SUD, Centre d'Orsay, F 91405 Orsay, France.*

\* *Station de Bioclimatologie I.N.R.A., Route de Saint-Cyr, F 78000 Versailles, France.*

*Abstract*

Biophysical models which describe photosynthetic gas exchanges lead to the definition of 4 resistances to $CO_2$ transfer: $r_a$, $r_s$, $r_m$ and $r_x$. The intracellular resistance ($r_i = r_m + r_x$) presents the most difficult interpretational problems. The separation of $r_i$ into a transport ($r_m$) and carboxylation ($r_x$) component is dependent upon the model used. For example, the high $r_m/r_x$ ratio observed in $C_3$ plants may be partly due to the form of the equation used to summarize the Calvin Cycle.

In contrast, good correlations observed between RuDP carboxylase activity and net maximum photosynthesis would suggest great importance of carboxylation processes.

Using *Lolium multiflorum* as material, the variation of $r_m$ and $r_x$ was compared with structural and metabolic parameters which may be involved in the resistances namely the number of chloroplasts on leaf area basis, the plastid ultrastructure, the activities of carbonic anhydrase, RuDP carboxylase and PEP carboxylase. High significant correlation was observed between nearly all parameters and the net maximum photosynthesis or the resistances. The limitations of the different kinds of approaches are discussed.

## Introduction

From the air outside the leaf to its reduction into glucides, $CO_2$ must cross a lot of physical and biochemical barriers. Usually the $CO_2$ pathway is divided into four parts, each part being associated with one barrier or resistance, namely:

— $r_a$ in the air boundary-layer of the leaf.

— $r_s$ between the surface of the leaf and the intercellular spaces. This resistance is mainly controlled by stomatal openings.

— $r_m$, the mesophyll resistance, between the intercellular spaces and the carboxylation sites where the RuDP carboxylase is located.

— $r_x$ the so-called carboxylation resistance which expresses the ability of the Calvin Cycle, and especially of the RuDP carboxylase, to fix the $CO_2$.

The sum of $r_m$ and $r_x$ is often named $r_i$ or intracellular resistance.

It is only near light saturation that resistances play an important role in the photosynthetic exchanges. In light saturated conditions almost any model yields the following equation proposed by Lake in 1967.

$$P_N = \frac{C - \Gamma}{r_a + r_s + r_i}$$

where $P_N$ is the net photosynthesis, C the $CO_2$ concentration in the air outside the leaf, $\Gamma$ the $CO_2$ compensation point and $r_a$, $r_s$ and $r_i$ the resistance defined above.

For $r_a$ and $r_s$ which are true diffusional resistances, Gaastra's method (1959) of determination is used. The two resistances are obtained from water vapour exchanges. The origins of inaccuracies are mainly:
— the anisolaterality in stomatal distribution
— the incomplete water saturation inside the leaf
— the values of the ratio of the diffusivity coefficients for water vapour and $CO_2$.

When $r_a$ and $r_s$ are known it is easy to obtain $r_i$. In comparing the $r_i$ of $C_3$ and $C_4$ plants an important result, already noted by Ludlow in 1971, arises (table 1). The intracellular resistance $r_i$ is low in $C_4$ plants which mean that one of the most important parameters controlling maximum photosynthetic rate in these plants is the stomata. On the other hand, in $C_3$ the intracellular resistance is by far the most important resistance.

To explain the better assimilation rate of $C_4$ plants it is often assumed that the carboxylation processes in $C_4$ plants are most efficient because:
— the PEP carboxylase located in the mesophyll cells cytoplasm is very active.
— the malate or aspartate formed is quickly moved into the bundle sheath chloroplasts providing the RuDP carboxylase with $CO_2$ under the best conditions.

One may consider that the processes located before the RuDP carboxylation step are essentially transfer processes, Chartier (1975).

As in $C_3$ and $C_4$ plants the terminal carboxylation is the same, i.e. the Calvin Cycle, one may suppose that the difference in $r_i$ lies in the transfer processes.

In $C_4$ plants, the transfer processes are known (PEP carboxylase system) and very efficient. In $C_3$ plants the nature and the importance of transfer processes are still questionable. Because of the large value of $r_i$, the problem of the partitioning of the intracellular resistance into transfer and carboxylation components is of great importance.

*Table 1.* Comparison of resistances to carbon dioxide transfer between $C_3$ and $C_4$ plants.

| Resistances | C3 PLANTS | | | C4 PLANTS | |
|---|---|---|---|---|---|
| | *Phaseolus vulgaris* CHARTIER 1969 | *Lolium multiflorum* $I_{growth}$=110 W.m$^{-2}$ PRIOUL 1971 | *Calopogonium mucunoides* LUDLOW 1971 | *Zea mays* CHARTIER | *Sorghum almum* LUDLOW 1971 |
| $r_a$ (m$^{-1}$.s) | 42 | 71 | 56 | 60 | 76 |
| $r_s$ (m$^{-1}$.s) | 109 | 248 | 84 | 250 | 94 |
| $r_i$ ($= r_m + r_x$) (m$^{-1}$.s) | 715 | 668 | 300 | 135 | 110 |

From models, when we subdivide $r_i$ into $r_m$ and $r_x$, the value for $r_m$ is always much greater than $r_x$, indicating that the main resistance to $CO_2$ assimilation is that of $CO_2$ transfer between the intercellular spaces and the carboxylation sites.

In contrast, several authors, Björkman (1968), Neales and his coworkers (1971), Treharne & Cooper (1969) measuring RuDP carboxylase activity under various conditions found a good correlation between carboxylation and net $CO_2$ assimilation rate at light saturation which might suggest that $r_x$ is the greatest resistance.

In order to have a better understanding of these problems, we have studied on plants grown under different light intensities the resistance variation and the variation of some metabolic and structural parameters. The material used was Italian rye-grass (*Lolium multiflorum*).

## Methods and Techniques

Choosing the parameters associated with intercellular resistances:

Since the carboxylation resistance $r_x$ is dependent on the ability of Calvin Cycle to fix $CO_2$, it is logical to compare $r_x$ variation with that of RuDP carboxylase. Simultaneously it is easy to measure PEP carboxylase activity, a carboxylation enzyme not directly involved in the Calvin Cycle.

For mesophyll resistance ($r_m$), it must be emphasized that it is certainly not a purely diffusive resistance. The number of elementary processes (physical and biochemical) which may be involved is indeed very high. If we consider the $CO_2$ pathway from the intercellular spaces to the carboxylation sites we can see that $CO_2$ must cross:

– the cell wall in a dissolved state.
– the cell membrane which seems more permeable for $CO_2$ than for bicarbonate.
– a thin layer of cytoplasm where $CO_2$ is dissolved and where pH dependent equilibrium occurs between $CO_2$ and bicarbonate.
– the chloroplast envelope, double lipoprotein membrane which is again more permeable for $CO_2$ than for bicarbonate ions. At this point the $CO_2$ is redissolved in chloroplast stroma with a new $CO_2$-bicarbonate equilibrium and finally $CO_2$ migrates to the carboxylation sites which seem to be located partly in the stroma and partly on the membranes. If we look at the diffusion distance it is evident that average intraplastidial diffusion length is many times greater than the extra-chloroplastidial pathway. So the importance of the intrachloroplastic membranes as barriers for $CO_2$ or bicarbonate diffusion must be emphasized. One other important element for intraplastidal $CO_2$ migration is the pH and its correlative action on $CO_2$-bicarbonate equilibrium. Werdan & Heldt (1972) have shown that in the light an important pH gradient appears between the cytoplasm and the stroma, so that the stroma pH is about 8. But at this pH the ratio of $CO_2$ to bicarbonate is only 1%. As $CO_2$ has been shown by Cooper T.G. et al. (1969) to be the substrate for RuDP carboxylase, a possible barrier to $CO_2$ fixation may be the low availability of free $CO_2$ at the carboxylation sites. In this regard the role of carbonic anhydrase present in large amount in the chloroplast may be important (Everson 1970 and Graham & Reed 1971).

For all these reasons the 3 following parameters were measured:

- carbonic anhydrase activity
- number of chloroplasts on leaf area basis
- variation in the lenght of grana and intergrana membranes per chloroplast

It should be emphasized that these 3 parameters represent only some elements among all the possible components involved in the mesophyll resistance.

## Experimental

Plants of *Lolium multiflorum* (var. Westerwold Barenza) were grown in controlled environment under a range of light intensity (16, 45, 85 and 110 W.m$^{-2}$). The third leaf which appeared on the main shoot was sampled randomly a few days after full expansion. For photosynthetic measurements the leaves were still attached to the shoot (Prioul, 1971 and 1973).

The water vapour and $CO_2$ fluxes were measured and resistances calculated by methods detailed elsewhere (Prioul, 1971) and similar to those of Chartier (1969).

Carboxylation activities: the preparation of crude enzyme extracts and reaction mixtures were similar to the procedures described by Björkman (1968) and Treharne & Cooper (1969). Concentrations of ribulose diphosphate or phosphoenolpyruvate and bicarbonate were adjusted to avoid substrate limitations and the linearity of the relation between the enzyme activity and the time of incubation was checked.

The carbonic anhydrase activity was measured from enzyme extracts prepared according to Poincelot (1972). A reaction mixture similar to that described by Everson (1970) was used, but instead of the colorimetric method, the pH variations were measured with a glass electrode and recorded. All the enzyme activities are expressed on a leaf area basis.

Number of chloroplasts per unit of leaf area: after the measurement of photosynthetic gaseous exchanges, small parts of the leaves were sampled for subsequent observation by light microscopy. On photographs of a complete section in the median part of the leaf the total number of chloroplasts was counted.

For investigations on the chloroplast ultrastructure small pieces of leaves tissues were prepared by conventional techniques detailed elsewhere (Prioul, 1973). A biometric method was used to describe quantitatively the ultrastructural modifications.

## Results and Discussion

Net photosynthetic rate at light saturation ($P_N$ max) increases linearly with irradiance during growth (fig. 1). Under the same conditions the reciprocal of resistances varies in the same way, the better linearity being obtained for the mesophyll resistance. On the other hand it appears using Chartier's model that $r_m$ is always the most important resistance whatever irradiance received during growth may have been, and that $r_m$, $r_x$ ratio is very high (Table 2) (Prioul, 1971).

The variation of RuDP carboxylase as a function of irradiance during growth (Fig. 2) gives also a straight line. Therefore the correlation coefficient between $P_N$ maximum and RuDP carboxylase activity is about 0.99. We find the same value

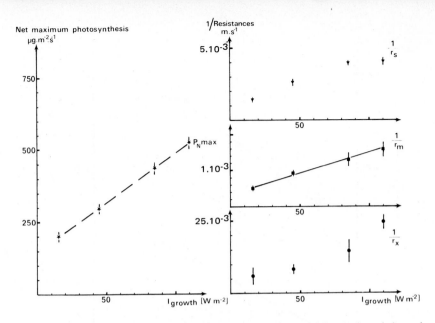

*Fig. 1.* Variation in *Lolium* leaves in net maximum photosynthesis ($P_N$ max), and the reciprocals of resistances $r_s$, $r_m$ and $r_x$ as a function of the irradiance during growth. (Each point with the standard error of mean, for Fig. 1, 2, 3, 4, 5.)

*Table 2.* Resistance to carbon dioxide transfer from *Lolium* plants grown under 4 different levels of irradiance (Prioul, 1971).

| | 16 W.m$^{-2}$ | 45 W.m$^{-2}$ | 85 W.m$^{-2}$ | 110 W.m$^{-2}$ |
|---|---|---|---|---|
| $r_a$ (m$^{-1}$.s) | 71 | 71 | 71 | 71 |
| $r_s$ (m$^{-1}$.s) | 707 | 386 | 252 | 249 |
| $r_m$ (m$^{-1}$.s) | 1964 | 1101 | 768 | 627 |
| $r_x$ (m$^{-1}$.s) | 167 | 123 | 69 | 41 |

for the correlation coefficient between the reciprocal of the carboxylation resistance and the RuDP carboxylase activity. These results are in good agreement with those of Björkman (1968), Treharne & Cooper (1969), Neales and coworkers (1971), etc. But the consistency between RuDP carboxylase activity and $1/r_x$ value must be noticed.

If we consider now the PEP carboxylase activity (Fig. 2) the same excellent correlation between $P_N$ max and PEP carboxylase on one hand, and $1/r_x$ and PEP

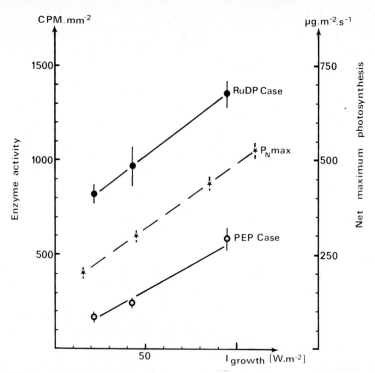

*Fig. 2.* Comparison between the variation in RuDP carboxylase or PEP carboxylase and the net maximum photosynthesis ($P_N$ max) for *Lolium* plants grown under different irradiances.

carboxylase on the other hand, is observed. Nevertheless in $C_3$ plants there is no evidence for a direct role of this enzyme in Calvin Cycle.

A linear relationship is also observed between carbonic anhydrase activity and irradiance during growth (Fig. 3). An excellent correlation between this activity and net assimilation or $1/r_m$ is also found. Downton & Slatyer (1972) studying temperature dependance of photosynthesis have noticed similar relationships between carbonic anhydrase activity and photosynthetic rates.

The number of chloroplasts per unit of leaf area (Fig. 4) increases steadily from low light to a maximum at 85 W.m$^{-2}$ and decreases slightly but significantly for the highest irradiance. In spite of this non linear variation the correlation coefficient is still significant at the 0.1% level. However the magnitude of the variation in the number of plastids is too low to account for the variation in the assimilation rate. Kariya & Tsunoda (1972) have obtained similar results from 9 strains of *Brassica.* If the chloroplast envelope would provide a resistance to $CO_2$ transfer, a better index for comparison with assimilation rate would be the ratio of total chloroplast area to leaf area called by Kariya & Tsunoda (1972) the chloroplast area index (C.A.I.). This value is obtained by multiplication of the chloroplast number per leaf area and the chloroplast size. As in *Lolium* the chloroplast sizes do not seem to vary significantly the C.A.I. varies in the same way as the chloroplast number.

Plastids from plants grown under the lower light intensity consist of numerous

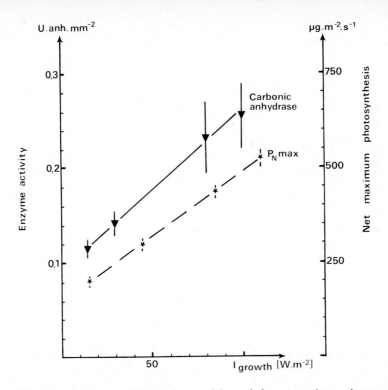

*Fig. 3.* Variation in the carbonic anhydrase activity and the net maximum photosynthesis ($P_N$ max) against irradiance during growth.

*Table 3.* Intraplastidal membranes as a function of irradiance under which *Lolium* plants where grown.

|  | 16 W.m$^{-2}$ | 45 W.m$^{-2}$ | 110 W.m$^{-2}$ |
|---|---|---|---|
| Total length of stroma lamellae ($\mu$m/plastid) | 63 | 40 | 47 |
| Total length of grana ($\mu$m/plastid) | 29 | 15 | 19 |
| Total length of stroma lamellae and grana ($\Sigma$L) | 92 | 55 | 66 |
| $\dfrac{\Sigma L}{N}$ ($10^6$ mm$^3$) N = Number of plastids on leaf area basis | 1.12 | 0.575 | 0.613 |

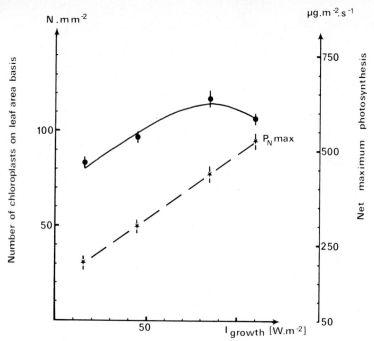

*Fig. 4.* Variation in the number of chloroplasts on a leaf area basis and the net maximum photosynthesis ($P_N$max) against irradiance during growth ($N = 10^3$ chloroplasts).

grana containing a lot of appressed membranes and between the grana a great number of long stroma lamellae is found. The general aspect of these plastids gives an impression of great abundance of the lamellae.

In contrast the lamellar content of plastids from plants grown under the higher irradiances is less important: the number of grana and the thickness of grana is lower and stroma lamellae are shorter (Prioul, 1973). The biometric method shows that there is a decrease of about 50% of the internal membrane system (Table 3). A similar reduction of plastidial structure when light intensity during growth increases was observed by Ballantine & Forde (1970) in Soybean. Björkman (1974) mentions also a great increase in the grana stacking in plants from extremely shaded habitats.

If we return to the idea expressed above that chloroplast lamellae are barriers for intraplastidial $CO_2$ migration, it is possible to suppose that the ratio of the plastid membrane length to the number of chloroplasts per leaf area is related to mesophyll resistance. If we compare the variation of this ratio with $r_m$ (Fig. 5) we can observe a correlative variation between these 2 variables from 16 to 45 W.m$^{-2}$. However, the variation of this parameter is unable to account for the total variation of $r_m$ or of the assimilation rate.

If we consider the parameter which may be involved in $CO_2$ transfer, a good correlation between nearly all components and the overall process can be seen. It is clear, however, that these kinds of results are unable to give a conclusive argument on the relative importance of each elementary process and especially about the ratio of transfer to carboxylation barriers. Furthermore, the correlation

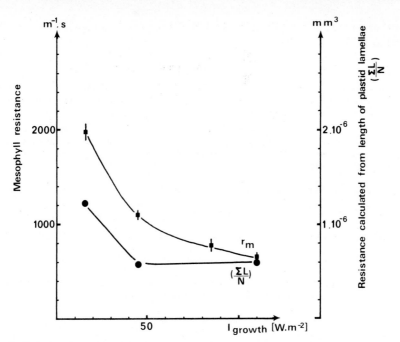

*Fig. 5.* Variation in mesophyll resistance and the resistance calculated from the length of intraplastidial lamellae against irradiance during growth.

coefficient technique does not allow us to demonstrate any causal relationship. PEP carboxylase variation may illustrate this fact.

A possible solution to this methodological dilemma is to find conditions where no correlation may be observed between one component and the overall process. A first example is provided by Randall & Bouma (1973) who studied zinc deficiency in Spinach leaves. They have observed considerable decrease in carbonic anhydrase activity with only a small lowering of net photosynthesis which provides evidence against a close relationship between this activity and net photosynthesis. But even this experiment does not exclude the possibility that very small amounts of carbonic anhydrase may be sufficient to facilitate $CO_2$ transfer. A second example is given in this volume by Oben & Marcelle on *Phaseolus* and Michel in *Euglena*. They have observed from material grown with CCC a decrease in net photosynthetic exchange, but no significant changes in electron transport or RuDP carboxylase activities with *Phaseolus* extracts; an increase in electron transport is even observed in *Euglena*. If we assume that in vitro measurements are representative of in vivo behaviour, which is always questionable, this fact would provide direct evidence that the carboxylation and photochemical reactions might not be the main restraints in leaf photosynthetic $CO_2$ exchanges. Furthermore, it is often admitted that CCC acts upon membrane permeability which is consistent with the view of an important value of $r_m$.

From theoretical point of view a general solution to the problem would be the use of adequate models. To do so it is necessary to check the most important assumptions underlying the models and to improve them to get a better fitting with experimental data. For Chartier's model the main assumptions are summarized in Table 4.

*Table 4.* List of assumption for model.

1 — Steady-state conditions are obtained in all stages of assimilatory process when constant exchange of $CO_2$ at leaf surface is observed.

2 — The resistance $r_s$ can be obtained from the transpiration measurement (e.g. difference in membrane diffusion for $CO_2$ and water, complete saturation of water vapour in intercellular spaces, negligible effect of anisolaterality of stomata).

3 — $r_m$ and $r_x$ are constant in the range used for calculation.

4 — No influence of $CO_2$ gradient inside the intercellular spaces.

5 — No influence of light gradient inside the mesophyll.

6 — No significant influence of the imprecision of respiration rate R.

7 — Linear relationship between net assimilation rate and external $CO_2$ concentration from $\Gamma$ to 300 v.p.m. (i.e. measured photosynthesis far from $CO_2$ saturation).

8 — Rectangular — hyperbolic equation of photosynthetic curve at carboxylation.

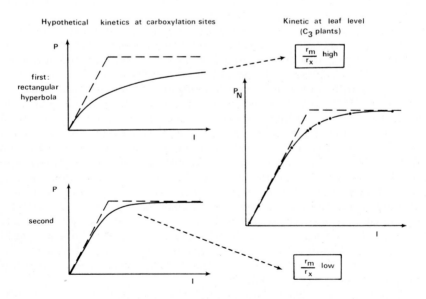

*Fig. 6.* Illustration of the effect of the selected hypothesis at the carboxylation sites upon $r_m, r_x$ ratio in $C_3$ plants.

The $r_m/r_x$ ratio is dependent upon the hypothesis 5, 6 and 8. For hypothesis 5, Laisk (1969) has shown that neglecting light gradient inside the leaf led to underestimation of $r_m$. Comparing the influence of various assumptions for photorespiration Chartier (1971) demonstrated the $r_m/r_x$ ratio was not very much affected.

From the theoretical investigations it appears now that $r_m$, $r_x$ ratio is strongly dependent upon the shape of the equation used for photosynthesis at the carboxylation sites. For $C_3$ plants, the shape of photosynthetic light curves for leaf is near that of Blackman's curve (Fig. 6). If the equation which represents the relation between photosynthesis and irradiance at carboxylation sites is that of a rectangular hyperbola $r_m$, $r_x$ ratio must be high in order to match Blackman's curve at leaf level. In contrast, if the shape of the equation at the carboxylation sites is similar to Blackman's curve $r_m$,$r_x$ ratio should be low.

In the absence of information on global kinetics at the carboxylation sites the authors, following Rabinowitch (1951) or using a Michaelis Menten equation (Jones & Slatyer, 1972), have used a rectangular hyperbolic relation to summarize Calvin Cycle, so high values of $r_m/r_x$ are necessarily obtained.

The need for experiments to check this fundamental assumption is clear. And more generally the study of the uncertainties associated with the relative importance of transfer to carboxylation processes supposes a better coordination between investigations on structure and function on one hand, and model and experiment on the other hand.

# REFERENCES

Ballantine, J.E.M. & B.J. Forde (1970): The effect of light intensity and temperature on plant growth and chloroplast ultrastructure in Soybean. *Amer. J. Bot.*, 57: 1150-1159.

Björkman, O. (1968): Carboxydismutase activity in shade-adapted and sun adapted species of higher plants. *Physiol. Plant*, 21: 1-11.

Björkman, O. (1974): this volume.

Chartier, P. (1969): Assimilation nette d'une culture couvrante. II. La réponse de l'unité de surface de feuille. *Ann. Physiol. Vég.*, 11: 221-263.

Chartier, P. (1970): A model of $CO_2$ assimilation in the leaf. In Prediction and Measurement of Photosynthetic Productivity. Pp 307-315. Pudoc, Wageningen.

Chartier, P. (1975): La productivité primaire à l'échelle foliaire. *Physiol. Vég.* (in press).

Cooper, T.G., R.D. Filmer, M. Wishnick & M.D. Lane (1969): The active species of $CO_2$ utilized by ribulose diphosphate carboxylase. *J. biol. Chem.*, 244: 1081-1083.

Downton, J. & R.O. Slatyer (1972): Temperature dependence of photosynthesis in Cotton. *Plant Physiol.*, 50: 518-522.

Everson, R.G. (1970): Carbonic anhydrase and $CO_2$ fixation in isolated chloroplasts. *Phytochemistry*, 9: 25-32.

Gaastra, P. (1959): Photosynthesis of crop plants as influenced by light, carbon dioxide, temperature and stomatal diffusion resistance. *Meded. Landbouwhogesch. (Wageningen)*, 59 (13): 1-68.

Graham, D. & M.L. Reed (1971): Carbonic anhydrase and the regulation of photosynthesis. *Nature New Biol.*, 231: 81-84.

Jones, H.G. & R.O. Slatyer (1972): Estimation of the transport and carboxylation components of the intracellular limitation to leaf photosynthesis. *Plant Physiol.*, 50: 283-289.

Kariya, K. & S. Tsunoda (1972): Relationship of chlorophyll content, chloroplast area index and leaf photosynthesis rate in *Brassica. Tohoku J. Agr. Res.*, 23: 1-14.

Laisk, A. (1969): Light curves of photosynthesis considering light profile in the leaf. In Voprosy effektivnosti fotosinteza (Tartu). Pp 64-92.

Lake, J.V. (1967): Respiration of leavers during photosynthesis. II. Effects on the estimation of mesophyll resistance. *Aust. J. Biol. Sci.,* 20: 495-499.

Ludlow, M.M. (1971): Analysis of the difference between maximum leaf net photosynthetic rates of $C_4$ grasses and $C_3$ legumes. In Hatch, M.D., Osmond, C.B. & Slatyer, R.O. (ed): Photosynthesis and Photorespiration. Pp 63-67. Wiley-Interscience.

Michel, J.M. (1974): This volume.

Neales, T.F., K.J. Treharne & P.F. Wareing (1971): A relationship between net photosynthesis, diffusive resistance, and carboxylating enzyme activity in Bean leaves. In Hatch, M.D., Osmond, C.B. & Slatyer, R.O. (ed): Photosynthesis and Photorespiration. Pp 89-96. Wiley-Interscience.

Oben, G. & R. Marcelle (1974): This volume.

R.P. Poincelot (1972): The distribution of carbonic anhydrase and ribulose diphosphate carboxylase in Maize leaves. *Plant Physiol.,* 50: 336-340.

Prioul, J.L. (1971): Réactions des feuilles de *Lolium multiflorum* à l'éclairement pendant la croissance et variation des résistances aux échanges gazeux photosynthétiques. *Photosynthetica,* 5: 364-375.

Prioul, J.L. (1973): Eclairement de croissance et infrastructure des chloroplastes de *Lolium multiflorum Lam.* Relation avec les résistances au transfert de $CO_2$. *Photosynthetica,* 7: 373-381.

Rabinowitch, E.I. (1951): Photosynthesis and related processes. II (1) Interscience, New-York.

Randall, P.J. & D. Bouma (1973): Zinc deficiency, carbonic anhydrase and photosynthesis in leaves of spinach. *Plant Physiol.,* 52: 229-232.

Treharne, K.J. (1972): Biochemical limitations to photosynthetic rates. In Crop Processes in Controlled Environments, Rees, A.R., Cockshull, K.E., Hand, D.W., Hurd, R.G., ed. Academic Press New-York and London. Pp 285-303.

Treharne, K.J. & J.P. Cooper (1969): Effect of temperature on the activity of carboxylases in tropical and temperate gramineae. *J. Exp. Botany,* 20: 170-175.

Werdan, K. & H.W. Heldt (1972): Accumulation of bicarbonate in intact chloroplasts following a pH gradient. *Biochim. Biophys. Acta,* 283: 430-441.

# TOMATO LEAF PHOTOSYNTHESIS AND RESPIRATION IN VARIOUS LIGHT AND CARBON DIOXIDE ENVIRONMENTS

L.J. LUDWIG, D.A. CHARLES-EDWARDS & A.C. WITHERS

*Glasshouse Crops Research Institute, Littlehampton, Sussex, U.K.*

*Abstract*

The carbon dioxide exchange rates of single attached tomato leaves were continuously measured over a 24 h period using IRGA techniques. Net photosynthetic rates were measured in a constant light and carbon dioxide environment for 8 h. The leaf was then darkened and the rate of $CO_2$ evolution measured for a further 16 h. Different light levels or carbon dioxide concentrations were used in these assays each day.

A positive linear correlation was found between the integrated rate of dark respiration and the integrated rate of net photosynthesis in the preceding light period when the photosynthetic rate was varied by changing the light level. A different positive linear correlation was observed when photosynthesis was varied by changing the carbon dioxide concentration. In both 'high' and 'low-light'-adapted leaves, the integrated rate of dark respiration was more sensitive to changes in the light level during the preceding light period than to changes in the carbon dioxide concentration, even when similar amounts of carbon dioxide were assimilated during the light period.

## Introduction

The effect of carbon dioxide enrichment on photosynthesis and plant growth is well established and is commercially exploited by the glasshouse industry. However, surprisingly little information is available on the physiological basis of the response of photosynthesis and growth to carbon dioxide, and there is little quantitative information on its interaction with other environmental factors such as light and temperature.

An experimental programme was initiated to study the effects of light, carbon dioxide and temperature on photosynthesis, photorespiration and dark respiration in relation to growth of young tomato plants. In this paper we report some preliminary results which suggest that the leaf environment during photosynthesis affects not only the amount of carbon dioxide assimilated during the light period but also the respiratory activity of the leaf during the subsequent dark period.

## Materials and Methods

Tomato plants, *Lycopersicon esculentum* L. (c.v. Kingley Cross) were grown in nutrient culture in a controlled-environment cabinet at $20°C$, 0.7 kPa water vapour pressure deficit, 300 vpm $CO_2$ and either 80 $Wm^{-2}$ PAR (400-700 nm) during a 16 h light period or 20 $Wm^{-2}$ during an 8 h light period. Plants were also grown in a glasshouse in pots of peat-sand containing base fertilizer. The minimum glasshouse day and night temperature was set at $20°C$, with ventilators opening at $24°C$. The mean daily integral of PAR inside the glasshouse was 5.8 $MJm^{-2}$.

Six week old plants were moved to the laboratory where the $CO_2$ exchange rates of either the fifth or sixth leaf were measured for one day under standard conditions using IRGA techniques. The chosen leaf was enclosed in a perspex assimilation chamber, illuminated on the upper surface by a 250 W high-pressure mercury halide lamp, and the net $CO_2$ exchange rate continuously monitored at a constant light flux density. After 8 h the light was switched off and the rate of $CO_2$ evolution in the dark was measured for a further 16 h. The $CO_2$ concentration in the air entering the assimilation chamber was maintained at 300 vpm and the leaf temperature at $20°C$. Assays were conducted under a different light flux density each day. In a second series of experiments a constant light level was maintained throughout, but each day a different $CO_2$ concentration was used in the chamber during the light period. In these experiments the $CO_2$ concentration was returned to 300 vpm for the dark period.

## Results

The results of experiments on the glasshouse grown plants are shown in Fig. 1. During each experiment the light flux density was held constant throughout the 8 h light period, but was changed between experiments. The $CO_2$ concentration in the air entering the chamber was maintained at 300 vpm. The rate of dark respiration integrated over the dark period ($R_D$) was found to be linearly related to the

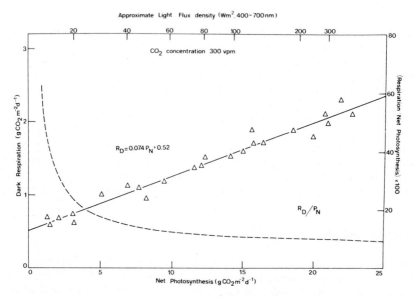

*Fig. 1.* Dark respiration integrated over 16 h ($R_D$) plotted against the previous net photosynthesis integrated over 8 h ($P_N$) for leaves of 'glasshouse' plants. The $CO_2$ concentration was 300 vpm, the leaf temperature $20°C$, and a constant light level was given for 8 h. A different leaf, and light level, was used in each experiment. $R_D$ is also expressed as a percentage of $P_N$.

rate of net photosynthesis integrated over the light period ($P_N$), the relationship being described by the equation

$$R_D = aP_N + b$$

where $a$ is a dimensionless constant and $b$ a constant with dimensions $(gCO_2)m^{-2}d^{-1}$. A regression analysis of the data shown in Fig. 1 gave values for these constants of 0.074 and 0.52 $(gCO_2)m^{-2}d^{-1}$ respectively; that is, in one day the leaves lost through respiration at night 0.52 $(gCO_2)m^{-2}$ plus 7.4% of the carbon dioxide assimilated during the light period. It should be noted that respiratory losses during the light period, whether by dark respiration or photorespiration, have been accounted for by the use of net photosynthetic rates.

The integrated night respiration rate expressed as a percentage of the integrated net photosynthetic rate is also shown in Fig. 1. At a light flux density of approximately 6 $Wm^{-2}$ (about twice the leaf light compensation value) $R_D$ was equal to $P_N$ (100%), but at higher light flux densities the percentage dropped rapidly until at light saturation $R_D$ was only about 10% of $P_N$.

Similar data were obtained for leaves of plants grown in a controlled environment cabinet at 80 $Wm^{-2}$ PAR during a 16 h light period. A linear relationship between $R_D$ and $P_N$ was again observed (Fig. 2) when the rate of net photosynthesis was varied by changing the light level, and a regression analysis of these data gave values of $a = 0.071$ and $b = 0.97$ $(gCO_2)m^{-2}d^{-1}$. A comparison of these values with those derived from Fig. 1 shows that $a$ was similar for the two groups of plants. However, $b$ was considerably higher for leaves of plant grown in the controlled environment cabinets.

Further experiments were made to determine whether the effect of the light level on subsequent dark respiration rate was specific, or whether any environmental factor which affected the leaf photosynthetic rate and thereby the amount of assimilate at the end of the light period, had a similar effect on subsequent dark respiration.

The experimental procedures were similar to those already described. Instead of varying the light flux density each day, however, it was maintained throughout at 80 $Wm^{-2}$ PAR and the net rate of photosynthesis was altered by changing the $CO_2$ concentration in the leaf chamber. For the subsequent dark period the $CO_2$ concentration was returned to 300 vpm. $R_D$ was plotted against $P_N$ and these results are also shown in Fig. 2.

A linear relationship was again observed between $R_D$ and $P_N$, but the slope and intercept of the regression line were different from those determined in the previous experiment. The analysis of these latter data gave values of $a = 0.026$ and $b = 1.55$ $(gCO_2)m^{-2}d^{-1}$, which were significantly different ($P = 0.01$) from those derived from the 'light flux density' experiments.

The two regression lines relating $R_D$ and $P_N$ should cross at the value of $P_N$ corresponding to a light flux density of 80 $Wm^{-2}$ and a $CO_2$ concentration of 300 vpm. Consequently, at low $P_N$ values, $R_D$ was greater when photosynthesis was $CO_2$ limited. When photosynthetic rates were high, subsequent dark respiration was greater when photosynthesis was light saturated. For these 'high-light' grown leaves, $P_N$ at 300 vpm was increasingly $CO_2$ limited at light flux densities above 120 $Wm^{-2}$ (i.e. as light saturation was approached). Thus the relationships present-

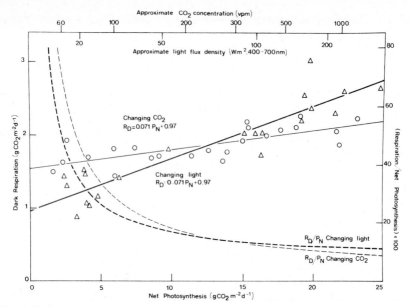

*Fig. 2.* Dark respiration integrated over 16 h ($R_D$) plotted against the previous net photosynthesis integrated over 8 h ($P_N$) for leaves of controlled environment plants grown in 'high' light. $P_N$ was changed by changing either the light level at 300 vpm $CO_2$ ($\triangle$), or the $CO_2$ concentration at a light level of 80 $Wm^{-2}$ PAR ($\bigcirc$). In the dark $CO_2$ concentration was 300 vpm in all experiments. $R_D$ is also expressed as a percentage of $P_N$.

ed in Fig. 2 show that for equivalent amounts of $CO_2$ fixed during the light period, subsequent dark respiration was greater following $CO_2$-limited rather than light-limited photosynthesis.

This effect of light and $CO_2$ on subsequent dark respiration is clearly demonstrated in Fig. 3 which shows the actual $CO_2$ exchange rates for representative leaves over the whole 24 h period. Typical data with environmental combinations of light and $CO_2$ resulting in similar net photosynthetic rates, were selected from the extremes of the two linear relationships shown in Fig. 2.

The pattern of respiration was similar following each of the four treatments. When the leaf was darkened there was a high rate of $CO_2$ evolution during the first minute. This 'burst', presumably a remnant of photorespiration, has been omitted from Fig. 3. Following this 'burst' there was a temporary depression of the respiration rate followed after 20-30 min by a second period of higher respiration. Thereafter the respiration rate slowly declined to a minimum and then increased to reach another small peak about 2 h before the beginning of the next light period. The factors responsible for these transients and diurnal changes are not known. A general comparison of these four examples shows that night respiration was more sensitive to changes in the light flux density during the light period than to changes in the $CO_2$ concentration. The differential effect of light and $CO_2$ on subsequent dark respiration was apparent throughout the whole of the dark period.

In Fig. 4 both the initial and subsequent 'basal' rates of night respiration from

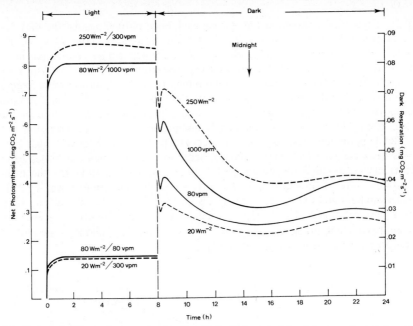

*Fig. 3.* Actual $CO_2$ exchange rates for the whole 24 h period for 4 leaves selected from the extremes of the two linear $R_D$:$P_N$ relationships shown in Fig. 2. Solid lines: $CO_2$ level; broken lines: changing light level.

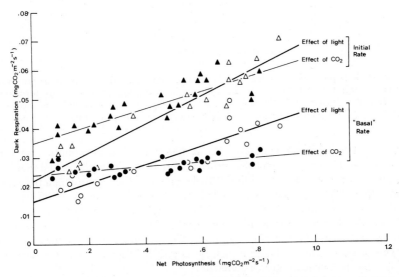

*Fig. 4.* The initial 'maximum' (△,▲) and subsequent 'basal' (○,●) rates of dark respiration each plotted against the previous photosynthetic rate for both series of experiments shown in Fig. 2. Open symbols – changing light level; closed symbols – changing $CO_2$ level.

all the experiments shown in Fig. 2 have each been plotted against the previous net photosynthetic rate. Both initial and 'basal' respiration rates were linearly related to the previous photosynthetic rate, the relationships responding more markedly to the light level than to the $CO_2$ concentration (cf. Fig. 2). Whilst the marked effect of the previous light environment on dark respiration was apparent in both initial and 'basal' respiration rates the effect of the $CO_2$ concentration during the light period was primarily on the initial respiration rates although the 'basal' respiration rate was not completely independent of $CO_2$ concentration.

The experiments described so far used plants grown at relatively high light levels during a 16 h light period. Experiments were also carried out on leaves of plants grown at 20 $Wm^{-2}$ during an 8 h period to establish whether the night respiration rate of 'low-light' grown plants responded in a similar way to daily changes in light and $CO_2$ levels. The effects of three $CO_2$ levels (60, 300 and 1000 vpm) were examined at two light flux densities (20 and 200 $Wm^{-2}$) to establish the approximate limits of the light-$CO_2$ response surface. $R_D$ and $P_N$ were plotted as before and the response is shown in Fig. 5.

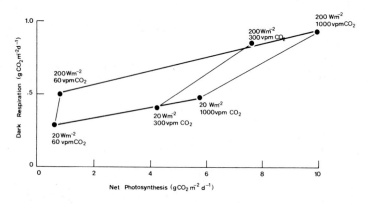

*Fig. 5.* Dark respiration integrated over 16 h ($R_D$) plotted against the previous net photosynthesis integrated over 8 h ($P_N$) for leaves of controlled environment plants grown in 'low' light. $P_N$ was changed by changing the light and $CO_2$ levels as indicated. The $CO_2$ concentration was 300 vpm in the dark. Each point is a mean of 5 experiments.

In the 'low-light' adapted leaves, $P_N$ at 300 vpm and light saturation was only one third of that of the 'high-light' grown material (shown in Fig. 2) and there was a similar reduction in $R_D$. In both the 'high-' and 'low-light' grown material at 300 vpm a change in light level from 200 $Wm^{-2}$ to 20 $Wm^{-2}$ resulted in a 50% decrease in $R_D$. However, whilst the decrease in $P_N$ in 'high-light' grown leaves was 84% it was only 45% in the 'low-light' grown leaves, indicating that respiration is not directly related to the leaf photosynthesis. This was more clearly demonstrated at 60 vpm where a similar change in light level reduced $P_N$ by 25% but reduced $R_D$ by 42% in the 'low-light' grown material.

As in the 'high-light' adapted leaves, $R_D$ was more sensitive to changes in the previous light level than to changes in the $CO_2$ concentration during the light period, this sensitivity being most marked at low $CO_2$ concentrations.

## Discussion

A number of workers (Ludwig et al., 1965; McCree & Troughton, 1966; McCree, 1970; Heichel, 1970; Penning de Vries, 1972) have observed that when the rate of photosynthesis of a single leaf is altered by changing the light level the subsequent rate of dark respiration changes concomitantly with the photosynthetic rate. McCree (1970) and Penning de Vries (1972) have both demonstrated a positive linear relationship between net photosynthesis and the subsequent dark respiration of the whole plant. The respiration rate appeared to have two components, one component being proportional to the preceding rate of photosynthesis and the other to the total dry weight of the plant (a maintenance component).

Heichel (1970) has reported that in maize leaves an effect of light on subsequent dark respiration can be demonstrated in $CO_2$-free air, and has suggested that respiration was quantitatively related to the prior illumination but independent of the rate of net $CO_2$ fixation.

The results presented here show that there is a positive linear correlation between the integrated rate of dark respiration and the integrated rate of net photosynthesis when photosynthesis is varied by changing the light level. A different positive linear correlation is shown when photosynthesis is altered by changing the $CO_2$ concentration during the light period. These results suggest that dark respiration is not simply related to the amount of $CO_2$ assimilated during photosynthesis and also raises doubts about the simple partitioning of respiration into growth and maintenance components in the present situation.

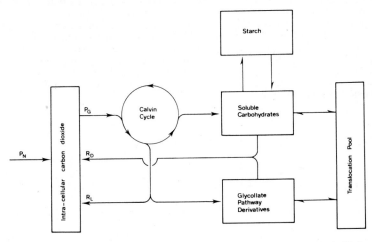

*Fig. 6.* Scheme showing carbon flow and relationships between photosynthesis ($P_G$), photorespiration ($R_L$), dark respiration ($R_D$) and translocation which forms the basis of a model simulating the responses observed in Fig. 5. (Charles-Edwards & Ludwig, to be published).

A simple model describing photosynthesis and photorespiration has been described elsewhere (Charles-Edwards & Ludwig, 1974), and it has been extended to examine some aspects of the carbon metabolism of the leaf (Charles-Edwards

& Ludwig, to be published). The scheme on which the model is based is shown in Fig. 6. Changes in the ambient $CO_2$ concentration and light level affect the net $CO_2$ exchange rate by changing both absolute and relative rates of photosynthesis and photorespiration. It follows that both the absolute amount of carbon dioxide assimilated and its partitioning between the products of the Calvin cycle and the glycollate pathway are affected. As a consequence of this asymmetry in partitioning with changing light and $CO_2$ levels the model predicts a light-$CO_2$ response for the $R_D/P_N$ relationship of the type shown in Fig. 5.

The data presented here, and also the model, are concerned with the response of the leaf to a constant light environment during the light period. Under natural light conditions a more complex relationship between $R_D$ and $P_N$ might be expected. Before these results can be extended with confidence to a field situation, further work is needed. In particular it is important to establish how these complex patterns of respiratory activity are related to both dry-matter production and the maintenance of leaf metabolism.

## REFERENCES

Charles-Edwards, D.A. & L.J. Ludwig (1974): A model for leaf photosynthesis by C3 plant species. *Ann. Bot.,* 38.

Heichel, G.H. (1970): Prior illumination and the respiration of maize leaves in the dark. *Pl. Physiol.,* 46: 359-362.

Ludwig, L.J., Saeki, T. & Evans, L.T. (1965): Photosynthesis in artificial communities of cotton plants in relation to leaf area. I. Experiments with progressive defoliation of mature plants. *Aust. J. Biol. Sci.,* 18: 1103-1118.

McCree, K.J. & J.H. Troughton (1966): Prediction of growth rate at different light levels from measured photosynthesis and respiration rates. *Pl. Physiol.,* 41: 559-566.

McCree, K.J. (1970): An equation for the rate of respiration of white clover plants grown under controlled conditions. In Setlik, I. (ed.): Prediction and measurement of photosynthetic productivity. Proc. IBP/PP Technical Meeting, Trebon, 14-21 September 1969 Pp. 221-229. Centre for Agricultural Publishing & Documentation, Wageningen.

Penning de Vries, F.W.T. (1972): Respiration and Growth. In Rees, A.R., Cockshull, K.E., Hand, D.W. & Hurd, R.G. (ed.): Crop processes in controlled Environments. Pp. 327-347 Academic Press, London and New York.

# THE BASIS OF EXPRESSION OF LEAF PHOTOSYNTHETIC ACTIVITIES

D.A. CHARLES-EDWARDS & L.J. LUDWIG

*Glasshouse Crops Research Institute, Littlehampton, Sussex, U.K.*

*Abstract*

Since the leaf is a heterogeneous assemblage of tissues, only a part of which are photosynthetically active, the anatomy and structure of the leaf may affect its photosynthetic performance. In order to examine this problem a simple analysis has been made of the role of leaf structure in determining the rates of transfer of carbon dioxide from the air to the assimilatory sites within the leaf.

The analysis is supported by published experimental data. An investigation of leaf photosynthesis by *Lycopersicon esculentum* grown at three light levels suggests that despite large differences in the thickness of leaves from the different growth environments the rate of transfer of carbon dioxide from the air to the assimilatory sites was in no case a limiting factor to photosynthesis.

The basis of expression of leaf photosynthetic data is discussed in relation to the physiological and mathematical interpretation of the response of leaf photosynthesis to the plant growth environment.

## Introduction

The pathway for movement of carbon dioxide from the air to the assimilatory sites within the leaf, and the numbers and distribution of those sites, are determined by the anatomy and structure of the leaf. Some of the variation in leaf photosynthetic activities can be attributed to differences in leaf structure (e.g. Charles-Edwards, Charles-Edwards & Sant, 1974), and when seeking to optimise the photosynthetic capabilities of a plant by either genetic or environmental manipulation of the leaf photosynthetic apparatus it is important to be able to identify the role played by the leaf structure in that apparatus.

The problem can be investigated at a number of levels. In the present communication we first consider the role of leaf structure in determining the rate of transfer of carbon dioxide from the air to the assimilatory sites, and compare some of the relationships predicted by our analysis with published data. Second, we examine in greater detail the response of leaf photosynthesis in tomato to different growth light environments to assess the quantitative effects of leaf structural changes on photosynthesis.

## A Simple Mathematical Analysis

A mathematical model for photosynthesis and photorespiration by leaves of

*Acknowledgments:* We wish to thank Mr. A.P. Gay for providing us with the data on stomatal frequencies, and Dr. J.H.M. Thornley for advice during the preparation of the manuscript.

$C_3$-plant species has recently been proposed (Charles-Edwards & Ludwig, 1974). The model relates the net rate of carbon dioxide exchange per unit leaf area to the ambient carbon dioxide and oxygen concentrations and the light flux density on the leaf surface. For the present purpose the model can be simplified, and the relation of the net exchange rate per unit leaf area, F, to ambient carbon dioxide concentrations and light flux densities can be described by three equations

$$F = h\phi_1(C_a - C_b), \tag{1}$$

$$F = h\phi_2(C_b - C_c), \tag{2}$$

and

$$F = \frac{h(s_1 I C_c - s_2 I)}{s_3 I + s_4 C_c} - R_d, \tag{3}$$

where $C_a$, $C_b$ and $C_c$ are the ambient, inter-cellular and intra-cellular concentrations of carbon dioxide respectively, I is the incident light flux density, $R_d$ is the dark respiration rate per unit leaf area, h is the leaf thickness, $\phi_1$ and $\phi_2$ are stomatal and mesophyll diffusion constants and $s_1...s_4$ are constants describing the biochemical apparatus. If it is assumed that the transfer of carbon dioxide from the air to the assimilatory sites is a passive diffusive process, the constants $\phi_1$ and $\phi_2$ can be written in the form

$$\phi = \frac{D'A}{V\Delta x}, \tag{4}$$

where D' is an effective diffusion coefficient, A the leaf area, V the leaf volume and $\Delta x$ the mean diffusion pathlength.

Movement of carbon dioxide from the air to the inter-cellular air spaces occurs via the stomata. The actual area across which the carbon dioxide is diffusing can be written as the product of the stomatal density (number per unit leaf area), n, the mean stomatal cross-sectional area, a, and the leaf area, so that Eqn.(4) becomes

$$\phi_1 = \frac{DAna}{V\Delta x} = \frac{Dna}{h\Delta x}, \tag{5}$$

where D is the actual diffusion coefficient, and substituting for $\phi_1$ in Eqn.(1),

$$F = \frac{Dna}{\Delta x}(C_a - C_b). \tag{6}$$

Eqn.(6) can be written as

$$F = \frac{(C_a - C_b)}{r_s}$$

where $r_s$ is the stomatal resistance, and is equal to $\frac{\Delta x}{Dna}$ .

Eqn.(6) relates the net carbon dioxide exchange rate per unit leaf area to both the stomatal density and the mean stomatal cross-sectional area.

Movement of carbon dioxide from the inter-cellular spaces to the intra-cellular assimilatory sites occurs across the mesophyll cell surface. If it is assumed that mesophyll cells are spherical (Charles-Edwards, Charles-Edwards & Sant, 1972) with a mean radius r and a density m (number of cells under unit leaf area), $\phi_2$ can be written approximately as

$$\phi_2 = \frac{DAm(4\pi r^2)}{V\Delta x} , \tag{7}$$

and assuming also that the mean diffusion pathlength, $\Delta x$, depends on the volume to surface area ratio (radius) of the mesophyll cell (i.e. $\Delta x = br$, where b is a constant)

$$\phi_2 = \frac{4D\pi mr}{bh} ; \tag{8}$$

substituting for $\phi_2$ in Eqn.(2),

$$F = \frac{4D\pi mr}{b}(C_b - C_c) . \tag{9}$$

Eqn.(9) relates the net exchange rate per unit leaf area to both the mesophyll cell density and the mean cell radius, the former relationship having been observed in *Lolium* (Wilson & Cooper, 1969a).

Leaf thickness is often found to be correlated with mean mesophyll cell size (e.g. Charles-Edwards, Charles-Edwards & Sant, 1974; Pieters, 1974), that is h $\alpha$ r. If it is assumed that h = gr, where g is a constant, Eqn.(9) can be re-written as

$$F = \frac{4D\pi mh}{bg}(C_b - C_c) , \tag{10}$$

which relates the net exchange rate per unit leaf area to the product of the leaf thickness and mesophyll cell density. This relationship has been reported for *Lolium* (Wilson & Cooper, 1967, 1969a) where the exchange rate was assayed at high carbon dioxide concentrations and light saturation, a situation where ($C_b - C_c$) might be expected to be similar for a range of leaf material.

The net carbon dioxide exchange rate per unit volume of leaf mesophyll tissue,

F*, can be written as

$$F* = \frac{4D\pi r^2(C_b - C_c)}{br(4/3\pi r^3)} = \frac{3D(C_b - C_c)}{br^2},$$ (11)

relating the net exchange rate per unit leaf mesophyll tissue volume inversely to the mean mesophyll cell cross-sectional area. This relationship has also been observed in *Lolium* (Wilson & Cooper, 1969b).

## Leaf Photosynthesis in Tomato

Tomato plants (cv. Minibellella) were grown in controlled environment cabinets at $20°C$ at three light levels (20, 50 and 80 $Wm^{-2}$ visible radiation) during a 16 h light period. Light response curves for net carbon dioxide exchange per unit leaf area were obtained for leaves of plants from each of the growth environments. The response curves were obtained at $20°C$ and at an ambient carbon dioxide concentration of 600 $mg(CO_2)m^{-3}$ using an infra-red gas analyzer. Transpiration rates were measured simultaneously. Stomatal conductances ($h\phi_1$) were calculated from the transpiration rates, enabling the inter-cellular carbon dioxide concentration to be calculated at each of the light levels used in obtaining the light response curves. With the experimentally determined inter-cellular carbon dioxide concentration and light flux density as the two independent variables, net carbon dioxide exchange rates were calculated from the model using $Eq^{ns.}(2)$ and (3), the parameters of the model being adjusted to obtain best fit with the experimentally determined exchange rates. The fitting was done by computer using a surface fitting routine developed at this Institute by Dr. J.H.M. Thornley.

Both the experimental values and the curves fitted using the model are shown in Fig. 1. A more detailed report of the experiment and fitting procedures will be given elsewhere. We are concerned here only with the response of the stomatal and mesophyll conductances to the different growth environments.

The stomatal conductances (calculated as means over the range of light levels used in the light response assays), the mesophyll conductances, and the rates of light-saturated photosynthesis per unit leaf area are shown in Table 1. All three measurements increased with the increasing growth light levels, and it might be concluded that the rate of light-saturated photosynthesis per unit leaf area was determined by the conductances to carbon dioxide transfer. However, the diffusion constants $\phi_1$ and $\phi_2$, obtained by dividing the conductances by the estimated leaf thickness, do not alter appreciably between the growth environments: thus, the time for transfer of unit amount of carbon dioxide from the air to the assimilatory sites is roughly the same for leaves from the three environments. If expressed on a leaf volume basis the rate of light-saturated photosynthesis also remains the same in all growth environments, and the data support the view that the apparent enhancement of leaf photosynthetic activity on a leaf area basis is the result of the increased size of the biochemical apparatus under unit leaf area in material grown at the higher light levels. The estimated leaf thickness, diffusion constants and rates of light-saturated photosynthesis per unit leaf volume are

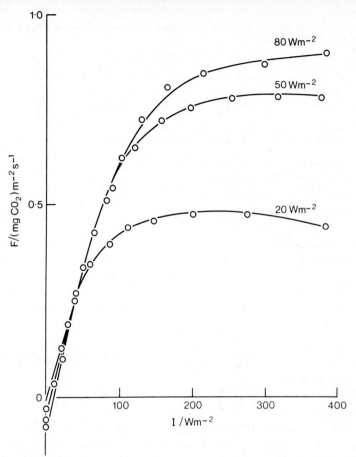

*Fig. 1.* Experimental values (o) and fitted curves showing the response of the net $CO_2$ exchange rate per unit leaf area to changing light flux densities for leaves from three growth light environments.

shown in Table 1.

If the stomatal diffusion constant, $\phi_1$, is independent of leaf thickness we can write from Eq$^n$.(5) that

$$h\alpha\frac{na}{\Delta x}. \tag{12}$$

The mean diffusion pathlength, $\Delta x$, might be expected to increase with the leaf thickness, but for the material studied it was observed that there was an unsymmetrical increase in upper and lower surface stomatal densities with increasing leaf thickness, a result that might be expected to keep the mean diffusion pathlength effectively constant. At the same time the total stomatal density increased with increasing leaf thickness, an observation consistent with Eq$^n$.(12) if a is constant.

*Table 1.* Some parameters of leaf photosynthesis and structure of tomato leaves from plants grown at three light levels.

| Grown light level | 20 | 50 | 80 | Wm$^{-2}$ |
|---|---|---|---|---|
| stomatal conductance: $h\phi_1$ | $3.13 \times 10^{-3}$ | $5.21 \times 10^{-3}$ | $5.68 \times 10^{-3}$ | ms$^{-1}$ |
| mesophyll conductance: $h\phi_2$ | $1.28 \times 10^{-3}$ | $2.26 \times 10^{-3}$ | $2.72 \times 10^{-3}$ | ms$^{-1}$. |
| light-saturated photosynthesis per unit leaf area: F | 0.475 | 0.785 | 0.890 | (mgCO$_2$)m$^{-2}$s$^{-1}$ |
| estimated leaf thickness: h | $1.52 \times 10^{-4}$ | $1.97 \times 10^{-4}$ | $2.78 \times 10^{-4}$ | m |
| stomatal diffusion constant: $\phi_1$ | 20.6 | 26.4 | 20.4 | s$^{-1}$ |
| mesophyll diffusion constant: $\phi_2$ | 8.4 | 11.5 | 9.8 | s$^{-1}$ |
| light-saturated photosynthesis per unit leaf volume: F/h | $3.13 \times 10^3$ | $3.98 \times 10^3$ | $3.20 \times 10^3$ | (mgCO$_2$)m$^{-3}$s$^{-1}$ |
| stomatal density (upper surface: $n_u$ | $42 \times 10^6$ | $117 \times 10^6$ | $152 \times 10^6$ | m$^{-2}$ |
| stomatal density (lower surface): $n_l$ | $173 \times 10^6$ | $262 \times 10^6$ | $269 \times 10^6$ | m$^{-2}$ |

If the mesophyll diffusion constant, $\phi_2$, is independent of leaf thickness Eq$^n$.(10) requires that the number of mesophyll cells under unit leaf area should decrease with increasing leaf thickness. This is consistent if the increase in leaf thickness at the higher growth light levels is the result of mesophyll cell expansion rather than an increase in the number of layers of cells in the leaf.

## Discussion

Our analysis of the role of leaf structure in determining leaf photosynthetic activities is very simple, but it does provide an initial basis for the interpretation of published data.

The consideration of leaf photosynthesis by tomato plants grown in different light environments illustrates the difficulties which can arise in interpretation if photosynthetic activity is expressed on a leaf area basis alone. Our results indicate that leaves can adapt during growth to minimise the effects of changes in leaf thickness on the rate of carbon dioxide transfer. The correlation that exists between the rate of light-saturated photosynthesis per unit leaf area and the conductances does not necessarily mean that the conductances are themselves the limiting factors in leaf photosynthesis, and in the present case it can be seen that the increase in the size of the assimilatory apparatus is the determining factor. A similar interpretation has been made of the apparent stimulation of photosynthesis per unit leaf area at low growth temperature in *Lolium* (Charles-Edwards, Charles-Edwards & Sant, 1974).

The structure of the leaf should be examined simultaneously with its photosynthetic activity when investigating environmental or genetic variation in leaf photosynthetic activity. By so doing, it becomes possible to discriminate between the contributions of gaseous diffusion and chemical fixation to photosynthesis, and to identify the primary cause of any variation. Consideration of photosynthetic rates expressed on a leaf area basis alone can be misleading.

## Symbols

| | | |
|---|---|---|
| F | net $CO_2$ exchange rate per unit leaf area | $(mgCO_2)m^{-2} s^{-1}$ |
| Rd | dark respiration rate per unit leaf area | $(mgCO_2)m^{-2} s^{-1}$ |
| C | carbon dioxide concentration | $(mgCO_2)m^{-3}$ |

subscripts a, b and c denote ambient, inter-cellular and intra-cellular concentrations respectively

| | | |
|---|---|---|
| I | light flux density | $Wm^{-2}$ or $Jm^{-2} s^{-1}$ |
| $\phi$ | diffusion constant | $s^{-1}$ |

subscripts 1 and 2 denote stomatal and mesophyll diffusion constants respectively

| | | |
|---|---|---|
| $s_1$ | constant | $m^2 W^{-1} s^{-1}$ |
| $s_2$ | constant | $(mgCO_2) m^{-1} W^{-1} s^{-1}$ |
| $s_3$ | constant | $m^2 W$ or $sm^2 J^{-1}$ |
| $s_4$ | constant | $m^3 (mgCO_2)^{-1}$ |
| A | leaf area | $m^2$ |
| V | leaf volume | $m^3$ |
| h | leaf thickness | $m$ |
| D | diffusion coefficient | $m^2 s^{-1}$ |
| a | mean stomatal cross-sectional area | $m^2$ |
| n | stomatal density | $m^{-2}$ |
| r | mean mesophyll cell radius | $m$ |
| m | mesophyll cell density | $m^{-2}$ |
| $\Delta x$ | diffusion pathlength | $m$ |

## REFERENCES

Charles-Edwards, D.A., J. Charles-Edwards & F.I. Sant (1972): Models for mesophyll cell arrangement in leaves of ryegrass (*Lolium perenne* L.). *Planta (Berl.)*, 104: 297-305.

Charles-Edwards, D.A., J. Charles-Edwards & F.I. Sant (1974): Leaf photosynthetic activity in six temperate grass varieties grown in contrasting light and temperature environments. *J. exp. bot.*, 25: 86.

Charles-Edwards, D.A., & L.J. Ludwig (1974): A model for leaf photosynthesis by $C_3$ plant species. *Ann. Bot.*, 38.

Pieters, G.A. (1974): The growth of sun and shade leaves of *Populus euramericana* 'robusta' in relation to age, light intensity and temperature. *Meded. Landb. Hoogesch. Wageningen*, 71, 11.

Wilson, D., & J.P. Cooper (1967): Assimilation of *Lolium* in relation to leaf mesophyll. *Nature*, 214: 989-992.

Wilson, D. & J.P. Cooper (1969a): Effects of light intensity and $CO_2$ on apparent photosynthesis and its relation with leaf anatomy in genotypes of *Lolium perenne* L. *New Phytol.*, 68: 627-644.

Wilson, D. & J.P. Cooper (1969b): Effect of temperature during growth on leaf anatomy and subsequent light-saturated photosynthesis among contrasting *Lolium* populations. *New Phytol.*, 68: 1115-1123.

# CONTROL OF PHOTORESPIRATION IN SOYBEAN AND MAIZE

WILLIAM L. OGREN

*United States Regional Soybean Laboratory, Agricultural Research Service, United States Department of Agriculture, Urbana, Illinois 61801, U.S.A.*

## Abstract

The relative rates of photosynthesis and photorespiration in soybean are determined by the kinetic properties of ribulose 1,5-diphosphate (RuDP) carboxylase. Carbon dioxide and oxygen compete for RuDP at the carboxylase. High $CO_2$ and low $O_2$ favor carboxylation of RuDP to 3-phosphoglycerate, thereby promoting photosynthesis. Low $CO_2$ and high $O_2$ concentrations favor oxygenation of RuDP to P-glycolate, thereby promoting photorespiration. In maize, the phosphopyruvate carboxylase – malic enzyme couple acts to increase the $CO_2$ concentration in the bundle sheath cells, site of RuDP carboxylase in $C_4$ species. Maize RuDP carboxylase is also sensitive to $O_2$, so this increased $CO_2$ concentration in the bundle sheath stimulates photosynthesis and reduces photorespiration. Since photorespiration and an associated oxygen inhibition of photosynthesis markedly inhibit net photosynthesis in soybean under standard atmospheric conditions of 300 ppm $CO_2$ and 21% $O_2$, and since no required role for photorespiration has been demonstrated, attempts are being made to reduce or eliminate photorespiration in the soybean and other $C_3$ crop species.

Atmospheric oxygen is a potent inhibitor of photosynthetic $CO_2$ uptake in leaves of soybean [*Glycine max* (L.) Merr.] and other $C_3$ species, but has little effect on photosynthesis in leaves of maize (*Zea mays* L.) and other $C_4$ species (Forrester, Krotov & Nelson, 1966a, 1966b; Downes & Hesketh, 1968). Since all major crop species except maize, sorghum [*Sorghum bicolor* (L.) Moench], and sugar cane (*Saccharum officinarum* L.) are $C_3$ plants, the oxygen sensitivity of $C_3$ photosynthesis represents a substantial loss of potential agricultural productivity. An understanding of the regulatory processes of photorespiration and of an associated direct oxygen inhibition of photosynthesis may help to determine whether the $O_2$ sensitivity of $C_3$ photosynthesis can be reduced or eliminated, and whether a reduction in $O_2$ sensitivity will lead to increased dry matter production and agronomic yield in these crop species.

*Abbreviations:* RuDP, ribulose 1,5-diphosphate; PEP, phospho(enol)pyruvate; 3-PGA, 3-phosphoglycerate; $C_3$, Calvin cycle photosynthesis; $C_4$, $C_4$-dicarboxylic acid photosynthesis; APS, apparent (net) photosynthesis; $CE_O$ carboxylation efficiency in the absence of oxygen; exp, base of natural logarithms (*l*); C, carbon dioxide concentration; O, oxygen concentration; v, enzyme velocity; V, maximal enzyme velocity; S, substrate concentration; I, inhibitor concentration; Km, Michaelis constant; Ki, inhibition constant; $v_c$, velocity of RuDP carboxylase reaction; $v_o$, velocity of RuDP oxygenase reaction; $V_c$, maximal velocity of RuDP carboxylase reaction; $V_o$, maximal velocity of RuDP oxygenase reaction; $K_c$, Michaelis (or inhibition) constant for $CO_2$; $K_O$ Michaelis (or inhibition) constant for $O_2$; k, constant relating $CO_2$ compensation concentration to $O_2$ concentration; t, proportion of P-glycolate carbon ultimately released as $CO_2$; $\alpha$-HPMS, $\alpha$-hydroxy-2-pyridinemethanesulfonate.

## Soybean

Oxygen sensitivity of soybean photosynthesis was first demonstrated by Forrester et al. (1966a). From the data provided by Forrester et al. (1966a), Ogren & Bowes (1971) developed an equation which mathematically characterized the nature of the $O_2$ inhibition (Equation 1),

$$APS = CE_o \exp(-O/K_o)C - CE_o \exp(-O/K_o)kO \qquad (1)$$
$$= (\text{true photosynthesis}) - (\text{photorespiration})$$

where APS is apparent (net) photosynthesis, $CE_o$ is the carboxylation efficiency of the leaf (as defined by Forrester et al., 1966a) in the absence of $O_2$, exp is the base of natural logarithms ($e$), C is the $CO_2$ concentration, O is the $O_2$ concentration, $K_o$ is the inhibition constant for $O_2$ on carboxylation efficiency, and k is a proportionality constant. Experimentally, carboxylation efficiency is determined as the slope of the $CO_2$ response curve of photosynthesis, $K_o$ is determined from the effect of $O_2$ on the $CO_2$ response curve of photosynthesis, and k is determined as the slope of a plot of $CO_2$ compensation concentration versus $O_2$ concentration. Equation 1 is valid only in the linear portion of the $CO_2$ response curve of photosynthesis.

Equation 1 indicates that carboxylation efficiency, $CE_o$, regulates both the rate of photosynthesis and the rate of photorespiration, and that carboxylation is inhibited by $O_2$ in a definite and regular manner. Ogren & Bowes (1971) and Bowes & Ogren (1972) observed that RuDP carboxylase, the first enzymatic step in $C_3$ photosynthetic $CO_2$ fixation, is competitively inhibited by $O_2$ with respect to $CO_2$. Bowes, Ogren & Hageman (1971) further observed that RuDP carboxylase catalyzes the oxidation of RuDP by $O_2$ to yield P-glycolate as a product. P-glycolate is a presumed intermediate in photorespiration (Tolbert, 1973). The occurrence of RuDP oxygenase activity was subsequently confirmed by Lorimer, Andrews & Tolbert (1973) and by Takabe & Akazawa (1973). Thus RuDP carboxylase exhibits the characteristics of carboxylation efficiency as deduced from Equation 1. RuDP carboxylase is inhibited in a definite and regular manner by $O_2$, namely, $O_2$ is a competitive inhibitor of the enzyme with respect to $CO_2$. Further, since RuDP carboxylase catalyzes the first step in photosynthesis, it may regulate photosynthesis, and since this enzyme catalyzes the first step in photorespiration, P-glycolate synthesis, it may regulate photorespiration.

More recently, Laing, Ogren & Hageman (1974) observed that $CO_2$ is a competitive inhibitor of oxygenase activity with respect to $O_2$, and that Equation 1 can be approximated by an equation based on the enzyme kinetics of the carboxylase and oxygenase activities of RuDP carboxylase. The velocity, v, of an enzyme reaction in the presence of a competitive inhibitor is given by Equation 2,

$$v = VS/[S + Km(1 + I/Ki)] \qquad (2)$$

which expands to,

$$v = VSKi/(KmKi + KiS + KmI) \qquad (3)$$

where S is substrate concentration, I is inhibitor concentration, V is maximal velocity, Km is the Michaelis constant, and Ki is the inhibition constant. Since $CO_2$ and $O_2$ compete with each other for the available RuDP through RuDP carboxylase, and since $Km(CO_2)$ in the carboxylase reaction equals $Ki(CO_2)$ in the oxygenase reaction and $Km(O_2)$ in the oxygenase reaction equals $Ki(O_2)$ in the carboxylase reaction (Laing et al., 1974), the velocity of the carboxylase and oxygenase reactions in the presence of both $CO_2$ and $O_2$ can be expressed,

$$v_c = V_c C K_o / (K_c K_o + K_c O + K_o C) \tag{4}$$

and

$$v_o = V_o O K_c / (K_c K_o + K_c O + K_o C) \tag{5}$$

where $v_c$ and $v_o$ are the velocities of the carboxylase and oxygenase reactions, respectively, $V_c$ is the maximal velocity of the carboxylase reaction, $V_o$ is the maximal velocity of the oxygenase reaction, $K_c$ is the Michaelis (or inhibition) constant for $CO_2$, $K_o$ is the Michaelis (or inhibition) constant for $O_2$, C is the $CO_2$ concentration, and O is the $O_2$ concentration.

If apparent (net) photosynthesis is considered to be equal to the difference between true photosynthesis and photorespiration, then,

$$APS = v_c - t v_o \tag{6}$$

where t is the fraction of P-glycolate carbon ultimately released as $CO_2$. According to the photorespiratory scheme of Tolbert (1973), $t = 0.25$. If Equations 4 and 5 are considered in the range of $CO_2$ concentrations where $v_c$ is proportional to C, $K_o C$ is negligible and can be dropped from the equations. Substituting Equations 4 and 5 into Equation 6, followed by expanding and rearranging, gives Equation 7,

$$APS = (V_c/K_c)[K_o/(K_o + O)]C \\ - (V_c/K_c)[K_o/(K_o + O)](t V_o K_c / V_c K_o)O \tag{7}$$

Comparing Equations 1 and 7, if $CE_o$ is defined as $V_c/K_c$, if k is defined as $t V_o K_c / V_c K_o$, and since $K_o/(K_o + O)$ is an approximation of $\exp(-O/K_o)$, it is evident that Equation 7, derived from the kinetics of soybean RuDP carboxylase, is identical in form to Equation 1, developed from $CO_2$ exchange characteristics of soybean leaves. The agreement between Equations 1 and 7 supports the concept set forth by Ogren & Bowes (1971) that photorespiration and $O_2$ sensitivity of photosynthesis in soybean leaves is regulated by the enzyme RuDP carboxylase.

From a study of the kinetics of soybean leaf photosynthesis and of purified soybean RuDP carboxylase, and from information in the literature, Laing et al. (1974) concluded that the ratio of the two activities of RuDP carboxylase in vivo at standard atmospheric conditions (300 ppm $CO_2$ 21% $O_2$) at $25°$ C was one molecule of $O_2$ fixed for every four molecules of $CO_2$ fixed (Fig. 1). Fig. 1 indicates that the rate of photorespiration under the conditions stated above is

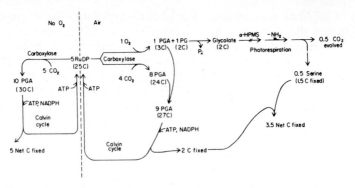

*Fig. 1.* Suggested stoichiometry of photosynthesis and photorespiration in soybean leaves at 300 ppm $CO_2$, 21% $O_2$ and 25°C. The stoichiometry will be altered in favor of photorespiration and $O_2$ inhibition of photosynthesis by higher temperatures, lower $CO_2$ concentrations, and higher $O_2$ concentrations. (From Laing et al., 1974).

about one-sixth the rate of net photosynthesis. This rate is consistent with rates of photorespiration as determined by 1) measuring the rate of $CO_2$ evolution into $CO_2$-free air, 2) measuring differential $^{12}CO_2$ and $^{14}CO_2$ uptake, and 3) extrapolating the $CO_2$ response curve for photosynthesis to zero $CO_2$ (Forrester et al., 1966a; Ludwig & Canvin, 1971; Canvin & Fock, 1972). Fig. 1 suggests that, in the absence of $O_2$, the rate of photosynthesis will increase by about 43% if the $CO_2$ concentration is held constant (5 net C fixed in low $O_2$/3.5 net C fixed in air). An increase in photosynthesis of this magnitude has been observed with soybean (Forrester et al., 1966a) and with several other $C_3$ species (Downes & Hesketh, 1968). Also, the scheme indicates that in presence of an inhibitor of glycolate oxidase, α-HPMS, 50% of the radiocarbon incorporated will accumulate as glycolate, as was observed by Zelitch (1959). Finally, Fig. 1 indicates that photorespiratory $CO_2$ evolution accounts for one-third of the $O_2$ inhibition of photosynthesis in air, while direct inhibition accounts for the remaining two-thirds of the total $O_2$ inhibition. This conclusion is consistent with direct measurement, by differential uptake of $^{12}CO_2$ and $^{14}CO_2$, of the magnitude of the two $O_2$ effects (Ludwig & Canvin, 1971).

It should be noted that the stoichiometry in Fig. 1 is suggested to occur at 300 ppm $CO_2$, 21% $O_2$, and 25°C. At other $CO_2$ or $O_2$ concentrations, or at other temperatures, the ratio of activities will change in a manner dictated by the environmental conditions and by the relevant kinetic constants of RuDP carboxylase. For example, at 75 ppm $CO_2$, 21% $O_2$, and 25°C, the $CO_2$ concentration is one-fourth that assumed in Fig. 1. Therefore the rate of $CO_2$ fixation will be one-fourth as great at 75 ppm $CO_2$ as at 300 ppm $CO_2$, while the rate of $O_2$ fixation will be essentially unchanged. Thus the ratio of $CO_2/O_2$ fixed will be approximately one, and net photosynthesis will be about one-sixth the rate at 300 ppm $CO_2$, as was observed by Forrester et al. (1966a). The $CO_2$ compensation concentration will be reached when the rate of $CO_2$ uptake equals the rate of $CO_2$ release. This will occur when the ratio of $CO_2/O_2$ fixed is 0.5. If the rate of photosynthesis is a linear function of $CO_2$ concentration between the compensation concentration and 300 ppm $CO_2$, the compensation concentration should be equal to one-eighth of 300 ppm $CO_2$ (corresponding to a reduction in the ratio of $CO_2/O_2$ fixed from 4 at 300 ppm to a ratio of 0.5 at the compensation concentration), or about 38 ppm $CO_2$. This is the approximate $CO_2$ compensation concentration observed with soybean leaves at $25°C$ (Forrester et al., 1966a; Curtis, Ogren & Hageman, 1969).

## Maize

Photorespiration in maize, a $C_4$ plant, is absent or greatly reduced, as indicated by an absence of inhibition of photosynthesis by 21% $O_2$ (Forrester et al., 1966b), a compensation concentration of approximately zero ppm $CO_2$ (Moss, Willmer & Crookston, 1971), and lack of $CO_2$ evolution into $CO_2$-free air (Laing & Forde, 1971). According to the scheme of maize photosynthesis as outlined by Hatch (1971a),$CO_2$ first reacts via PEP carboxylase in mesophyll cells to yield oxalacetate, which is subsequently reduced to malate by $NADP^+$-malate dehydrogenase and transported to the bundle sheath cells. In the bundle sheath, malate is decarboxylated by malic enzyme to produce $CO_2$ and pyruvate. Pyruvate is transported back to the mesophyll, while $CO_2$ serves as a substrate for RuDP carboxylase in the bundle sheath. Following fixation of $CO_2$ into 3-PGA, the carbon then proceeds through the Calvin cycle reactions in the bundle sheath. The distribution of enzymes in maize as suggested by Hatch (1971a) has been disputed by Bucke & Long (1972), who concluded that all enzymes of maize photosynthesis are localized in the mesophyll cells.

Chollet & Ogren (1972a, 1972b, 1973) isolated strands of maize bundle sheath cells and determined that these preparations contained most of the leaf RuDP carboxylase and malic enzyme, but very little PEP carboxylase or $NADP^+$-malate dehydrogenase. The isolated bundle sheath cells were photosynthetically active, and the pathway of photosynthesis appeared to be similar to that found in $C_3$ species such as soybean. Photosynthesis was inhibited by 21% and 100% $O_2$, $O_2$ stimulated glycolate synthesis, $O_2$ inhibition of photosynthesis and stimulation of glycolate synthesis were both reversed by high $CO_2$ concentrations, and the initial products of photosynthetic $CO_2$ fixation were sugar phosphates, not dicarboxylic acids. These results support the distribution of photosynthesis enzymes in maize as proposed by Hatch (1971a).

PEP carboxylase activity is considerably greater than RuDP carboxylase activity in maize leaf extracts (Bowes & Ogren, 1972; Chollet & Ogren, 1972a, 1972c), so the PEP carboxylase — malic enzyme couple may act as a $CO_2$ pump, increasing the $CO_2$ concentration in the bundle sheath above ambient. Hatch (1971b) has presented evidence to support the notion of an increased $CO_2$ concentration in the bundle sheath in the light. Bowes & Ogren (1972) and Chollet & Ogren (1972b) suggested that an increased $CO_2$ concentration in the bundle sheath could reduce photorespiration and relieve $O_2$ inhibition of photosynthesis by virtue of the competition between $CO_2$ and $O_2$ for RuDP, a competition mediated by RuDP carboxylase. The $O_2$ sensitivity of maize RuDP carboxylase is similar to the $O_2$ sensitivity of soybean enzyme, since $O_2$ inhibits maize RuDP carboxylase (Bowes & Ogren, 1972) and substitutes for $CO_2$ to yield P-glycolate (Bowes et al., 1971). An increased $CO_2$ concentration in the bundle sheath will

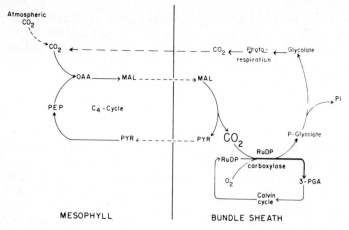

*Fig. 2.* Suggested mechanism of inhibition of photorespiration in maize. The PEP carboxylase – malic enzyme couple acts to concentrate $CO_2$ in the bundle sheath, increasing the $CO_2/O_2$ ratio over ambient and thereby reducing P-glycolate synthesis and photorespiration. Any photorespiratory $CO_2$ released in the bundle sheath cells may be recaptured by PEP carboxylase in the mesophyll cells before it can escape the leaf. (From Bowes & Ogren, 1972; Chollet & Ogren, 1972b).

permit $CO_2$ to compete more effectively with $O_2$ for the active site(s) of RuDP carboxylase, thereby stimulating photosynthesis and reducing P-glycolate synthesis and photorespiration (Fig. 2). If any photorespiratory activity does occur, and it is likely that some photorespiratory $CO_2$ is produced, the $CO_2$ released by this process will be recaptured by PEP carboxylase in the mesophyll before it can escape from the leaf. Thus the specialized leaf anatomy and enzyme complement of maize, permitting an increase in the $CO_2$ concentration at the site of RuDP carboxylase and an ability to recapture photorespiratory $CO_2$, is responsible for the reduced, and not readily detectable, photorespiratory activity in this species.

## Agricultural interest

No obligatory function has yet been determined for photorespiration, and it may be agronomically advantageous to eliminate this process from $C_3$ crop species. Of the major crop plants, only maize, sugar cane, and sorghum have overcome the problem of photorespiration. The $C_4$ solution is not applicable to the present $C_3$ plants, for a new anatomy and new enzymes would have to be introduced into these plants. Successful conversion of a $C_3$ species into a $C_4$ species has not been achieved even where the appropriate genetic material is available (Björkman & Berry, 1973). In the case of $C_3$ crop plants, $C_4$ relatives sufficiently similar to permit cross breeding have not even been found. Thus if photorespiration and $O_2$ sensitivity of photosynthesis in $C_3$ plants is to be controlled and reduced, a novel solution to the problem will have to be found.

As discussed earlier, the relative rates of photosynthesis and photorespiration are determined by the kinetic properties of RuDP carboxylase. Thus to decrease photorespiration, it will be necessary either to increase the affinity or maximal velocity of RuDP carboxylase with respect to $CO_2$ or to reduce the affinity or maximal velocity of the enzyme with respect to $O_2$. For example, if the affinity or maximal velocity of RuDP carboxylase with respect to $CO_2$ could be increased so that nine $CO_2$ molecules are fixed to every one $O_2$ molecule, then, from Fig. 1, net photosynthesis would increase by about 20% (from 3.5 net C fixed to 4.25 net C fixed). Such a change in the kinetic properties of RuDP carboxylase might be achieved by mutagenic or chemical alteration of the enzyme.

One manifestation of an increased sensitivity of the enzyme toward $CO_2$ or a reduced sensitivity toward $O_2$ would be a reduced $CO_2$ compensation concentration. Using the example in the previous paragraph, a ratio of nine $CO_2$ per one $O_2$ fixed would give a compensation concentration of approximately 17 ppm $CO_2$, compared to about 38 ppm $CO_2$ observed in normal soybean leaves. Widholm & Ogren (1969) described a system to screen for photorespiratory deficiency based on the reduction in $CO_2$ compensation concentration which would accompany reduced photorespiration. Widholm & Ogren (1969) observed that $C_3$ plants, when maintained at a $CO_2$ concentration below compensation, experienced a loss in chlorophyll, protein, soluble sugars, and RNA, and that after a week or so under these conditions, the plants senesced. The rate of senescence was stimulated by factors which stimulate photorespiration, namely high $O_2$ concentration, high temperature, and high light intensity. It was concluded that the senescence observed was the result of photorespiratory-induced loss of energy reserves.

This screening system, which can be used to test for both photorespiratory-deficient mutants and for chemical inhibitors of photorespiration, has not located either a desired mutant or chemical. Widholm & Ogren (unpublished) have located no photorespiration mutants among 200,000 M2 generation soybean seed, nor did they find any photorespiration deficiency in more than 6,000 cultivars of oat (*Avena sativa* L.). Cannell, Brun & Moss (1969) reported no differences in 2,458 soybean cultivars. No chemicals which reduce the $CO_2$ compensation concentration have been reported. Nonetheless, since net photosynthesis in soybean and other $C_3$ crop species can be increased by about 45% in the absence of $O_2$, and since increases in photosynthetic productivity can be translated into seed yield increases (Hardy & Havelka, 1974), the search for mutant or chemical reduction control of photorespiration must continue until mutants or chemicals are found or until an obligatory function for photorespiration is determined.

# REFERENCES

Björkman, O. & J. Berry (1973): High-efficiency photosynthesis. *Sci. Amer.*, 229: 80-93.

Bowes, G. & W.L. Ogren (1972): Oxygen inhibition and other properties of soybean ribulose 1,5-diphosphate carboxylase. *J. biol. Chem.*, 247: 2171-2176.

Bowes, G., W.L. Ogren & R.H. Hageman (1971): Phosphoglycolate production catalyzed by ribulose diphosphate carboxylase. *Biochem. Biophys. Res. Commun.*, 45: 716-722.

Bucke, C. & S.P. Long (1972): Location of enzyme assimilating and shuttling carbon dioxide in sugarcane and maize leaf tissue. In Forti, G., Avron, M. & Melandri, A. (eds.) Proc. IInd Int. Cong. Photosynthesis Res., Vol. 3, pp. 1935-1941. Junk, The Hague.

Cannell, R.Q., W.A. Brun & D.N. Moss, (1969): A search for high net photosynthetic rate among soybean genotypes. *Crop Sci.*, 9: 840-841.

Canvin, D.T. & Fock, H. (1972): Measurement of photorespiration. *Methods in Enzymol.*, 24: 246-260.

Chollet, R. & Ogren, W.L. (1972a): Oxygen inhibits maize bundle sheath photosynthesis. *Biochem. Biophys. Res. Commun.*, 46: 2062-2066.

Chollet, R. & W.L. Ogren (1972b): The Warbug effect in maize bundle sheath photosynthesis. *Biochem. Biophys. Res. Commun.*, 48: 684-688.

Chollet, R. & Ogren, W.L. (1972c): Greening in a virescent mutant of maize. II. Enzyme studies. *Z. Pflanzenphysiol.*, 68: 45-54.

Chollet, R. & Ogren, W.L. (1973): Photosynthetic carbon metabolism in isolated maize bundle sheath strands. *Plant Physiol.*, 51: 787-792.

Curtis, P.E., W.L. Ogren & R.H. Hageman (1969): Varietal effects in soybean photosynthesis and photorespiration. *Crop Sci.*, 9: 323-327.

Downes, R.W. & J.D. Hesketh (1968): Enhanced photosynthesis at low $O_2$ concentrations: differential response of temperate and tropical grasses. *Planta (Berl.)*, 78: 79-84.

Forrester, M.L. G. Krotkov & C.D. Nelson (1966a): Effect of oxygen on photosynthesis, photorespiration, and respiration in detached leaves. I. Soybean. *Plant Physiol.*, 41, 422-427.

Forrester, M.L., G. Krotkov & C.D. Nelson (1966b): Effect of oxygen on photosynthesis, photorespiration, and respiration in detached leaves. II. Corn and other monocotyledons. *Plant Physiol.*, 41: 428-431.

Hardy, R.W.F. & U.D. Havelka (1974): The nitrogen barrier. *Crops & Soils Mag.*, February: 11-13.

Hatch, M.D. (1971a): Mechanism and function of the $C_4$-pathway of photosynthesis. In Hatch, M.D., Osmond, C.B. & Slatyer, R.O. (eds.): Photosynthesis and photorespiration, pp. 139-152. Wiley-Interscience, New York.

Hatch, M.D. (1971b): The $C_4$-pathway of photosynthesis. Evidence for an intermediate pool of $CO_2$ and the identity of the donor $C_4$dicarboxylic acid. *Biochem. J.*, 125: 425-432.

Laing, W.A. & B.J. Forde (1971): Comparative photorespiration in Amaranthus, soybean and corn, *Planta (Berl.)*, 98: 221-231.

Laing, W.A. W.L. Ogren & R.H. Hageman (1974): Regulation of soybean net photosynthetic $CO_2$ fixation by the interaction of $CO_2$, $O_2$, and ribulose 1,5-diphosphate carboxylase. *Plant Physiol.*, 54: 678-685.

Lorimer, G.H., T.J. Andrews & N.E. Tolbert (1973): Ribulose diphosphate oxygenase. II. Further proof of reaction products and mechanism of action. *Biochemistry* 12: 18-23.

Ludwig, L.J. & D.T. Canvin (1971): The rate of photorespiration during photosynthesis and the relationship of the substrate of light respiration to the products of photosynthesis in sunflower leaves. *Plant Physiol.*, 48: 712-719.

Moss, D.N., C.M. Willmer & R.K. Crookston (1971): $CO_2$ compensation concentration in maize (*Zea mays* L.) genotypes. *Plant Physiol.*, 47: 847-848.

Ogren, W.L. & G. Bowes (1971): Ribulose diphosphate carboxylase regulates soybean photorespiration. *Nature New Biol.*, 230: 159-160.

Takabe, T. & T. Akazawa (1973): Oxidative formation of phosphoglycolate from ribulose-1,5-diphosphate carboxylase catalyzed by *Chromatium* ribulose-1,5-diphosphate carboxylase. *Biochem. Biophys. Res. Commun.*, 53: 1173-1179.

Tolbert, N.E. (1973): Compartmentation and control in microbodies. *Symp. Soc. Exp. Biol.*, 27: 215-239.

Widholm, J.M. & W.L. Ogren (1969): Photorespiratory-induced senescence of plants under conditions of low carbon dioxide. *Proc. Nat. Acad. Sci. U.S.*, 63: 668-675.

Zelitch, I. (1959): The relationship of glycolic acid to respiration and photosynthesis in tobacco leaves. *J. biol. Chem.*, 234: 3077-3081.

# PROBLEMS IN BUILDING COMPUTER MODELS FOR PHOTO-SYNTHESIS AND RESPIRATION

J.D. HESKETH, J.M. McKINION, J.W. JONES, D.N. BAKER, H.C. LANE, A.C. THOMPSON & R.F. COLWICK

*Cotton Production Research, Agricultural Research Service, U.S. Department of Agriculture, P.O. Box 5367, Mississippi State, Mississippi 39762 U.S.A.*

*Abstract*

We present a set of differential equations that tie together changes in dry weight, photosynthesis, and respiration in the plant, and which follow the status of the carbohydrate reserve. We discuss some of our efforts to measure components of these equations, emphasizing difficulties in estimating gross photosynthesis. We need quantitative estimates of respiration in the light caused by the accumulation of photosynthate, to add to measurements of net $CO_2$ exchange. Also needed are quantitative estimates of reductant produced and used in metabolism not associated with respiration or the carbon balance. The role of photorespiration is still confusing.

We offer here only a simple summary of the status of the problem of determining gross photosynthesis, and plead with those involved in photosynthetic and respiratory research to address more directly some of these problems.

In recent years our group has developed a set of differential equations for handling photosynthesis, respiration and growth (cf. Jones et al., 1974a, b; McKinion et al., 1975). Here we will discuss how we estimate experimentally various parts of the system of equations to depict a carbon balance for the plant. The discussion will focus on the importance of measuring and predicting the effect of environment on gross photosynthesis and experimental problems involved. We hope to stimulate further discussion of these problems here and in future review papers by others, and thereby encourage photosynthetic experimentalists to contribute to a general effort to construct production models.

The equations are as follows:

$$dW/dt = dP/dt - RW \tag{1}$$

$$W = Q + S, \quad dW = dQ + dS \tag{2}$$

$$RW = R_o Q + G_R \, dQ/dt + R_s \, dS/dt; \tag{3}$$

where $dQ/dt$ and $S \geqslant O$, and

$$dS/dt = [dP/dt - (dQ/dt)(1 + G_R) - R_o Q]/(1 + R_s) \tag{4}$$

where all the terms are as defined in Table 1.

We feel that these equations, or some modification thereof, have great potential for experimentally and theoretically studying these relationships between dry-weight changes ($dW/dt$), reserve carbohydrate changes ($dS/dt$), various kinds of respiration ($R$), and gross photosynthesis ($dP/dt$). These equations were de-

*Table 1.* Components of a photosynthesis: growth: respiration model describing the carbon balance in a plant

| | | |
|---|---|---|
| $W^*$ | = | dry weight in g/plant (oven dry weight) |
| $P$ | = | gross photosynthate in g $CH_2O$/plant |
| $R$ | = | overall respiration rate per g dry weight/plant (g $CH_2O$/g oven dry weight plant-time) |
| $S$ | = | starch and soluble sugars, or carbohydrate reserve in g $CH_2O$/plant |
| $Q^*$ | = | plant tissue weight in g oven dry weight/plant excluding S or carbohydrate reserve |
| $R_s$ | = | respiration rate associated with storage of photosynthate into reserve S (g $CH_2O$/g $CH_2O$) |
| $R_0$ | = | maintenance respiration rate (g $CH_2O$/g $Q^*$-time) |
| $G_R^*$ | = | growth respiration rate (g$CH_2O$/g Q), including adjustments for losses of O, H and gains of N, S and other atoms in transforming $CH_2O$ to Q. |
| $d(\cdot)/dt$ | = | time derivative or change in $(\cdot)$ per unit time |

\* We switch from g $CH_2O$ to g plant dry weight in these equations by allowing $G_R$ to correct for more than growth respiration.

veloped into a computer program for plant growth (cf. McKinion et al., 1975), after several of the components were described as functions of the biological status of the plant or the environment, and instances were considered when photosynthate demand, supply and storage rate or capacity were out of balance. We also developed a similar set of equations for nitrogen metabolism in the soil-plant system, which at times affect the potential demand value (dW/dt) in the plant for photosynthate (Jones et al., 1974).

Here we would like to discuss some experimental aspects of equations (1)-(4). The maintenance respiration ($R_o$) factor is affected by temperature; gross photosynthetic rate (dP/dt) is affected by biological and environmental factors discussed at this symposium. The carbohydrate reserve S has a maximum capacity, and the rate of depletion (when dS/dt is negative) and increase (when dS/dt is positive) have maximum limits depending on the reserve level and temperature. The maximum tissue growth rate (dQ/dt) is limited by nutritional (N, P, K, etc.), morphogenetic, and environmental parameters. As a result of these relationships, there is a feedback control of the photosynthetic rate dP/dt such that no excess photosynthate can accumulate at any time because equation (4) must balance. The capacity of reserves, the maximum rate of increase in reserves, the tissue biomass Q, and the maximum growth rate for tissue (dQ/dt) would determine if the photosynthetic rate (dP/dt) were out of balance with the rest of the system. In other words, we can study the possibility of feedback control of photosynthesis and growth using these techniques. Or, we can study the effect of sink size (potential maximum tissue growth rate (dQ/dt) for the environmental, nutritional and morphogentic conditions prevailing in the system) on photosynthesis. Such studies would require experimental estimates of the terms in Table 1, as well as maximum rates of increase of tissue growth (dQ/dt), reserves (dS/dt), and the maximum rate of depletion of reserves (−dS/dt).

# Photorespiration[*]

Rabinowitch (1945 - p. 567) defines true photorespiration as 'a photochemical activation of the respiratory system which sets in immediately upon illumination and disappears abruptly upon return to darkness,' or as an 'indirect stimulation which sets in slowly in the light and persists for some time in the dark." Rabinowitch would not confuse photorespiration with the stimulation of respiration caused by the accumulation of photosynthate. Such a stimulatory effect was reported by Borodin (1881), Palladin (1893), and by many subsequent scientists (cf. Weintraub, 1944, for a thorough discussion of this effect). The indirect definition of photorespiration might best describe the hypothesis of Bowes & Ogren, where C-2 fragments are produced from C-5 sugars as a result of $O_2$ competing with $CO_2$ for the carboxylating enzyme (as hinted at by Franck & French, cf. Rabinowitch, p. 329, 1944), resulting in possible $CO_2$ production from further metabolism of the C-2 sugar. Rabinowitch's reviews (1944-1956) of photorespiratory phenomena are interesting in view of what we now know (photorespiration, light respiration, the oxygen effect, the C-4 pathway, etc.). As is well known, the C-3: C-4 plant phenomenon has tied together lots of bits and pieces from past photosynthetic and related research.

If photosynthate or photosynthetic energy is available for plant growth as a consequence of the photorespiratory reactions, then we must estimate and add this contribution to our estimate of gross P. For example, if some biochemical intermediates are synthesized more efficiently using photorespiratory products instead of those from nonphotosynthetic intermediary metabolism, then this must be accounted for. Conversely, if some of our gross photosynthate is being used up in the photorespiratory reactions, then we must add a $-R_p(dP/dt)$ term with $R_p$ a function of the environment independent of environmental effects on dP/dt.

The slowing of gross photosynthesis by $CO_2$ from photorespiratory processes and other respiratory processes interacting with $CO_2$ diffusion into the leaf need

* Photorespiration was used for several years to describe the difference in $CO_2$ fixation between C-3 and C-4 plants in light and $CO_2$-free air, a rather puzzling phenomenon then and now. It was suggested that C-4 plants had no photorespiration and the word was used to describe the C-3:C-4 photosynthetic phenomenon. Some of us never really believed this after studying the differences in anatomy between C-3 and C-4 plants, but we still used the word photorespiration to describe everything we were seeing. There was considerable discussion at the time about what to call it, but most of us were so enthralled with our new results that we couldn't be bothered by strict definitions. It also seemed appropriate at the time to wait until we knew better what was causing the results we were measuring, before we settled upon a better word. The younger generation of experimentalists seem to get quite upset by the words we used, but we only remind them that we were distracted by far more important things at the time.

The C-3:C-4 phenomenon was discovered independently among three groups of scientists at about the same time. Parts of the phenomenon were available to two of the groups who independently determined the gas-exchange aspects of the phenomenon, and certainly the biochemical differences discovered by the third group would have led quickly to the gas-exchange differences. However, communications between the three groups were also good, which hastened the pace at which discoveries were made. The availability of the infra-red gas analyzer to a large number of scientists in the early 1960's was perhaps most responsible for the necessary extensive comparisons of gas exchange among genotypes.

to be accounted for in a carbon budget, as long as our estimates of gross P do not include this component of potential photosynthesis. However, we do need to include respiratory $CO_2$ refixed in the leaf chloroplasts that came from sugars metabolized to provide the plant with energy (ATP, etc.) for growth or maintenance.

## Maintenance Respiration ($R_0$)

Penning de Vries (1973) thoroughly discussed maintenance and growth respiration, with emphasis on biochemical estimates for rates of such processes. We will not attempt to review again that information here, but his research, calculations, and literature reviews are very timely in the context of subjects to be discussed at this symposium.

McCree & Troughton (1966) define maintenance respiration as that after 24 hrs in the dark. It, in turn, would be a function of temperature and the biological status of the plant. Their growth respiration was similar to the respiration stimulated by the accumulation of photosynthetic products for a 24-hr period. From our equations, some growth can be expected after 24 hrs in the dark; the amount depends on S and $-dS/dt$ in the plant at the time the measurement is taken (Equation (3). Depending upon $dS/dt$ and $dQ/dt$ for a particular time period, their estimate of maintenance respiration could vary considerably. One such time might be late in the growing season or at certain stages of plant maturity when considerable photosynthate is stored as sugars or starch in roots of legumes.

$R_0$ can be estimated from transformation of some of the growth equations given above (Hesketh et al., 1971; Thornley & Hesketh, 1973; Hughes, 1973) or from biochemical considerations (Penning de Vries, 1973). The state of the art is such that we need estimates using all available techniques. Part of Q, of course, can be dead or aged, with reduced respiration or energy requirement for maintenance. In a model, it is convenient to eliminate Q that is a certain age B:

$$\text{Maintenance respiration} = R_0[Q_t - Q_{t-B}] \tag{5}$$

## Growth Respiration ($G_R$)

Growth respiration consists of two factors: (1) The amount of $CH_2O$ used to form ATP and similar energy-rich intermediates to meet the energy requirements for conversion of photosynthate to plant tissue (fats, proteins, cellulose, membranes, etc.), and (2) the losses of oxygen and hydrogen, as well as the gains in nitrogen, sulfur, and other minerals as a consequence of converting photosynthetic $CH_2O$ to plant tissue and plant uptake of ions.

We estimated $G_R$ using the same techniques as for estimating $R_0$ (see above), and Penning de Vries (1973) made 'paper biochemical' estimates of growth respiration. Changes in plant composition would affect all the above methods of estimating $G_R$ and corrections for such changes would depend on frequency, accuracy, and numbers of metabolites monitored. Presumably, growth respiration would include the losses in energy associated with translocation and ion uptake as needed for growth; similarly, maintenance respiration would require estimates for associated translocation and ion uptake.

## Gross Photosynthesis (dP/dt)

Here we define gross photosynthesis as the photosynthetically fixed energy that is available for use in plant growth or maintenance. By combining all the methods listed previously with growth analysis and $CO_2$-exchange measurements of a plant, there are four methods for estimating gross photosynthesis. All of these must be corrected for respiration during the photosynthetic measurement, Table 2 (cf. Alberte et al., 1974). So far we have added respiration measurements taken in the dark shortly after a period of photosynthesis to the net flux of $CO_2$ into the crop canopy. We realize all the changes in respiration that can occur when the lights go out; however, our techniques were crude enough so that we felt we were averaging out some of this behavior. The above estimate of gross photosynthesis is based on $CO_2$-exchange and would not include photosynthetic reductant used directly in plant metabolism. Our objective here is to bring this problem up once again for discussion, and plead for better reviews of this problem.

*Table 2.* Methods for estimating plant photosynthesis production with the corresponding respiratory and synthesis corrections. (Adapted from Alberte et al., 1974).

| Method | Respiratory Correction |
|---|---|
| (1) Chamber ($CO_2$ exchange) | Measure light, then dark for enclosed system. |
| (2) Flux ($CO_2$ profiles) | Measure same time as $CO_2$ profile determined, for plant and soil. |
| (3) Leaf model, canopy model | Measure light, then dark for leaf. |
| (4) Dry weight sampling | Develop a respiration model for the whole plant with corrections for losses of $H_2O$ and gains in N or S. |

## Plant Reserve (S, dS/dt)

We have to assume a reserve capacity for the plant as a fraction of its dry weight or a function of its leaf area. Also we assumed a maximum rate at which the reserve could be built up or used, with temperature corrections. Hofstra (personal communication) has measured reductions in photosynthesis in leaves containing considerable starch; in our model, dP/dt that cannot be instantly converted into dQ/dt, dS/dt or respiration has to be eliminated.

# The Morphogenetic Demand for Photosynthate, or Maximum dQ/dt

Hackett (1973) outlined the logic for a carbon economy of the plant to determine if photosynthetic supply or demand limits growth; he referred to our earlier attempts (Duncan, 1972) to generate a carbon economy for the cotton plant to show that photosynthate supply can limit boll set and growth. Later, however, we had difficulties after running a nitrogen economy concurrent with a carbon economy to determine which was limiting growth. If nitrogen supply limits growth or the plant's demand for $CH_2O$, then it follows from rather simple logic that photosynthate probably is adequate for maximum possible growth.

For nitrogen and carbon budgets, one needs to know the time course of both N and C in growing and mature plant parts under optimum and stress or severe stress conditions for a range of temperatures. We have made progress in obtaining maximum dQ/dt values for bolls by enriching the air with $CO_2$ under light-rich conditions (cf. McArthur et al., 1974), but have only begun to make progress in repeating this work with nitrogen (cf. Jones et al., 1974a). We have discussed methods for estimating the maximum dQ/dt possible under a particular set of growing conditions and morphological status of a plant. Information such as we describe is needed to evaluate relationships between photosynthetic demand and supply. It also seems likely that supply and demand will often be out of balance either way during the life cycle of the plant. Sometimes the weather may limit supply; at other times the stage of plant growth or other factors may affect demand. Removal of sinks from the plant by natural or artificial means may create a big imbalance, such as the practice of harvesting immature fruit from horticultural plants. A plant geneticist could select better morphological strategies for flowering in some of these species once we better define what is going on in a particular management situation.

## Leaf Respiration

Leaf photosynthesis-canopy models do not need to consider root, stem, or fruit respiration, but stem and fruit photosynthesis needs to be accounted for independently of the leaf models. Raven (1972a,b), Ludlow & Jarvis (1971), and Alberte et al. (1974) discussed sources of respiration in the leaf or in photosynthetic cells, and how such respiration might be altered by photophosphorylation in the chloroplast and by mechanisms of transport of such energy in and out of the chloroplast to the rest of the cell without involving $CO_2$ production. Chapman and co-workers (1974) recently published a series of research papers indicating that the tricarboxylic acid cycle operates in the light at a rate comparable to that in darkness.

Particularly frustrating in all this work are possible fluxes of energy-rich chemical bonds out of chloroplasts active in intermediary metabolism but not associated with $CO_2$ production, as well as the direct reduction of nitrogeneous and sulfurous intermediates and other biochemical intermediates near chloroplasts.

Resistance analyses of $CO_2$ flux into leaves and photosynthetic cells, as well as from respiratory sources in photosynthetic cells and other places inside the leaf, could profit from better analysis of all possible respiratory sources in the light (cf. Alberte et al., 1974).

## Photosynthesis Models

Rabinowitch (1945-56) and Clayton (1965) reviewed the status of photosynthesis models. Clayton emphasized the role of the experimentalist in learning about photosynthetic systems (p. 192). Gaastra's model (1959) for $CO_2$ and $H_2O$ flux in and out of the leaf, Table 3, represented an important innovation in modeling because of its simplicity and because the various components of his model could be estimated experimentally. Models for the fluxes of gases and energy in and out of the crop have taken on some significance as better techniques have been developed for estimating various parts of these models (cf. Inoui et al., 1958; Lemon, 1965). We hope that our system of equations relating photosynthate supply to carbohydrate reserves, respiratory needs, and growth demands will soon develop as strong an experimental base as these other models. We are working on aspects of the reserve, morphogenetic demand, and on a similar set of equations and research for an N balance which will interact with the C balance.

*Table 3.* An example of a useful photosynthetic model, with an indication of disciplines integrated (Gaastra, 1959).

| Needed Component | Discipline(s) Involved |
|---|---|
| $\gamma_a$, air boundary resistance to gas flux | Micrometeorology |
| | Transpiration |
| $\gamma_s$, stomatal resistance to gas flux | Stomatal guard cell physiology and anatomy |
| $\gamma_i$, leaf air space – photosynthetic cell surface resistance | Leaf anatomy (cf. C. Edwards, these Proceedings). |
| $\gamma_x$, photosynthetic reactions | Photosynthetic light and dark reactions, chemical kinetics, physical chemistry. |

## REFERENCES

Alberte, R.S., J.D. Hesketh & D.N. Baker (1974): Aspects of predicting gross photosynthesis (net photosynthesis plus light and dark respiration) for an energy-metabolic balance in the plant. In Gates, D., (Ed.): Ecological Studies: Analysis and Synthesis. Vol. 6, Biophysical Ecology. Springer-Verlag, Berlin.

Chapman, E.A. & D. Graham (1974): The effect of light on the tricarboxylic acid cycle in green leaves. I. Relative rates of the cycle in the dark and the light. II. Intermediary metabolism and the location of control points. *Plant Physiol.*, 53: 879-892.

Clayton, R.K. (1965): Molecular physics in photosynthesis. Blaesdell Publ. Co., N.Y. 205 pp.

Duncan, W.G. (1972): SIMCOT: A simulator of cotton growth and yield. In Murphy, C., et al. (Ed.): Proc. Workshop on Tree Growth Dynamics and Modeling. Pp. 115-118. Duke University, October 1971. Oak Ridge Nat. Lab., Tenn.

Gaastra, P. (1959): Photosynthesis of crop plants as influenced by light, carbon dioxide, temperature, and stomatal diffusion resistance. *Meded. Landb.hoogesch. Wageningen,* 59: 1-68.

Hackett, C. (1973): An exploration of the carbon economy of the tobacco plant. I. Inferences from a simulation. *Aust. J. Biol. Sci.,* 26: 1057-1071.

Hesketh, J.D., D.N. Baker & W.G. Duncan (1971): Simulation of growth and yield in cotton: Respiration and the carbon balance. *Crop Sci,* 11: 394-398.

Hughes, A.P. (1973): A comparison of the effects of light intensity and duration on *chrysanthemum morifolium* cv. Bright Golden Anne in controlled environments. II. Ontogenetic changes in respiration. *Ann. Bot.,* 37: 275-286.

Inoui, E., N. Tani, K. Imar & S. Isole (1958): The aerodynamic measurement of photosynthesis over the wheat field. *J. Agr. Meterol.,* (Tokyo) 13: 121-125.

Jones, J.W., A.C. Thompson & J.D. Hesketh (1974): Analysis of SIMCOT: Nitrogen and growth. Proc. Beltwide Cott. Prod. Res. Confs., Nat'l. Cott. Council Amer., Memphis, Tenn. USA. *pp.* 111-118.

Jones, J.W., R.W. Wensink, R.S. Sowell & J.D. Hesketh (1974): Formulation of a decision model to select optimum crop production practices. ASAE paper No. 74-4023.

Lemon, E. (1965): Micrometeorology and the physiology of plant in their natural environments. In Steward, F.C. (Ed.): Plant Physiology, A Treatise. Pp. 203-227. Academic Press, New York.

Ludlow, M.M. & P.G. Jarvis (1971): Methods for measuring photorespiration in leaves. In Sestak, Z., Catsky, J., & Jarvis, P.G. (Eds.): Plant Photosynthetic Production, Manual of Methods. Pp. 294-315. Dr. W. Junk N.V.-Publishers, The Hague.

McArthur, J.A. J.D. Hesketh & D.N. Baker (1974): Cotton. In Evans, L.T. (Ed.): Crop Physiology: Some case histories. Cambridge Univ. Press. In press.

McCree, K.J. & J.H. Troughton (1966): Prediction of growth rate at different light levels from measured photosynthesis and respiration rates. *Plant Physiol.,* 41: 559-566.

McKinion, J.M., J.W. Jones & J.D. Hesketh (1975): A system of growth equations.

Penning de Vries, F. (1973): Substrate utilization and respiration in relation to growth and maintenance in higher plants. PH.D. Thesis. Landbouwhogeschool TC. Wageningen.

Rabinowitch, E.I. (1945-1956): Photosynthesis and Related Processes. Interscience, New York. 2088 pp.

Raven, J.A. (1972): Endogeneous inorganic carbon sources in plant photosynthesis. I. Occurrence of the dark respiratory pathways in illuminated green cells. II. Comparison of total $CO_2$ production in the light. *New Phytol.,* 71: 227-247; 995-1014.

Thornley, J.H.M. & J.D. Hesketh (1972): Growth and respiration of cotton bolls. *J. Appl. Ecol.,* 9: 315-317.

Weintraub, R.L. (1944): Radiation and plant respiration. *Bot. Rev.,* 10: 383-459.

# EFFECT OF GROWTH TEMPERATURE ON PHOTOSYNTHETIC AND PHOTO–RESPIRATORY ACTIVITY IN TALL FESCUE.

K.J. TREHARNE[1] & C.J. NELSON[2]

*Welsh Plant Breeding Station, Aberystwyth*

*Abstract*

Whilst the effect of temperature on photosynthesis is well established, the influence of growth temperature on net carbon exchange (NCE) and component processes is less well documented. In the present study the degree of temperature adaptation on plant growth in relation to NCE and photo-respiratory activity and selected *in vitro* enzyme activities has been determined in Tall Fescue grown at $10^\circ$ and $25^\circ$. Different optima for NCE have been shown to be related to effect of temperature on photo-respiratory and dark respiratory activity. No environmental influence on qualitative characteristics of enzymes was observed but quantitative differences in enzyme activities can explain the different pattern of temperature response. Activity of RuDPc was similar in both environments but activities of respiratory enzymes MDH and ICDH were significantly higher at $10^\circ$ growth. Stomatal conductances were markedly reduced at $30^\circ$ and caused a sharp decline in NCE in plants from both growth temperatures. Thus in different environmental conditions different components act as rate controlling factors.

## Introduction

The temperate species are well documented as showing optima in their growth characteristics and rates of photosynthesis at temperatures of $20\text{-}25^\circ$ C in contrast to tropical species which exhibit optima around $35^\circ$ C. The lack of response in photosynthesis or net carbon exchange (NCE) to temperatures above $25^\circ$ C in Festucoid grasses may involve several physiological characteristics including gaseous diffusion, carboxylation and respiratory components. Moreover, the influence of growth temperature on photosynthesis and related biochemical parameters is not well established.

Previous studies have shown a positive relationship between NCE and *in vitro* activity of the carboxylating enzyme ribulose diphosphate carboxylase (RuDPc) in several species (Björkman, 1968; Eagles & Treharne, 1969), and activity of RuDPc has been implicated as an important rate-determinant in photosynthesis (Wareing *et al* 1968). Furthermore, the *in vitro* temperature response of crude extracts of RuDPc showed maximal activity at about $25^\circ$ C in some temperate *Gramineae* (Treharne & Cooper, 1969) and a degree of adaptation to growth temperature, related to temperature response in NCE, in *Dactylis glomerata* L.(Treharne & Eagles, 1970).

1. Present address: International Institute of Tropical Agriculture, Ibadan, Nigeria.
2. On study leave from University of Missouri.

*Acknowledgements:* One of us (2) acknowledges the support of a N.A.T.O. fellowship. The authors wish to thank Dr J.P. Cooper for valuable advice, Dr D. Wilson for conducting the stomatal conductance measurements and excellent technical assistance of E. Lloyd and F.I. Sant.

Thus the question is posed as to whether carboxylating activity is the underlying factor determining low temperature optima in NCE or is this phenomenon related to other factors such as enchanced rates of dark and photo-respiration with increase in temperature, or anatomical-morphological features affecting $CO_2$ diffusion.

In the present study the degree of temperature adaptation on plant growth in relation to NCE and photo-respiratory activity and selected *in vitro* enzyme activities has been determined in *Fectuca arundinaceae,* Schreb, (Tall Fescue). Here we report temperature response data for two growth temperatures of one variety, which formed a part of a study of eight diverse populations of Tall Fescue grown in a range of controlled environments which will be reported in detail elsewhere.

## Materials and Methods

Seedlings of Tall Fescue (var. S170) were grown in pots of John Innes No.I in controlled environments of 16 hr photoperiod with an incident irradiance of 70 $Wm^{-2}$ provided by mercury fluorescent lamps, at constant $10°$ and $25°$, respectively. Growth analysis was conducted between the 4th and 5th leaf stage in the different growth regimes and the parameters calculated from modified classical equations of Briggs *et al* (1920).

Photosynthetic rates of young fully expanded 5th leaves were determined at 200 $Wm^{-2}$ incident irradiation in a 10-channels infrared gas analyser starting 2 hr after commencement of the light period in the growth room, as described by Eagles & Treharne (1969). Leaf temperatures were measured with copper-constantan thermocouples on the undersurfaces of the leaves. Temperature response of NCE at 300 ppm $CO_2$ was conducted from $10°$ to $35°C$. Measurements were routinely started at the low temperature and after equilibration to achieve stable readings, the air (21% $O_2$) was replaced with a mixture of air (10% v/v) and $N_2$ containing 300 ppm $CO_2$ (90% v/v) giving ca 2% $O_2$. The enchancement in rate of NCE at 2% $O_2$ over the rate in 21% $O_2$ was used as an estimate of photo-respiration. Following measurement in the gas mixture, the temperature was increased by $5°$ intervals and the above procedure repeated at each respective temperature.

Dark respiration was measured on the 6th leaf in the gas analyser at 21% $O_2$ 300 ppm $CO_2$ commencing readings at $10°$ and in subsequent $5°$ intervals up to $35°C$.

After gas-exchange measurements, the leaves were removed for area measurements, part of the leaf used for stomatal count and anatomical observation and the remainder dried in a forced-draft oven at $80°C$ allowing determination of specific leaf weight.

Leaf diffusive resistances were measured with a diffusion porometer (Lamba Instruments mod. LI. 60) using young fully expanded leaves, fully exposed to the light source, of plants transferred from each growth environment to other controlled growth rooms at different temperatures. Plants were allowed to 'equilibrate' at the different temperature for 5 hrs before porometer measurements.

Enzyme extraction from 100 mg fresh weight of leaf tissue and assay for ribulose diphosphate carboxylase activity was conducted as described by Treharne & Cooper (1969). Suitable aliquots of extract were used for measurements of glycolate oxidase by monitoring change in absorbance of phenylhydrazine at 340 nm in an LKB model 8600 reaction rate analyser. Activities of malate and isocitrate dehydrogenases were also determined in the LKB analyser at 35° following oxidation of NADH or reduction of NADP, respectively. Temperature response characteristics of the enzymes were measured in a Unicam SP800 spectrophotometer.

## Results and Discussion

Plants of Tall Fescue S170 showed a marked influence of growth temperature on leaf area ratio (L) which was reflected in a lower relative growth rate (R) at 10° than at 25°C (Table I). The lower shoot: root ratio, also shown in Table I,

Table 1. Growth characteristics of Tall Fescue (var. S170) at two temperatures

| Growth temp °C | R mg mg$^{-1}$d$^{-1}$ | L mm$^2$mg$^{-1}$ | E mg mm$^{-2}$d$^{-1}$10$^3$ | SLW mg cm$^{-2}$ | LA increase mm$^2$d$^{-1}$ | Shoot/Root |
|---|---|---|---|---|---|---|
| 10 | 0.106 | 16.0 | 6.67 | 3.10 | 66 | 3.58 |
| 25 | 0.123 | 22.7 | 5.63 | 2.46 | 259 | 4.56 |

Abbreviations see text. Growth analysis 4th-5th leaf stage.

corroborated the fact that a lower portion of aerial growth occurred at low temperature. Leaf growth, expressed as area per day, was approximately 25% of that at 25°C and the reduced rate of leaf expansion resulted in leaves with higher specific leaf weight at 10°C. Net assimilation rate (E) was 18% higher in the 10°-grown plants but the influence of temperature on L, being 42% higher at 25°C, was the major factor affecting R.

Similar effects were previously reported (Treharne & Eagles, 1970) for *Dactylis* where growth temperature had no effect on E but resulted in a 3-4 fold increase in L and R from 5-25°C. Thus it is apparent that leaf growth attributes of temperate grasses are more affected by temperature than are physiological parameters such as assimilation rate. In this respect, the ability of Tall Fescue to exploit the growth environment is affected to a large extent by the leaf area and distribution of dry matter as well as the functioning of the photosynthetic surface.

The ability of leaves grown in both regimes to respond to a range of measuring temperatures in terms of net carbon exchange (NCE), was tested at 5° intervals, sequentially, from 10° to 35°C. Plants grown at 10°C showed a steady decline in NCE above 15°C (see Fig. 1a) whilst those grown at 25°C had an optimum temperature about 25°C when measured in air (21% $O_2$). The response curves obtained from leaves measured in 2% $O_2$ (Fig. 1b) showed no effect of temperature on NCE of 10°-grown plants between 10° and 30°, whereas the

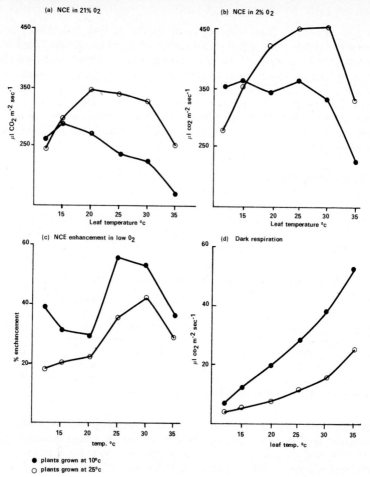

(a) NCE in 21% $O_2$

(b) NCE in 2% $O_2$

(c) NCE enhancement in low $O_2$

(d) Dark respiration

● plants grown at 10°c
○ plants grown at 25°c

*Fig. 1.* Temperature response of photosynthesis and respiration in Tall Fesque grown at 10° and 25° C.

25°-grown plants showed a continuing increase in rate up to 30° C. A sharp decline in rate of NCE was observed above 30° in plants from both growth environments, in both 21% and 2% $O_2$. The difference between NCE in 2% and 21% $O_2$ is expressed in Fig. 1c as % enchancement and can be considered as an estimate of photo-respiratory rate and it is clear that this is significantly greater in plants grown at 10° C than at 25° C. The pronounced effect of temperature increase on % enhancement from 20° to 25° is in part responsible for the low temperature maximum of NCE in 21% $O_2$ in the 10° plants. Although a similar increase in photo-respiratory rate with temperature was observed in the 25° plants, the rate of increase was lower and, moreover, was quantitatively less, thus producing an effect on the NCE response curve in 21% $O_2$ only above 25° C. However, in view of the fact that the response curve in all cases showed a marked decline above 30°, some phenomenon other than promoted rate of photo-respiration must be involved.

The temperature effect on rate of dark respiration (Fig. 1d) indicated a linear response up to $35°C$ in plants from both growth temperatures. A significantly higher rate of dark respiration was observed in the $10°$-grown plants and may also contribute to the difference in NCE response between the two environments, particularly if photo-respiration and dark respiration are considered additive in their effect on gas exchange with temperature increase. It is feasible that the continued response in dark respiratory rate above $30°$ may contribute in some measure to the sharp decline in NCE above $30°$, but from the carbon budget cannot be invoked as the full quantitative cause.

It has been suggested previously that the temperature response characteristics of whole leaves may in part be determined by the thermal properties of the carboxylating enzyme (Treharne & Cooper, 1969). The decline in photosynthesis at high temperature in *Caltha intraloba* was associated (Philips & McWilliam, 1971) with thermolability of ribulose diphosphate carboxylase (RuDPc). In contrast, Tieszen & Sigurdson (1973) have reported that crude preparations of RuDPc from several temperate and tropical species have optima around $50°C$, a stability well beyond the maxima in NCE.

Temperature response characteristics of crude extracts and of preparations partially purified by Sephadex G-200 separation were obtained from plants grown in both temperature regimes. The response curves for respective enzymes were identical from both growth regimes and effect of growth temperature is neglected in the ensuing consideration of the enzyme characteristics. Activity of RuDPc, representative of photosynthesis, increased linearly with temperature up to $35°C$ in crude preparation, but showed no rate increase with higher temperature. This pattern of temperature response is in contrast to those previously reported for *Lolium* and *Dactylis* (Treharne & Cooper, 1969; Treharne & Eagles, 1969) which showed maximal activity between 20 and $25°C$, and in marked contrast to the reported maximum around $50°C$ (Tieszen & Sigurdson, 1973). Since in all cases crude extracts of enzyme and very similar assay procedures were employed, it would appear that species differences exist, either in enzyme properties *per se* or in some factor associated with a crude extract, such as the presence of interfering components in a protein mixture or partial inhibitors. The temperature response curve of RuDPc after G-200 purification is shown with that of the crude extract in Fig. 2a, the curves being means of individual replicates expressed as % of maximum activity. It is clear that removal of low molecular weight components, or some other factor, by G-200 chromatography, resulted in a marked difference in temperature response, in that rate of carboxylation continued to increase up to $50°C$, with a sharp decline at $60°C$ probably associated with thermal denaturation. The nature of the component/s affecting activity of the carboxylase above $35°C$, and removed by G-200 chromatography, remains unknown but is/are possibly present in only very low concentration in the 3 arctic grasses and *Sorgum* and *Andropogon* in the studies of Tieszen & Sigurdson (1973). This would explain the anomalies in the temperature responses reported by the different studies. It is evident that a great deal of caution needs to be exercised in attempting to determine and interpret characteristics of enzyme components *in vitro*.

Temperature response curves of malate and isocitrate dehydrogenases, taken as biochemical indices of dark respiration, are shown in Fig. 2c and 2d, respectively. No effect of growth environment was evident, the enzyme activities in-

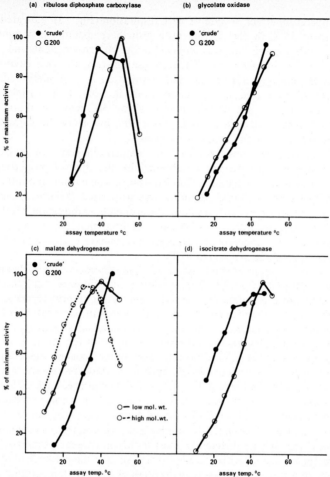

(a) ribulose diphosphate carboxylase      (b) glycolate oxidase

(c) malate dehydrogenase      (d) isocitrate dehydrogenase

*Fig. 2.* Temperature response of photosynthetic and respiratory enzymes isolated from leaves of Tall Fescue.

creasing linearly with temperature increase up to 45° in crude extracts. Two species of MDH were obtained on G-200 separation, one of high molecular weight and one of low molecular weight. The temperature response of the two species (**Fig. 2c**) showed different maxima but there was no difference between plants from the 2 growth regimes.

Activity of glycolate oxidase (GO), a component of the photo-respiratory pathway, showed progressive increase in activity with increase in assay temperature but again no effect of growth temperature of the plants was observed on the pattern of response.

The salient feature of the data is that the temperature response characteristics of RuDPc and the respiratory enzymes from Tall Fescue grown at either 10° or 25° C were identical, and cannot explain the difference in pattern of temperature response of NCE in the different growth regimes.

Quantitative determinations of enzyme ectivities from both growth regimes

*Table 2.* Enzyme activities of 5th leaf of Tall Fescue (var. S170) grown at two different temperatures.

| Growth temp. °C | RuDPc | GO | ICDH | MDH |
|---|---|---|---|---|
| 10 | 2.46 | 9.9 | 0.44 | 26.8 |
| 25 | 3.01 | 11.7 | 0.29 | 18.7 |
| Significance level | N.S. | N.S. | 0.01 | 0.01 |

Abbreviations see text. Activities as Units $min^{-1} cm^{-2}$ leaf area.

were made and were expressed in terms of unit leaf area (Table 2) to facilitate comparison with the gas exchange parameters. Activities of RuDPc and GO were not significantly different between leaves at different growth temperatures. However, activities of both MDH and ICDH were significantly higher on a leaf area basis in plants grown at $10°C$, as reflected in gas-exchange measurement of dark respiration. The qualitative characteristics of temperature response of the enzymes were same from both environments, but the effect of the quantitative differences in the respiratory components may be considered as significantly contributing to the different patterns of response and temperature optima of NCE in the two treatments.

The effect of increase in temperature from 30-35°, causing a sharp decline in NCE, cannot be interpreted by reduced carboxylating activity or by enhanced photorespiration, although dark respiration may be partly involved. It was, therefore, possible that the temperature effect above 30° was related to gas diffusion properties. Porometer measurements of leaf diffusive resistances of plants placed in 4 different temperature regimes were conducted on both abaxial and adaxial surfaces. The data shown as conductance values in Table 3, reflect mainly stomatal behaviour although a small boundary-layer component would be included. The measurements, related to leaf temperature, were taken after 5 hours

*Table 3.* Stomatal characteristics of leaves of Tall Fescue (S170) grown in 10° and 25°C temperature regimes

| Leaf surface | Plant growth temperature | Stomatal No. per unit area | Stomatal conductance (1/R) (x100) at mean leaf temperature | | | |
|---|---|---|---|---|---|---|
| | | | 13° | 24° | 29° | 34° |
| Adaxial | 10° | 53 | 38 ± 4.3 | 47 ± 3.6 | 47 ± 3.6 | 24 ± 5.9 |
| | 25° | 74 | 37 ± 5.3 | 49 ± 2.8 | 56 ± 3.1 | 25 ± 5.0 |
| Abaxial | 10° | 0 | 9 ± 0.4 | 4 ± 1.2 | 5 ± 4.0 | 4 ± 0.8 |
| | 25° | 21 | 23 ± 6.3 | 19 ± 5.1 | 40 ± 2.8 | 10 ± 3.5 |
| Total | 10° | 53 | 57 | 52 | 59 | 26 |
| | 25° | 95 | 62 | 73 | 97 | 41 |

equilibration in the different temperature regimes, at about 6 hours after the start of the light period when stomatal resistances were least (D. Wilson, personal communication). In the $10°$-grown plants, stomatal conductance was found to show no change from $10$-$25°$, increased between $25$-$30°$ with a sharp decline at $30°$. The adaxial conductance of the $10°$ plants accounted for the largest part of this effect, since abaxial conductance was very low and this was confirmed by stomatal counts, there being no stomata on the abaxial surface.

In the $25°$-grown plants, there was a large increase in stomatal number per unit area (see Table 3) with significant stomatal conductance from both leaf surfaces. The response of stomatal conductance to measuring temperature showed the same pattern for both surfaces, increasing from $10$-$30°$ with a very sharp decline at $30°C$. These data indicate stomatal closure at $30°$ for plants from both growth regimes which accounts for the sharp decline in photosynthetic rate at temperatures above $30°$.

Thus it would appear from the various parameters investigated, that the different patterns of response to temperature of Tall Fescue, grown at low and high temperature, can be explained in part, over the range $10$-$30°$, by differences between levels of carboxylating and respiratory activity, at the biochemical level. However, the higher rate of photosynthesis in the $25°C$ grown plants was also associated with higher stomatal conductance of the leaves, and unquestionably the response to temperature above $30°$ is directly related to a lower rate of gaseous diffusion through stomatal closure.

The degree of physiological adaptation to growth temperature in the response characteristics of photosynthesis in Tall Fescue is, therefore, controlled in some measure by the activity of the biochemical components and also by physical factors affecting $CO_2$ diffusion. The relative importance of the different rate-controlling processes is clearly a function of environmental changes imposed.

# REFERENCES

Björkman, O. (1968): Further studies on differentiation of photosynthetic properties in sun and shade ecotypes of *Solidago virgaurea*. *Physiol. Plant.,* 21: 84-99.

Briggs, G.E., F. Kidd & C. West (1920): A quantitative analysis of plant growth. *Ann. appl. Biol.,* 7: 103-123.

Eagles, C.F. & K.J. Treharne (1969): Photosynthetic activity of *Dactylis glomerata* L. in different light regimes. *Photosynthetica* 3: 29-38.

Phillips, P.J. & J.R. McWilliam (1971): Thermal responses of the primary carboxylating enzymes from $C_3$ and $C_4$ plants adapted to contrasting temperature environments. In Hatch, H.B., Osmond, C.B. & Slatyer, R.O. (Ed.): Photosynthesis and Photorespiration. Pp 97-104. Wiley-Publishers, New York.

Tieszen, L.L. & D.C. Sigurdson (1973): Effect of temperature on carboxylase activity and stability in some Calvin cycle grasses from the Arctic. *Arctic and Alpine Res.,* 5, No.1: 59-66.

Treharne, K.J. (1972): Biochemical limitations to photosynthetic rates. In Rees, A.R., Cockshull, K.E., Hand, D.W. & Hurd, R.G. (Ed.): Crop Processes in controlled environments. Pp 285-303. Acad. Press - Publishers, London and New York.

Treharne, K.J. & Cooper, J.P. (1969): Effect of temperature on the activity of carboxylases in tropical and temperate *Gramineae*. *J. exptl. Bot.,* 20: 170-175.

Treharne, K.J. & C.F. Eagles (1970): Effect of temperature on photosynthetic activity of climatic races of *Dactylis glomerata* L. *Photosynthetica,* 4: 107-117.

Wareing, P.F., M.M. Khalifa & K.J. Treharne (1968): Rate-limiting processes in photosynthesis at saturating light intensities, *Nature,* 220: 453-5.

# THE EFFECTS OF TEMPERATURE AND $CO_2$ ENRICHMENT ON PHOTOSYNTHESIS IN SOYBEAN

G. HOFSTRA & J.D. HESKETH

*Department of Environmental Biology, University of Guelph, Guelph, Ontario, Canada.*
*Cotton Production Research, Agricultural Research Service, U.S. Department of Agriculture, Mississippi State, Mississippi, U.S.A. 39762*

*Abstract*

The effects of temperature and $CO_2$ concentration on the growth and gas exchange of soybean plants were studied in phytotrons. Growth of soybean plants can be greatly enhanced by $CO_2$ enrichment when temperatures are optimum for growth. Photosynthesis was negatively correlated with mesophyll resistance, starch content, and specific weight of leaves. There was no correlation between photosynthesis and stomatal resistance. With suboptimal temperatures, leaf sections in $CO_2$-enriched air showed a more dense palisade layer and high starch accumulation.

## Introduction

Plant growth is the result of the complex interaction of many processes, any of which may limit the rate and all of which are influenced by various environmental factors either directly or indirectly. Which process is limiting growth is often difficult to determine; yet this information may greatly aid man's attempt to improve crop growth and yield.

The photosynthetic rates of soybean leaves are lower than those for many crops (El Sharkawy & Hesketh, 1965). That photosynthesis may be the limiting process in the growth of soybeans and other plants has been indicated. Hesketh (1963) reported that different species had different chlorophyll levels and that the limitations were in $CO_2$ conductivity in the mesophyll. He found that by increasing the $CO_2$ concentration to 800 ppm (v/v), castor bean and tobacco had photosynthetic rates similar to those of corn at 300 ppm. Brun & Cooper (1967) found that increasing the $CO_2$ concentration at high light intensities also greatly increased the photosynthetic rates of soybean leaves.

However, many reports indicate that photosynthetic rates may be limited by the size of the sinks or the level of assimilate concentration in the leaf (Neales & Incoll, 1968). Seemingly, this contradicts the indications that photosynthesis may be the limiting process in these or closely related species.

In many of these experiments the environmental conditions were not carefully controlled or the experiment was conducted under only one set of conditions. It seems feasible that under optimal conditions for growth photosynthesis may be the limiting process, whereas under suboptimal conditions growth through assimilate concentration may control the rate of photosynthesis.

The studies reported here were initiated at CERES phytotron at Canberra, Australia and continued at the SEPEL phytotron at Duke University, Durham, N.C., U.S.A.

# Materials and Methods

## Experiment I

Soybean plants (*Glycine max* L., var Biloxi) were seeded in the CERES greenhouses at a day/night temperature of 27/22°C (cf. Hofstra,1972). When the primary leaves had just unfolfed, plants were selected for uniformity and transferred to artificially lit cabinets for studies on the effect of $CO_2$ on vegetative growth. The plants were grown under a 16-hr day, a light intensity of about 30 klux, $CO_2$ concentrations of 300 ppm (normal) or 1000 ppm, and day/night temperatures of 36/31, 27/22 and 22/17°C. To obtain the 1000-ppm level, $CO_2$ was bled into the chamber and monitored daily with an infrared gas analyzer.

Seven days after the plants had been transferred to the chamber and every 2 days thereafter, 10 randomly selected plants were harvested. The leaf area and dry weight were determined for each plant.

## Experiment II

'Wayne' soybean plants were grown under a wide range of temperatures in the SEPEL growth rooms (Type B, see Kramer et al., 1970). At each temperature regime the $CO_2$ was either retained at ambient or was increased to between 800 and 1000 ppm.

The gas exchange was measured by using a Beckman[1] infrared gas analyzer and Hygrodynamics[1] electric hygrosensors, measuring both air temperature and relative humidity. Leaf temperatures were measured with a copper-constantan termocouple pressed against the underside of the leaf.

The leaf chamber was stationed in the middle of a growth cabinet (Type C, see Kramer et al., 1970) with the light supplied by cool white fluorescent tubes and incandescent bulbs giving a light intensity of 25 klux. Compressed air was fed into a 2000-liter drum and mixed with moistened air to give a relatively constant moisture content in the air passing over the leaf. Air was drawn from the drum and passed over the attached leaf in the leaf chamber. The air passing over the leaf was maintained at 30 ± 1°C.

Plants were harvested at about 0.2 g (dry weight) per plant and again at about 10 g per plant. The data were used to calculate the growth rate at the different temperatures and the two $CO_2$ concentrations.

After the photosynthetic measurements were completed, the leaflet was harvested, its leaf area determined, and the leaflet oven-dried at 80°C and weighed. The dried leaves and others collected from different temperature and $CO_2$ treatments were ground and extracted with 80% boiling ethanol and the total sugar content was determined using anthrone reagent. The residue was incubated with amylglucosidase (Sigma Chem. Co.) at 55°C for 4 hours, and analyzed for glucose using the Glucostat method (Worthington Biochem. Co.).

Leaf samples of plants grown at the different temperatures and the two $CO_2$ concentrations were collected just before the lights came on in the morning, at noon, and during the late afternoon. The samples were fixed in FAA, sectioned, stained with IKI, and photographed.

---

[1]Mention of a proprietary product does not necessarily imply endorsement of the product by the University of Guelph or the U.S. Department of Agriculture.

Correlation coefficients were calculated among the photosynthetic rates, stomatal resistances, mesophyll resistances, sugar contents, starch contents and specific leaf weights. The correlation coefficients were calculated for all treatments combined, and for normal and enriched $CO_2$ air separately.

## Results and Discussion

In the artificially lit chambers at CERES, Biloxi soybean produced much more axillary growth than when grown in the greenhouse. The axillary growth was further stimulated by increased $CO_2$ concentration. This was in marked contrast to the cultivar Wayne used in the SEPEL phytotron, which produced almost no axillary growth under any of the temperature and $CO_2$ combinations.

Hofstra (1972) reported earlier that Biloxi soybean produced maximum leaf area at around 27-30°C and maximum dry matter at 33-36°C. He also reported that specific leaf weights varied more or less inversely with the rate of leaf area increase (Fig. 1) which suggested that all assimilates were used at 27-30°C and that at lower temperatures assimilates might exceed what the plant could use for new production. The finding that increasing the $CO_2$ concentration from 300 to 1000 ppm greatly increased the growth rate at some temperatures but not others supported the above postulate.

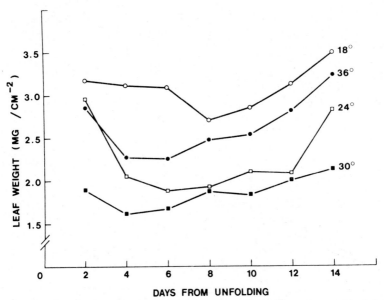

*Fig. 1.*Changes in specific weight of leaves (mg dry weight per cm[2]) with time after unfolding at different temperatures.

At 22/17°C, increasing the $CO_2$ concentration had no effect on rates of either leaf area (Fig. 2a) or dry weight production (Fig. 2b). At 27/22°C the rate of leaf area production was doubled (Fig. 2c) and, initially, the rate of dry matter

73

production was also doubled (Fig. 2d). The rates were also increased at 36/31°C, but much less so than at 27/22°C. The effects of increased $CO_2$ concentration thus depend greatly on temperature.

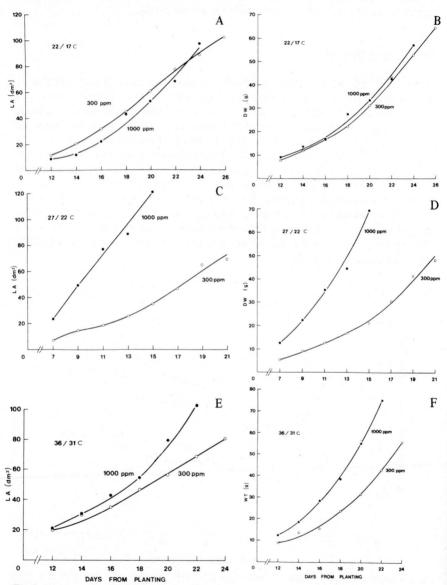

*Fig. 2.* The growth of soybean plants (Biloxi) at 22/17, 27/22 or 36/31°C and 300 ppm or 1000 ppm $CO_2$ expressed in terms of leaf area (LA) and dry weight (DW).

At 27/22°C, a large proportion of the increase in growth at 1000 ppm $CO_2$ was due to a threefold increase in axillary growth. At the other two temperatures, axillary growth was not significantly altered. The rate of leaf initiation was increased about 27% at 27/22°C, about 33% at 36/31°C and none at 22/17°C. Specific leaf weight increased only 2-3% at 27/22°C, but increased 23% at 22/17 and 37% at 36/31°C.

Thus an increase in $CO_2$ concentration greatly enhanced growth at 27/22°C in Biloxi soybeans, with most of the additional assimilate going into new leaf and stem tissue. At 22/17 and 36/31°C growth was stimulated little or none. Much of the weight increase was due to heavier leaves, particularly at 36/31°C.

A similar, but less striking response, was obtained with Wayne soybeans in the SEPEL phytotron. Growth was not stimulated until an average temperature of 25°C was reached (Fig. 3). This cultivar was affected considerably less than Biloxi, apparently because of their different growth habits. The main effect of increased $CO_2$ levels appeared to be an increase in the rate of leaf emergence; almost no axillary growth was produced (Hesketh et al., 1973). At lower temperatures, leaves in the enriched $CO_2$ atmosphere had a higher specific leaf weight.

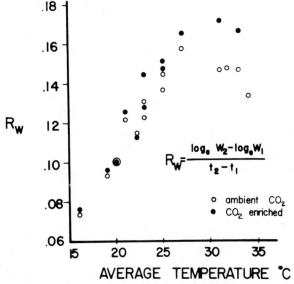

Fig. 3. The growth of soybean plants (Wayne) at 300 or 1000 ppm $CO_2$ at different temperatures expressed in terms of the growth rate.

Plants grown in the enriched $CO_2$ atmosphere had a lower photosynthetic rate than those grown in ambient air. A representative set of data is presented in Table 1. A decrease in the photosynthetic rate was evident in all but the youngest leaf on the plants grown in $CO_2$-enriched air. The leaves also had a higher specific leaf weight. The stomatal resistance of the leaves from the $CO_2$-enriched air was consistently higher than that of the leaves grown in ambient air, but most of the suppression in the photosynthetic rate was accounted for by the higher mesophyll resistance of the high $CO_2$ plants.

Generally, starch content was higher in the leaves of the $CO_2$-enriched plant, particularly in the youngest leaves (Table 2). The sugar content showed a more

*Table 1.* The effects of leaf age and $CO_2$ concentration on the photosynthetic rate, specific leaf weight, stomatal resistance and mesophyll resistance of soybean leaves (growth temperature 32/26/23°C).

| | Leaf | PS<br>mg $CO_2$/dm$^2$/hr | Sp. lf. wt.<br>mg/cm$^2$ | $r_s$<br>sec/cm | $r_m$<br>sec/cm |
|---|---|---|---|---|---|
| $CO_2$<br>enriched | 1* | 4.7 | 5.37 | 2.7 | 32.1 |
| | 2 | 6.9 | 5.28 | 2.9 | 21.3 |
| | 3 | 12.1 | 5.46 | 2.2 | 10.4 |
| | 4 | 13.3 | 5.42 | 2.7 | 8.6 |
| | 5 | 15.4 | 3.84 | 2.0 | 7.6 |
| Normal<br>$CO_2$ | 1 | 13.7 | 4.58 | 1.2 | 9.8 |
| | 2 | 14.4 | 4.71 | 1.5 | 8.9 |
| | 3 | 17.8 | 4.06 | 1.5 | 6.6 |
| | 4 | 17.7 | 3.40 | 1.5 | 6.6 |
| | 5 | 15.2 | 2.46 | 1.3 | 8.4 |

n = 2

\* Leaf 1 is the middle leaflet of the first trifoliate above the paired unifoliates of the main shoot.

*Table 2.* The effect of leaf age and $CO_2$ concentration on the starch and sugar contents of the leaf (growth temperature 26/23°C).

| | Leaf | Starch content<br>mg per g dry wt | Sugar content<br>mg per g dry wt |
|---|---|---|---|
| $CO_2$<br>enriched | 1 | 276 | 65.5 |
| | 2 | 388 | 59.7 |
| | 3 | 359 | 71.4 |
| | 4 | 334 | 42.8 |
| | 5 | 390 | 47.5 |
| | 6 | 470 | 46.3 |
| | 7 | 389 | 34.9 |
| | 8 | 338 | 45.4 |
| Normal<br>$CO_2$ | 1 | - - - | - - - |
| | 2 | 339 | 63.2 |
| | 3 | 345 | 79.6 |
| | 4 | 446 | 86.0 |
| | 5 | 356 | 116.6 |
| | 6 | 388 | 94.4 |
| | 7 | 280 | 141.1 |
| | 8 | 185 | 125.7 |

n = 1

dramatic difference, with the plant in normal air having the consistently higher sugar content. This presumably reflects the higher photosynthetic rates of these leaves, as the samples were taken during the middle of the day.

That the difference in starch content would have been greater at temperatures close to optimum is reflected in the cross sections of leaves that were taken at higher temperatures. At 32/39 and ambient $CO_2$, the leaf contained virtually no starch by midafternoon (Fig. 4a); under high $CO_2$, starch was still present in the leaf at 5 AM (Fig.4b). When the plant was tranferred from high $CO_2$ to ambient

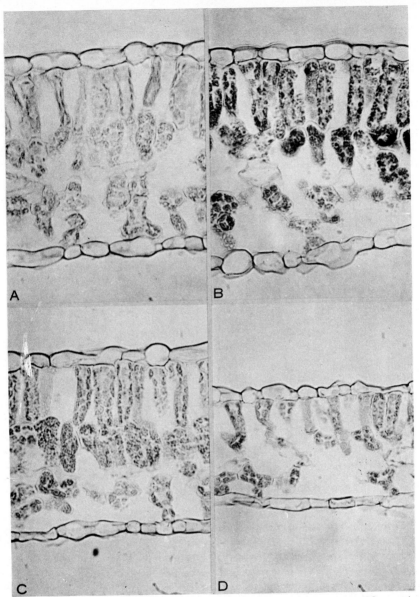

*Fig. 4.* Cross sections of soybean leaves grown at 32/29°C under (a) ambient $CO_2$, section taken midafternoon; (b) high $CO_2$, section taken 5 AM; (c) high $CO_2$ and transferred to low $CO_2$ 3 days previously, section taken 5 AM; (d) ambient $CO_2$ in the greenhouse, section taken midafternoon.

$CO_2$, the starch disappeared in about 3 days (Fig. 4c) and the photosynthetic rate increased to near the rate for a comparable leaf in ambient $CO_2$.

High $CO_2$ tended to increase the number of cells in the palisade layer and the development of the second layer of palisade cells. The greenhouse leaf grown during the late winter was thin with a single layer of small palisade cells (Fig. 4d) compared to the much larger palisade cells and the development of a second layer under artificial light.

At the lower temperature of 26/23/20°C (20°C during an 8-hr dark period, 23°C during a 4-hr light period, 26°C during an 8-hr light period with the $CO_2$ treatment, 23°C during another 4-hr light period, etc.) a slightly thicker leaf with better development of the second palisade layer and a more densely packed upper layer was produced under high $CO_2$ (Fig. 5a) than under normal $CO_2$ (Fig. 5b). Both sections were taken at 5 AM and showed the marked difference in starch content, although starch was present in both leaves at this time of day.

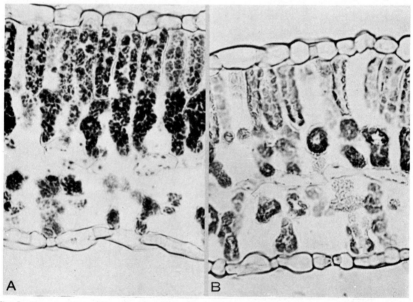

*Fig. 5.* Cross section of soybean leaves grown at 26/23/20°C under (a) high $CO_2$ and (b) ambient $CO_2$ both sections taken at 5 AM.

Correlation coefficients were calculated with all the data for different temperatures, age of leaf, $CO_2$ concentrations combined (Table 3). Stomatal resistance correlated very poorly with any of the parameters used. Correlation was highest between the photosynthetic rate and mesophyll resistance. This indicates that when leaves from different pretreatments are placed in the same environment, the stomata behave similarly and most of the difference in photosynthetic rates is due to changes in mesophyll resistance. Mesophyll resistance was significantly correlated with both starch levels and specific leaf weight when all the values were combined. As expected from these values, photosynthesis showed a significant negative correlation with starch content and specific leaf weight. Mesophyll resistance and sugar levels were weakly correlated, whereas photosynthesis and sugar level were positively correlated at the 5% level.

When the $CO_2$ treatments were separated (Tables 4 and 5) the photosynthetic

*Table 3.* Correlation coefficients among leaf photosynthetic rates, stomatal resistance, mesophyll resistance, sugar content, starch content and specific leaf weight. Values for different temperatures, age of leaf and two $CO_2$ concentrations combined.

| | Ps | $r_s$ | $r_m$ | Sug. | St. |
|---|---|---|---|---|---|
| $r_s$ | −.21 | | | | |
| $r_m$ | −.83** | −.14 | | | |
| Sug. | .25* | −.07 | −.22 | | |
| St. | −.32** | −.14 | .30** | −.43** | |
| Sp.lf.wt. | −.49** | −.01 | .40** | −.55** | 0.44** |

** Significant at 99% level.
* Significant at 95% level.

*Table 4.* Correlation coefficients among leaf photosynthetic rates, stomatal resistance, mesophyll resistance, sugar content, starch content and specific leaf weight. Values for different temperatures and age of leaf combined and for normal $CO_2$ concentration.

| | Ps | $r_s$ | $r_m$ | Sug. | St |
|---|---|---|---|---|---|
| $r_s$ | .08 | | | | |
| $r_m$ | −.77** | −.65* | | | |
| Sug. | .11 | .03 | −.09 | | |
| St. | −.02 | −.39 | .25 | −.38 | |
| Sp.lf.wt. | −.15 | −.32 | .25 | −.64** | .38 |

** Significant at 99% level.
* Significant at 95% level.

*Table 5.* Correlation coefficients among leaf photosynthetic rates, stomatal resistance, mesophyll resistance, sugar content, starch content and specific leaf weight. Values for different temperatures and age of leaf combined and for $CO_2$ enriched air.

| | Ps | $r_s$ | $r_m$ | Sug. | St. |
|---|---|---|---|---|---|
| $r_s$ | .04 | | | | |
| $r_m$ | −.90** | −.30 | | | |
| Sug. | .24 | .17 | −.31 | | |
| St. | −.43 | −.37 | .40 | −.40 | |
| Sp.lf.wt. | −.46* | −.06 | .41 | −.62** | .52* |

** Significant at 99% level.
* Significant at 95% level.

rate was strongly negatively correlated with mesophyll resistance, which was highest for the $CO_2$-enriched treatments. For $CO_2$-enriched air, weak correlations were also obtained for starch content and specific leaf weight. Stomatal resistance did not correlate with photosynthetic rates for either treatment.

Temperature, light, and $CO_2$ treatment affect the morphology of the leaf and greatly affect the starch content of soybean. A suppression of the photosynthetic rate as a result of various pretreatments is largely due to an increase in mesophyll resistance,[1] part of which seems to be attibutable to the starch content of the

[1] We are referring to the resistance that we calculated using the Gaastra method. It, of course, includes many things.

leaves. The sugar content did not appear to be important, but tended to increase as the rate of photosynthesis increased. In soybean the stomatal resistance appeared to be more or less independent of leaf morphology and starch content and more related to external conditions.

Photosynthesis appears to limit soybean growth at optimum temperatures, as indicated by the large growth response obtained by $CO_2$ enrichment. However, the indeterminate cultivar Biloxi was capable of a much greater growth response than Wayne. Both cultivars at 27-30°C had an increased rate of leaf initiation, which accounted for almost all the growth response in Wayne, but in Biloxi most of the growth response was due to increased axillary growth. A wide range of responses can be expected from $CO_2$ enrichment in different species. The response will depend on temperature, growth habit and other characteristics of the plant.

# REFERENCES

Brun, W.A. & R.L. Cooper (1967): Effects of light intensity and carbon dioxide concentration on photosynthetic rate of soybean. *Crop. Sci.,* 7: 451-454.

El-Sharkawy, M.A. & J. Hesketh (1965): Photosynthesis among species in relation to characteristics of leaf and $CO_2$ diffusion resistance. *Crop Sci.,* 5: 517-521.

Hesketh, J. (1963): Limitations of photosynthesis response for differences among species. *Crop Sci.,* 3: 493-496.

Hesketh, J.D., D.L. Myhre & C.R. Willey (1973): Temperature control of time intervals between vegetative and reproductive events in soybeans. *Crop Sci.,* 13: 250-254.

Hofstra, G. (1972): Response of soybeans to temperature under high light intensities. *Can. J. Plant Sci.,* 52: 535-543.

Ito, T. (1973): Plant growth and physiology of vegetable plants as influenced by carbon dioxide environment. Trans. Fac. Hort., Chiba Univ., no. 7. 134 p.

Kramer, P.J., H. Hellmers & R.J. Downs (1970): SEPEL: New Phytotrons for environmental research. *Bio Sci.,* 20: 1201-1208.

Neales, T.F. & L.D. Incoll (1968): The control of leaf photosynthesis rate by the level of assimilate concentration in the leaf: A review of the hypothesis. *Bot. Rev.,* 34: 107-125.

Madsen, E. (1975): Effect of $CO_2$-enrichment on growth, development, fruit production and fruit quality of tomato – from a physiological point of view. IN: Phytotrons in Horticultural Research, N. de Bilderberg & P. Chouard, Eds. In press.

Thomas, M.D. & G.R. Hill (1949): Photosynthesis under field conditions. IN: Photosynthesis in Plants, J. Franck & W.E. Loomis, Eds., The Iowa State College Press, Ames. p 19-52.

# Addendum

Erik Madsen, State Seed Testing Station, Skovbrynet 20, Denmark and T. Ito, Faculty of Agriculture, Chiba University, Matsudo, Chiba Prefecture, Japan have recently kindly supplied us with reprints and preprints describing their work on $CO_2$-enrichment in tomatoes. They both found photosynthesis decreased as starch increased up to 20-25% dry weight of the leaf. They both offer a comprehensive review of the literature related to $CO_2$ enrichment, photosynthesis and growth. Readers should also be aware of the experiments of Thomas and Hill in which they induced necrosis in tomato leaves at 0.3% $CO_2$ in a light-rich environment.

# RESPONSE OF STARCH SYNTHESIS TO TEMPERATURE IN CHILLING-SENSITIVE PLANTS

W.J.S. DOWNTON & J.S. HAWKER

*C.S.I.R.O. Division of Horticultural Research, Private Mailbag, Merbein, Vic. 3505 and P.O. Box 350, Adelaide, S.A. 5001, Australia*

*Abstract*

Temperatures in the chilling range (from about $12°$ to $0°C$) markedly inhibit photosynthetic electron transport, carboxylation and translocation in susceptible plants. A similar inhibition for starch synthesis is reported for the chilling-sensitive plants, avocado, maize and sweet potato. Arrhenius plots for soluble and starch grain-bound forms of starch synthetase from chilling-sensitive plants showed a change in slope (discontinuity) at about $12°C$. The chilling-resistant potato maintained a constant energy of activation from $23°C$ to $3°C$ for both forms of enzyme. The discontinuity in chilling-sensitive plants was abolished by high concentrations of *tert*-butyl alcohol, but not Triton X-100 or phospholipase A. The existence of an enzyme-lipid interaction within amylose molecules is suggested. The phospholipid, lysolecithin, may be uniquely involved in the control of starch biosynthesis and degradation.

## Introduction

Chilling temperatures in the range from $12°C$ down to freezing drastically disrupt normal physiological processes in many plants of tropical and subtropical origin. In this class are many important horticultural species such as citrus, avocado, banana, mango and pineapple as well as the commercially important tomato, sweet potato, bean, cucumber, corn, rice, millet, sorghum, sugarcane and cotton. Depending upon the conditions of exposure, susceptible plants may develop severe leaf chlorosis and their fruit may experience an abbreviated storage life. (Lyons, 1973).

The investigations of Raison (1973) and colleagues indicate that the primary response in chilling injury is the solidification of membrane lipids (a phase change in the state of lipid from a liquid-crystalline to a solid gel structure) at a critical temperature inducing conformational changes in membrane-associated enzymes. The $Q_{10}$ of enzyme activity, while approximately 2 at non-chilling temperatures, becomes much larger below the phase transition temperature in chilling-sensitive plants. Reactions catalyzed by nonmembrane-bound enzymes, on the other hand, decrease with a $Q_{10}$ of 2 through the chilling range. This differential effect of temperature in the chilling range on membrane-bound and 'soluble' enzymes leads to imbalances in metabolism at interfaces between nonmembrane-associated and membrane-associated processes, such as between glycolysis and the mitochondrial system, with the accumulation of metabolites such as ethanol and acetaldehyde which become toxic to the cell (Lyons, 1973). This explanation of chilling injury has been arrived at largely from mitochondrial studies.

*Acknowledgement:* Technical assistance by Mrs. V. Richards is gratefully acknowledged.

Chilling-resistant species show reduced growth and metabolism at low, but non-freezing temperatures, but otherwise are relatively unaffected. The lipids in these plants tend to contain a greater proportion of unsaturated fatty acids (Lyons et al., 1964) which solidify at a much lower temperature than the lipids in chilling-sensitive plants whose higher content of saturated fatty acids leads to a phase change at about 12° C.

More recently the effect of chilling temperatures on photosynthesis has received some attention. Shneyour et al. (1973) comparing the photochemical activities of chloroplasts isolated from chilling-sensitive and chilling-resistant plants found a large increase in the energy of activation ($E_{act}$) for the photoreduction of $NADP^+$ from water in chloroplasts of chilling-sensitive bean and tomato at about 12° C. The chilling-resistant lettuce and pea chloroplasts gave constant values of $E_{act}$ from non-chilling to chilling temperatures (25° C to 3° C). The effect of temperature in the chilling-sensitive chloroplasts was localized in photosystem I at the terminal electron transfer reaction to $NADP^+$ catalyzed by ferredoxin-$NADP^+$ reductase (E.C.1.6.99.4).

However chilling temperatures affect other components of photosynthesis besides the light reactions. McWilliam & Ferrar (1974) have observed that a range of chilling-sensitive $C_4$ grasses show large increases in $E_{act}$ below about 12° C for the primary carboxylating enzyme, phosphoenolpyruvate (PEP) carboxylase (E.C.4.1.1.31) both *in vitro* and *in vivo*. Chilling-resistant grasses do not display a discontinuity in the Arrhenius plot for PEP carboxylase. The abrupt decrease in the rate of photosynthesis seen in leaf slices of chilling-sensitive $C_4$ grasses at chilling temperatures (McWilliam & Ferrar, 1974) could be controlled either by photosynthetic electron transport, or carboxylation depending upon which process was rate-limiting under the experimental conditions chosen.

The translocation process is also subject to marked inhibition by chilling temperatures in susceptible species. Giaquinta & Geiger (1973) found a sudden increase in $E_{act}$ below 10° C for the translocation velocity of recently assimilated radiocarbon through bean petioles. Sugar beet, a chilling-resistant plant, did not show this response.

In this paper we report the effect of temperature from 23° C down to 3° C on starch synthesis by ADP-glucose: α-1,4-glucan α-4-glucosyltransferase (starch synthetase). It is evident that the synthesis of starch from recently assimilated carbon dioxide may also decrease abruptly at temperatures below about 12° C in chilling-sensitive plants.


## Material and Methods

*Chilling-sensitive:* Avocado cotyledons (*Persea americana* Mill. cv. Fuerte), maize kernels (*Zea mays* L. cv. NES 1002) and sweet potato tubers (*Ipomoea batatas* Lam.)

*Chilling-resistant:* Potato tubers (*Solanum tuberosum* L.)

*Extraction procedure:* Sweet potato tuber was ground in 50 mM Tris-HCl buffer pH 8.5 containing 10 mM DTT and 10 mM EDTA, or in 50 mM Tris-acetate buffer at pH 8.5 with 2 mM $Na_2S_2O_5$. The homogenate was filtered through Miracloth (Chicopee Mills, New York) and centrifuged at 30,000 x g for 15 min.

The supernatant was decanted and used as a source of soluble starch synthetase. The starch pellet was repeatedly washed and finally resuspended in a small quantity of buffer to provide a source of starch grain-bound starch synthetase. The other plant organs were processed similarly to sweet potato except that avocado cotyledon was ground in 350 mM Tris-acetate buffer pH 8.5 containing 20 mM EDTA, 11 mM diethyldithiocarbamate, 15 mM cysteine-HCl and 6% carbowax 4000. The buffer concentration was reduced to 20 mM during starch grain purification. Maize kernels were ground in 50 mM Tris-acetate buffer pH 8.5 containing 10 mM EDTA and 1 mM DTT. EDTA was omitted in experiments with phospholipase A.

*Assay of starch synthetase:* Transfer of $^{14}$C-glucose to a primer was measured as described by Ozbun et al. (1971). EDTA was omitted from reaction mixtures containing phospholipase A and replaced by 5 $\mu$mole $CaCl_2$. One mg rabbit liver glycogen in a final vol. of 0.2 ml provided a primer for soluble starch synthetase. For the assay of starch-bound starch synthetase, the grains themselves served as primer for the reaction. Reactions were terminated and the pellets handled by the procedures of Ghosh & Preiss (1965). Experiments with *tert*-butyl alcohol required further washing of the pellet to reduce the counts of the controls containing boiled enzyme.

Preparations treated with Triton X-100 were preincubated with 0.4 or 0.6% of the detergent on ice for 30 min prior to assay. Preparations treated with phospholipase A (*Crotalus terr. terr.* from Calbiochem, San Diego, Calif., U.S.A.) contained 2.4 I.U. of phospholipase with $CaCl_2$, 30 $\mu$l of starch synthetase preparation and in some cases Triton X. A pre-incubation period of 15 min at 23° was allowed prior to starting the starch synthetase assay with ADP$^{14}$C-glucose.

Reactions with *tert*-butyl alcohol contained 65 or 75% tert-butyl alcohol, 140 nmole ADP$^{14}$C-glucose and enzyme in a final vol. of 0.2 ml.

Assays were conducted at six temperatures between 3°C and 23°C ensuring that temperature variability throughout the experiment did not exceed $\pm 0.1$°C. The responses to temperature are presented as Arrhenius plots ($\log_{10}$ $^{14}$C-glucose incorporated into starch vs 1/temperature). Reaction rates were linear over the times used which were generally 10 min. No changes in the $K_m$ for ADP-glucose with temperature were apparent. Values for $^{14}$C-glucose incorporated into starch are means of triplicate assays at each temperature.

A more complete description of methodology used in these experiments can be found elsewhere (Downton & Hawker, 1975).


## Results and Discussion

Figure 1 illustrates the response of soluble and grain-bound starch synthetase from potato tuber to temperature from 23°C down through the chilling region to 3°C. Both synthetases in this chilling-resistant plant had a constant energy of activation. Soluble enzyme had a $K_m$ of 0.43 mM for ADP-glucose; starch-bound synthetase, a $K_m$ of 0.83 mM. The two starch synthetases in chilling-sensitive maize revealed a discontinuity in their Arrhenius plots at about 12°C (Figure 2). The $E_{act}$ values above the break were about half those below indicative of an abrupt decrease in the rate of starch synthesis below a critical temperature.

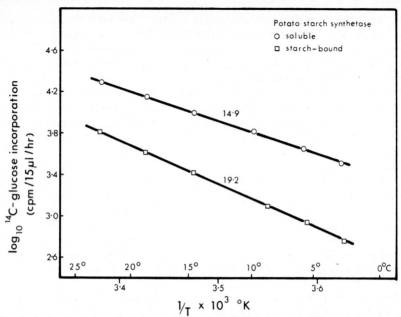

*Fig. 1.* Arrhenius plots of soluble and grain-bound ADPglucose-starch glucosyl transferase from potato tuber (chilling-resistant). The values beside each line are the calculated energies of activation (kcal/mole).

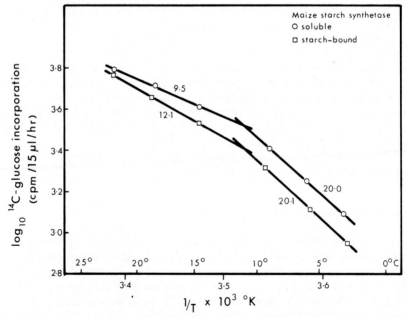

*Fig. 2.* Arrhenius plots of soluble and grain-bound ADPglucose-starch glucosyl transferase from maize kernels (chilling-sensitive). A discontinuity in the graph is apparent at about $12°$ for both enzymes.

*Table 1.* Effect of treatment on the energies of activation of starch synthetases from chilling-sensitive plants.

| Plant | Enzyme | Treatment | Energy of activation (kcal/mole) Above break | Below break |
|---|---|---|---|---|
| Sweet potato | Starch-bound | Control | 16.0 | 26.5 |
| | | Triton X (0.4%) | 14.6 | 23.8 |
| Avocado | Starch-bound | Expt. 1. | | |
| | | Control | 12.3 | 20.1 |
| | | Triton X (0.6%) | 12.8 | 18.3 |
| | | Expt. 2. | | |
| | | Control | 12.8 | 17.3 |
| | | *tert*-butyl alcohol (75%) | constant 23.9 | |
| | | *tert*-butyl alcohol (75%) | constant 29.7 | |
| | | Expt. 3. | | |
| | | Control | 6.4 | 17.4 |
| | | Phospholipase A | 7.8 | 14.9 |
| Maize | Soluble | Control | 9.1 | 19.6 |
| | | Triton X (0.4%) | 14.1 | 21.9 |
| | | Triton X + phospholipase A | 11.4 | 21.7 |
| | Starch-bound | Control | 11.4 | 20.2 |
| | | *tert*-butyl alcohol (65%) | constant 22.8 | |

The Table shows that the chilling-sensitive avocado and sweet potato responded like maize with respect to starch synthesis. Discontinuities in Arrhenius plots for some membrane-bound enzymes of chilling-sensitive plants can be abolished by low concentrations of detergents and phospholipase A (Raison, 1973; McGlasson & Raison, 1973). The application of Triton X-100 to sweet potato grain-bound synthetase, avocado grain-bound synthetase and maize soluble enzyme did not eliminate the discontinuity in the starch synthetase assay. Phospholipase A pre-treatment had no effect on avocado grain-bound enzyme and maize soluble enzyme.

Energies of activation for grain-bound enzyme from the cotyledons of avocado fruits harvested at different times over the season showed large differences. $E_{act}$ values above the transition temperature increased from 8 kcal/mole in early season fruit to 17 kcal/mole for late season fruit. $E_{act}$ values below the transition temperatures also increased as the season progressed. The values above and below 12° C came closer together later in the season. These changes correlated with germination of the seed within the fruit suggesting an altered metabolism within the cotyledon.

The Table shows that treatment with high concentrations of *tert*-butyl alcohol (65-75%) removed the discontinuity from avocado grain-bound synthetase and maize starch-bound enzyme. A constant $E_{act}$ was obtained which was somewhat greater than the $E_{act}$ in the chilling region of the control. *Tert*-butyl alcohol strongly inactivated maize soluble enzyme.

We consider it probable that a lipid interaction is responsible for the behaviour of starch synthetase at chilling temperatures in maize, avocado and sweet potato since the enzyme showed a break in Arrhenius plots at the same temperature that discontinuities have been reported for membrane-associated enzymes in these same species (Raison, 1973; Lyons, 1973).

Recent investigations have revealed that of the small lipid component in starch (about 1% of dry weight) up to 85% of this can be lysolecithin. (Acker & Schmitz, 1967a; Acker & Schmitz, 1967b; Wren & Merryfield, 1970). The lipid is specifically associated with the amylose of starch and is therefore absent from waxy varieties of maize and rice where starch consists almost entirely of amylopectin (Acker & Schmitz, 1967b; Nakamura et al., 1958a). Lysolecithin can be experimentally complexed to amylose. This reaction involves the entry of the lipid into the amylose helix as demonstrated by the displacement of iodine molecules from the helix (Nakamura et al., 1958b). We have found that *tert*-butyl alcohol can also displace iodine from amylose (Downton & Hawker, 1975). Starch synthetase is also considered to be bound to amylose in starch grains (Akazawa, 1965; Akazawa & Murata, 1965).

The observations that amylose is associated both with the enzyme starch synthetase and the lipid lysolecithin strengthen the suggestion of a lipid-protein interaction in starch grains which can be influenced by temperature and milieu, e.g. *tert*-butyl alcohol. A phase change in lysolecithin, or another lipid, at about 12° C could account for the discontinuity in the Arrhenius plot of grain-bound starch synthetase in chilling-sensitive plants. An environment of *tert*-butyl alcohol would be likely to change the type of lipid-enzyme interaction found in aqueous solution at chilling temperatures resulting in the elimination of a phase change at 12° C. This alcohol might also extract lipid from starch.

The break in the Arrhenius plot for maize soluble enzyme may be related to observations that small primers are bound to soluble starch synthetases (Schiefer, 1973; Hawker et al., 1974) to which a lipid could be bound. The constant values of $E_{act}$ found for the potato starch synthetases could result either from the reported absence of lysolecithin from potato starch (Acker & Schmitz, 1967b) or from the presence of a more unsaturated lipid whose phase transition temperature is much lower than $12°C$. Although phospholipase A can remove discontinuities in plots of certain membrane-associated enzymes, its lack of effect in the starch grain system may stem from its inability to penetrate the amylose helix. Also, lysolecithins of the $\beta$-isomer would not be susceptible to attack by phospholipase A (Acker & Schmitz, 1967a). Triton X has also been reported ineffective in eliminating the low temperature break in succinate oxidase activity of apple mitochondria (McGlasson & Raison, 1973).

If our hypothesis that lysolecithin or some other lipid within starch grains of chilling-sensitive plants can influence the activity of starch synthetase is correct, then the concept of enzyme control by lipid interaction as proposed by Raison and colleagues can be extended to include regulation by non-membrane lipids as well. PEP carboxylase probably provides another example of an enzyme controlled by some closely associated non-membrane lipid since most of the enzyme was not precipitated by high speed centrifugation (144,000 x g) suggesting no association with a membrane. This treatment did not eliminate the discontinuity in the Arrhenius plot (McWilliam & Ferrar, 1974).

No other biological material is known in which lysolecithin is present in such a high proportion of the total lipid (Wren & Merryfield, 1970). We suggest that lysolecithin may have a unique function in the biogenesis of starch as well as affect the degradation of starch. For example, the retarded degradation of starch seen in some $C_4$ grasses at $10°C$ could be controlled by the physical state of the lipid in starch grains (Garrard & West, 1972; Carter et al., 1972). Lysolecithin may further regulate starch synthesis by inhibiting the enzymatic conversion of all of the amylose molecules to amylopectin thereby permitting the simultaneous deposition of both amylose and amylopectin within the growing grain. The absence of lysolecithin from waxy starches supports this idea.

# REFERENCES

Acker, L. & H.J. Schmitz (1967a): Über die Lipide der Weizenstärke. 2. Mitt. Nachweis und Isolierung von Lysolecithin. *Die Stärke,* 19: 233-239.

Acker, L. & H.J. Schmitz (1967b): Über die Lipide der Weizenstärke. III. Mitt. Die übrigen Lipide der Weizenstärke sowie die Lipide anderer Stärkearten. *Die Stärke,* 19: 275-280.

Akazawa, T. (1965): Starch, inulin and other reserve polysaccharides. In Bonner, J. & Varner, J.E. (ed.): Plant Biochemistry. Pp 258-297. Academic Press, New York and London.

Akazawa, T. & T. Murata (1965): Adsorption of ADPG-starch transglucosylase by amylose. *Biochem. Biophys. Res. Comm.,* 19: 21-26.

Carter, J.L., L.A. Garrard & S.W. West (1972): Starch degrading enzymes of temperate and tropical species. *Phytochemistry,* 11: 2423-2428.

Downton, W.J.S. & J.S. Hawker (1975): Evidence for lipid-enzyme interaction in starch synthesis in chilling-sensitive plants. *Phytochemistry.*

Garrard, L.A. & S.H. West (1972): Suboptimal temperature and assimilate accumulation in leaves of 'Pangola' digitrass (*Digitaria decumbens* Stent.). *Crop Science,* 12: 621-623.

Giaquinta, R.T. & D.R. Geiger (1973): Mechanism of inhibition of translocation by localized chilling. *Plant Physiol.*, 51: 372-377.

Ghosh, H.P. & J. Preiss (1965): Biosynthesis of starch in spinach chloroplasts. *Biochemistry*, 4: 1354-1361.

Hawker, J.S., J.L. Ozbun, H. Ozaki, E. Greenberg & J. Preiss (1974): Interaction of spinach leaf adenosine diphosphate glucose α-1,4-glucan α-4-glucosyl transferase and α-1,4-glucan, α-1,4-glucan-6-glucosyl transferase in synthesis of branched α-glucan. *Arch. Biochem. Biophys.*, 160: 530-551.

Lyons, J.M. (1973): Chilling injury in plants. *Ann. Rev. Plant Physiol.*, 24: 445-466.

Lyons, J.M., T.A. Wheaton & H.K. Prat (1964): Relationship between the physical nature of mitochondrial membranes and chilling sensitivity in plants. *Plant Physiol.*, 39: 262-268.

McGlasson, W.B. & J.K. Raison (1973): Occurrence of a temperature-induced phase transition in mitochondria isolated from apple fruit. *Plant Physiol.*, 52: 390-392.

McWilliam, J.R. & P.J. Ferrar (1974): Photosynthetic adaptation of higher plants to thermal stress. In Bieleski, R.L., Ferguson, A.R. & Cresswell, M.M. (Ed.): Mechanisms of Regulation of Plant Growth. Pp 467-476. Published as Bulletin 12 by the Royal Society of New Zealand, Wellington, N.Z.

Nakamura, A., T. Kôno & S. Funahashi (1958a): Nature of lysolecithin in rice grains. Part I. Lysolecithin as a constituent of non-glutinous rice grains. *Bull. Agr. Chem. Soc. Japan*, 22: 320-324.

Nakamura, A., R. Shimizu, T. Kôno & S. Funahashi (1958b): Nature of lysolecithin in rice grains. Part II. Complex formation of lysolecithin with starch. *Bull. Agr. Chem. Soc. Japan*, 22: 324-329.

Ozbun, J.L. J.S. Hawker & J. Preiss (1971): Multiple forms of α-1,4 glucan synthetase from spinach leaves. *Biochem. Biophys. Res. Comm.*, 43: 631-636.

Raison, J.K. (1973): The influence of temperature-induced phase changes on the kinetics of respiratory and other membrane-associated enzyme systems. *Bioenergetics*, 4: 285-309.

Schiefer, S. (1973): The dependence of the 'unprimed' or 'de novo' reaction of starch synthetase on an endogenous primer. *Fed. Proc. Fed. Amer. Soc. Exp. Biol.*, 32: 603.

Shneyour, A., J.K. Raison & R.M. Smillie (1973): The effect of temperature on the rate of photosynthetic electron transport in chloroplasts of chilling-sensitive and chilling-resistant plants. *Biochem. Biophys. Acta*, 292: 152-161.

Wren, J.J. & D.S. Merryfield (1970): 'Firmly-bound' lysolecithin of wheat starch. *J. Sci. Fd. Agric.*, 21: 254-257.

# ENHANCED DARK $CO_2$ FIXATION BY MAIZE LEAVES IN RELATION TO PREVIOUS ILLUMINATION AND OXYGEN CONCENTRATION

J. POSKUTA & A. FRANKIEWICZ-JOZKO

*Laboratory of Plant Metabolism, Institute of Botany University of Warsaw, Poland*

## Abstract

The effect of blue, red and white light intensity and of oxygen concentration (21 or 100%) on the rates of $CO_2$ uptake by photosynthesis (PS) and by enhanced dark $CO_2$ fixation after preillumination ($En_{CO_2}$) in detached leaves of maize were investigated using an infra red $CO_2$ analyzer.

It has been found that the rates of $En_{CO_2}$ is a linear function of the rates of previous PS. For same rates of previous PS obtained with different light qualities, the blue light was always the most effective for $En_{CO_2}$ compared to red or white light in each $O_2$ concentration. Similar results were obtained after preillumination with light qualities and intensities required for the compensation of respiration. In these conditions, the high $O_2$ concentration also stimulates $En_{CO_2}$. It is proposed that light stimulation of $En_{CO_2}$ might result from an allosteric activation by some product(s) of PEP carboxylase activity, the most effective stimulation being obtained by blue light preillumination and high $O_2$ concentration.

## Introduction

The phenomenon of enhanced dark $CO_2$ fixation after preillumination presents a direct evidence of the occurrence of light and dark reactions in photosynthesis. Benson and Calvin (1947) have observed that $^{14}CO_2$ was picked-up in darkness after preillumination of *Chlorella* cells at rates several times faster than those of the ordinary dark $CO_2$ uptake. Thereafter a number of authors in the USA and in Japan obtained similar results with both algae and higher plants as well as with isolated chloroplasts systems (Gaffron et al., 1950; Gaffron & Fager, 1951; Tamiya or Miyachi & coworkers, 1955, 1957, 1958, 1959, 1960, 1962; Trebst et al., 1958; Bassham & Kirk, 1963; Togasaki & Gibbs, 1963, 1967; Hogetsu & Miyachi, 1970; Togasaki & Botos, 1971; Laber et al., 1971). The ability for $En_{CO_2}$ persisted for a few minutes in darkness. It is well established that in C-3 plants this process is mainly restricted to the carboxylation of ribulose-1,5-diphosphate accumulated during the preillumination to produce 3-phosphoglyceric acid. In C-4 plants the $En_{CO_2}$ was observed by Poskuta (1969) who showed that this process is related to $CO_2$ concentration. Samajima & Miyachi (1971), Stamieszkin (1972), Usuda et al. (1973) and Osmond (1974) have reported that in C-4 plants the dominant radioactive products of $En_{CO_2}$ were malate and aspartate.

They concluded that this process in C-4 plants is limited to the carboxylation of PEP to form C-4 acids. Therefore it is located in mesophyll cells only. It is reasonable to assume that the pool of PEP produced in excess during photosynthesis would be necessary to carry over the fixation of $CO_2$ from light to dark.

...arboxylation of PEP does not directly involves ATP. The synthesis of PEP however needs ATP. The conversion of oxaloacetate to malate requires $NADPH_2$. These compounds are produced during the light in the process of photophosphorylations. Therefore the conditions of preillumination are important for the process of $En_{CO_2}$. Moreover, Baldry et al. (1969) and Nagy et al. (1971) have reported that PEP carboxylase in C-4 plants is activated by light. Miyachi (1969) has observed the blue light stimulation of PEP carboxylase in *Chlorella* cells by a factor of 2-3 compared to red light; blue light had no effect on RuDP carboxylase. The $En_{CO_2}$ was reviewed by Frankiewicz-Jozko (1974). The blue light effect on photosynthesis was discussed by Voskresenskaya (1972).

The oxygen concentration is also an important factor influencing photosynthesis and carbon metabolism. Björkman (1966), Forrester et al. (1966), Poskuta (1969) have showed that leaves of C-4 plants did not exhibit the Warburg effect at $O_2$ concentrations below atmospheric. In our previous study (Stamieszkin et al., 1972) we noticed a Warburg effect for $En_{CO_2}$, but no change in the distribution of $^{14}C$ in the products of $En_{CO_2}$ was visible. The radioactivity of $En_{CO_2}$ was mainly located in malate and aspartate in either low or high $O_2$ concentration. It is postulated that oxygen inhibits photosynthesis in C-4 plants by inhibition of RuDP carboxylase in bundle sheath chloroplast without affecting PEP carboxylase in mesophyll cells (Chollet & Ogren, 1972 a, b; Bowes & Ogren, 1972). The accumulation of malate and aspartate in high $O_2$ concentration by maize leaves was observed in our previous work (Lewanty et al., 1971). It has been observed that these acids in turn inhibit PEP carboxylase (Lowe & Slack, 1971).

The objective of the present work was to investigate the interaction between conditions of preillumination (quality and intensity of light) and oxygen concentration on the $CO_2$ exchange rates in light and in darkness immediately after the lights were turned off in maize leaves.

## Material and Methods

Detached leaves of maize (*Zea mays* L., cv. 'Golden Bantham' or 'Konski zab') excised from plants of 3 to 9 weeks old were used as experimental material. Plants were grown in pots of garden soil in a growth chamber under controlled conditions: light intensity of 10.000 lux from fluorescent tubes during the 16 hours day at 28° C and 22° C during the night. Plants were watered daily to 70% of soil water capacity. The upper parts of leaves (about 30 cm$^2$) were placed in a polyethylene bag with their cut ends in water and were introduced in a plexiglass photosynthesis chamber connected by a closed system to an infra red $CO_2$ analyzer (Beckman 215 A). The volume of the system was 996 ml and the flow rate of air was 3 l per minute. The system was filled either with air (21% $O_2$) or with pure oxygen (100% $O_2$). The source of light was four 500 Watt photoflood lamps filtered through a 15 cm thick screen of running water to reduce infra red irradiation. Between the light source and the photosynthesis chamber, blue or red cellulose acetate filters were introduced to modify light quality. The transmission of these filters was described by Tregunna et al. (1962). The light intensity was measured by an YSI-65 radiometer. The experiments were carried out at 25° C. The following processes were measured according to the methods described earlier with some modifications (Poskuta, 1969): 1. rate of photosynthesis when $CO_2$ concentrations changed from 220 to 160 ppm. In our leaves we obtained a $CO_2$ compensation point equal to zero, indicating that the leaves appeared to lack in visible photorespiration. The initial concentration of $CO_2$ in the system was 300 ppm; 2. the rate of enhanced dark $CO_2$ fixation either after a period of photosynthesis during which $CO_2$ concentration in the system decreased from 300 ppm to 220 ppm or after a period of preillumination at the light intensity required for compensation of respiration.

In preliminary experiments, it has been observed that the $CO_2$ evolution in darkness after preillumination at light compensation point was retarded. This result was interpreted as the uptake of $CO_2$ arising from dark respiration, before it has a chance to diffuse to the outside atmosphere. Light compensation points were determined at $CO_2$ concentration of 220 ppm. The duration of illumination to obtain this equilibrium was about 2 minutes, after which it was kept constant for the next 2 minutes and then the light was turned off. The data described here are typical for number of experiments with both varieties of maize. Since the results were qualitatively similar only the data for maize var. 'Golden Bantham' of 4 weeks old plants are presented in this paper.

## Results and Discussion

Figure 1 a, b show the light curves of photosynthesis in 21 and 100% $O_2$ under white, red or blue light. In 21% $O_2$ and low light intensity, photosynthesis rates were similar for all light qualities but at high light intensity, they were markedly higher in white light compared to red or blue. No saturation was obtained at the used intensities. In 100% $O_2$, however, photosynthesis rates were saturated at relatively low light intensity; at the saturation level the $O_2$ inhibition was about 50-65% compared to 21% $O_2$. Figures 2 a, b, 3 a, b, 4 a, b, illustrate the effects of oxygen and light intensity on $En_{CO_2}$ rate. It is seen that an atmosphere of 100% $O_2$ inhibited $En_{CO_2}$ rates at all light intensities and qualities examined; the lowest rates were noticed after preillumination with blue light. The intensities of pre-illuminating lights were directly related with the $En_{CO_2}$ rates: the higher were the light intensities, the higher were the rates of $En_{CO_2}$. The life time of $En_{CO_2}$ was about 1 minute and showed no clear dependence on the previous intensity or quality of light.

The effects of light intensity during preillumination on $En_{CO_2}$ rates are shown on Figure 5 a, b; these curves can be considered as 'light curves' of $En_{CO_2}$. It is seen that $En_{CO_2}$ followed the similar pattern as previous photosynthesis (see Fig. 1 a, b); the oxygen inhibition was now about 30%. In contrast to photosynthesis, $En_{CO_2}$ was not saturated by previous illumination in both oxygen concentrations and was markedly higher after red light. When comparing the effect of light intensity and quality on the total $CO_2$ fixation in darkness after preillumination, the modified picture is visible (fig. 6 a, b). It is seen that in 21% $O_2$, total $En_{CO_2}$ was markedly lower after preillumination with white light compared to red or blue. In 100% $O_2$, $En_{CO_2}$ was markedly lower after preillumination with white light at low intensity compared to red or blue but very similar after preillumination at high light intensity. The data presented above clearly proved that the rate

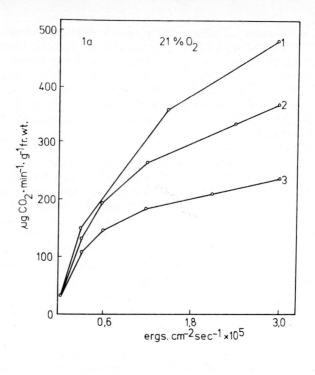

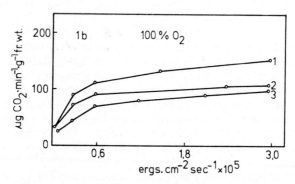

*Fig. 1 a,b.* Rates of $CO_2$ uptake in light by detached leaves of maize in relation to light intensity and quality in the atmosphere of 21 and 100% $O_2$. 1,2,3 white, red and blue light respectively.

of $En_{CO_2}$ as well as the total $CO_2$ uptake by $En_{CO_2}$ depended on the intensity and quality of preilluminating light and also on the concentration of oxygen in the atmosphere.

As mentioned in 'Introduction', $En_{CO_2}$ represents the carboxylation of PEP to form C-4 acids. Therefore the pool of PEP available to carry out this process should be produced during preillumination in some excess and not consumed in the current photosynthesis. The pool of PEP as well as the activity of PEP car-

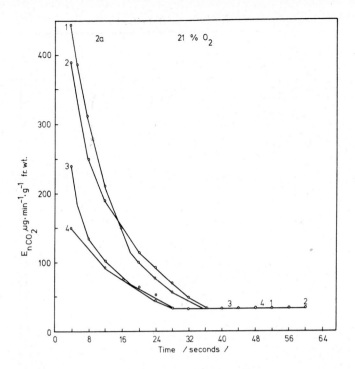

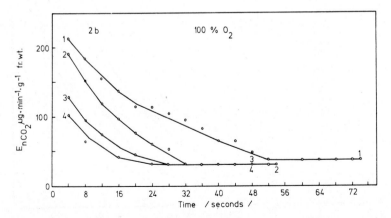

*Fig. 2 a,b, 3 a,b, 4 a,b.* Time courses of the dark decay of the rates of enhanced dark $CO_2$ fixation by detached leaves of maize after preillumination with white, red and blue light (Fig. 2-3-4) in the atmospheres of 21 and $100\%O_2$. Intensities of light: $1 : 3 \times 10^5$; $2 : 1.5 \times 10^5$; $3 : 0.6 \times 10^5$; $4 : 0.3 \times 10^5$ ergs $cm^{-2}$ $sec.^{-1}$.

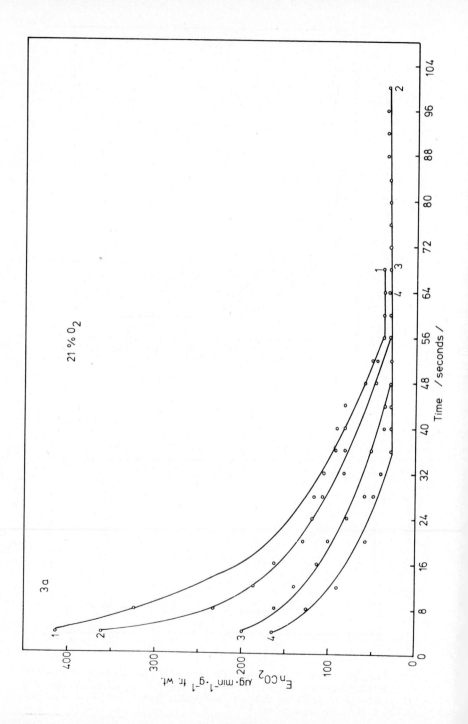

94

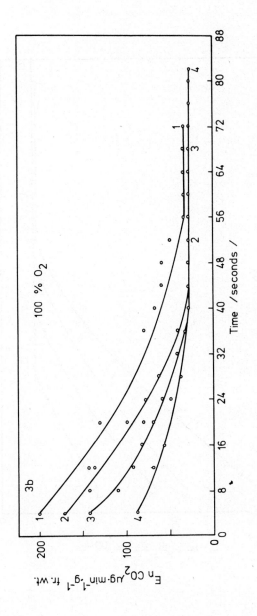

100 % $O_2$

3b

$E_n$ $CO_2$ $\mu g \cdot min^{-1} \cdot g^{-1}$ fr.wt.

Time /seconds /

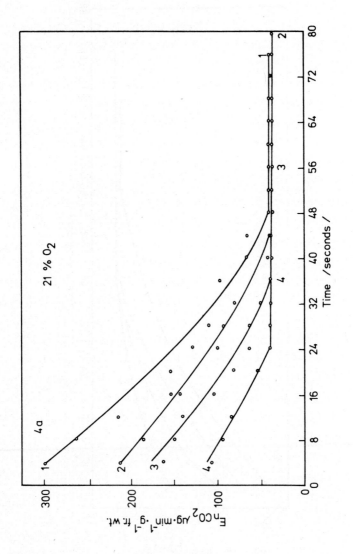

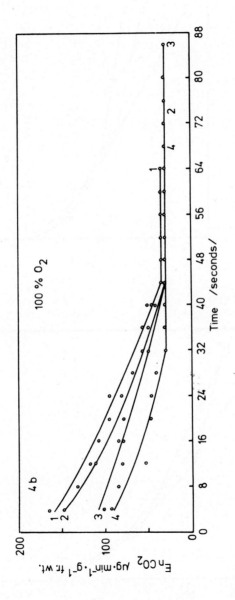

97

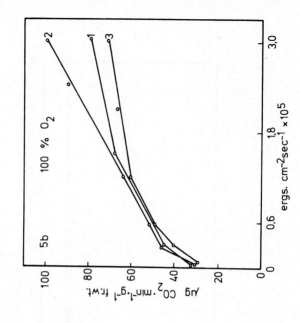

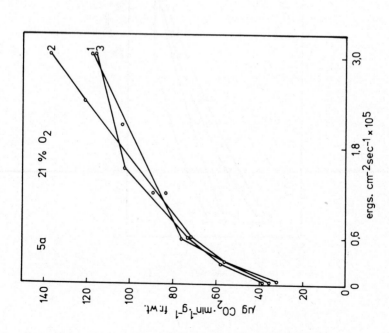

*Fig. 5 a,b.* The 'light curves' of the rates of enhanced dark $CO_2$ fixation by detached maize leaves in 21 and 100%$O_2$ after preillumination with either white (1), red (2) or blue light (3).

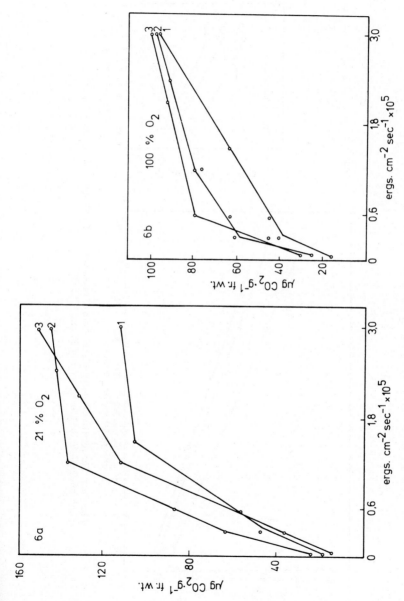

Fig. 6 a,b. The capacity for enhanced dark $CO_2$ fixation by detached maize leaves in 21 and 100% $O_2$ after preillumination with white (1), red (2) or blue light (3).

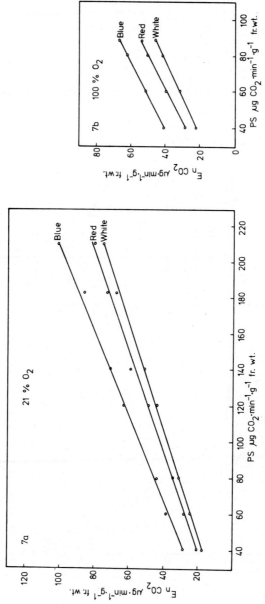

*Fig. 7 a,b.* The relationship between the rates of enhanced dark $CO_2$ fixation and the rates of photosynthesis in the atmospheres of 21 and 100%$O_2$

Table 1. The effect of light quality and $O_2$ concentration on light compensation point and enhanced dark $CO_2$ fixation of detached leaves of maize.

| Light quality | 21% $O_2$ | | | | 100% $O_2$ | | | |
|---|---|---|---|---|---|---|---|---|
| | Light intensity: $10^5$ Ergs cm$^{-2}$ sec$^{-1}$ | Time of apparent lack of $CO_2$ evolution in darkness (sec) | Dark respiration: $\mu g\,CO_2$ min$^{-1}$g$^{-1}$ fr. wt. | $EnCO_2$ total: $\mu g\,CO_2$ g$^{-1}$ fr. wt. | Light intensity: $10^5$ Ergs cm$^{-2}$ sec$^{-1}$ | Time of apparent lack of $CO_2$ evolution in darkness (sec) | Dark respiration: $\mu g\,CO_2$ min$^{-1}$g$^{-1}$ fr. wt. | $EnCO_2$ total: $\mu g\,CO_2$ g$^{-1}$ fr. wt. |
| White | 0.030 | 20.0 | 62.0 | 20.7 | 0.042 | 29.5 | 57.9 | 28.1 |
| Red | 0.036 | 23.4 | 61.4 | 24.0 | 0.048 | 30.0 | 62.0 | 29.9 |
| Blue | 0.042 | 29.7 | 69.4 | 36.3 | 0.090 | 40.9 | 64.6 | 44.2 |

boxylase would be responsible for the rates and the total $CO_2$ uptake in $En_{CO_2}$. In this sense the PEP and the activity of PEP carboxylase can be considered as photochemical products generated during the preillumination and utilized in darkness for the $CO_2$ fixation in dark reactions of photosynthesis. The rates of photosynthesis at a given intensity of illumination by white, red or blue light were different and consequently should produced different amounts of PEP available for $En_{CO_2}$. Therefore the important question arises whether similar or same rates of previous photosynthesis would produce similar or same rates of $En_{CO_2}$. A straight line relationship was found between the rate of $En_{CO_2}$ and the rate of previous photosynthesis for each light quality and $O_2$ concentration (Figure 7 a, b). It is also seen that same rates of photosynthesis induced by different light qualities produced different rates of $En_{CO_2}$. The established relation between photosynthetic rates and $En_{CO_2}$ rates is independent of the applied light quality. In an atmosphere of 100% $O_2$, the rates of $En_{CO_2}$ at a given photosynthesis rate could even be higher than those in air; blue light was then more effective for $En_{CO_2}$ than in air. Although photosynthesis and $En_{CO_2}$ rates at a given intensity of incident light were higher in white and red light compared to blue and that both photosynthesis and $En_{CO_2}$ were markedly lower in atmosphere of 100% $O_2$ compared to 21% $O_2$, blue light and 100% $O_2$ appeared to be the most effective for carrying over $En_{CO_2}$ when as comparison basis were taken the same previous rates of photosynthesis produced by the various conditions of preillumination. Same photosynthesis rates can also be obtained at light compensation points. We noticed that even after illumination of leaves with compensating light, $En_{CO_2}$ can be observed in maize (Table 1). The rates of dark respiration after preillumination were similar with white or red light but they were a little higher after preillumination with blue light in both $O_2$ concentrations. It is also seen on this table that higher intensities of blue and red lights were required to compensate respiration and for all light qualities, higher ones in 100% $O_2$ than in air. As mentioned in 'Material and Methods', the lack of net $CO_2$ exchange in darkness after illumination with compensating light means that $CO_2$ evolved by dark respiration was immediately fixed inside the leaves before it can escape to outside atmosphere. On table 1, the times of lack of $CO_2$ evolution due to $En_{CO_2}$ were much longer after preillumination with blue light compared to red and white light. It is also seen that these times were longer in 100% $O_2$ than in air. As a consequence, the total $En_{CO_2}$ uptake (time of lack of $CO_2$ evolution in darkness multiplied by the rate of dark respiration) was much higher after blue light than after red or white light. The total $CO_2$ uptake was considerably higher in 100% $O_2$ compared to air, and blue light again was the most effective light to carry over $En_{CO_2}$.

Recent data of Kagawa & Hatch (1974) showed that the crucial function of mesophyll chloroplast photoreactions would be the provision of ATP for conversion of pyruvate to PEP via pyruvate, Pi dikinase. These authors have shown that non-cyclic electron flow was responsible for providing ATP because PEP synthesis only occurred in the presence of Hill reagent. Moreover, they observed a light dependent PEP synthesis when pyruvate was added and the addition of ATP increased the formation of PEP. Earlier results of Hatch & Slack (1969) have shown that the synthesis of PEP in darkness was negligible and the production of malate by illuminating chloroplast was comparable to the NADP dependent malate dehydrogenase activity. On the basis of the biochemical data of the authors mentioned above it is reasonable to assume that the illumination conditions used in our experiments and the oxygen concentrations would affect either enzymes, the products of the $En_{CO_2}$ or both. We assume that the same rates of photosynthesis under different light quality and $O_2$ concentration would also produce the same magnitudes of PEP pool available for $En_{CO_2}$. If this assumption is valid, one can expect that variations of light quality and $O_2$ concentration which provide the same amount of PEP would not result in different $En_{CO_2}$ rates or total uptakes. The data however, indicate that $En_{CO_2}$ was stimulated by blue light and by 100% $O_2$. The tentative explanation of this finding implicates the activation of PEP carboxylase by blue light and high oxygen concentration. It is important that such activation of PEP carboxylase can be already observed at light compensation points which were attained in about 2 minutes. It is unlikely for such rapidity to postulate changes in the rates of synthesis or breakdown of some enzymes taking part in $En_{CO_2}$. A possibility exist that the stimulation of the PEP carboxylase activity by light and particularly by blue light and by high $O_2$ resulted from an allosteric activation by some products accumulating in the cells under the conditions applied. An allosteric nature of PEP carboxylase of C-4 plant *Pennisetum pupureum* was reported by Coombs et al. (1973). The second possibility can also be considered. Bell & Shuvalova (1971) have postulated a stimulatory effect of blue light on pseudoclyclic or cyclic photophosphorylation via photostructural changes produced by blue light. If blue light increases the production of ATP during preillumination, the pool of available ATP for the conversion of pyruvate to PEP can be a regulatory factor of the subsequent $En_{CO_2}$. The light curves of PS and $En_{CO_2}$ showed that in 21% $O_2$, PS was not saturated by the highest light intensities while in 100% $O_2$, saturation was obtained at relatively low light intensity. In contrast, $En_{CO_2}$ was not saturated in both $O_2$ concentrations. At high light intensity, 100% $O_2$ inhibited PS by about 50-65% but $En_{CO_2}$ by about 30%. Since the $En_{CO_2}$ is a mesophyll cell reaction this indicates that PS inhibition by 100% $O_2$ was partially due to the inhibition of photosynthesis in bundle sheath chloroplast and partially to the inhibition of $CO_2$ uptake in mesophyll cells due to accumulation of C-4 acids in high $O_2$ atmosphere as proposed by Lewanty et al. (1971) and Stamieszkin et al. (1972). The strong stimulatory effect on $En_{CO_2}$ of both blue light and high $O_2$ concentration due to the proposed activation of PEP carboxylase would be independent on the inhibitory effect of high $O_2$ concentration on $CO_2$ uptake by maize leaves produced by the feedback inhibitory mechanism proposed by Lowe & Slack (1971). Recently Chollet (1973) has shown that PEP-dependent $^{14}CO_2$ uptake in isolated crabgrass mesophyll cells is not inhibited by 100% oxygen. On the other hand 100% $O_2$ inhibited pyruvate – dependent $^{14}CO_2$ fixation in isolated mesophyll cells by 65% compared to air or 2% $O_2$. The author has proposed that one possible site of oxygen inhibition in the mesophyll could be the pyruvate-Pi dikinase reaction which converts pyruvate to PEP.

# REFERENCES

Baldry, C.W., C. Bucke & J. Coombs (1969): Light-phosphoenolpyruvate dependent carbon dioxide fixation by isolated sugar cane chloroplast. *Biochem. Biophys. Res. Commun.,* 37: 828-832.

Bassham, J.A. & M. Kirk (1963): Synthesis of compounds from $^{14}CO_2$ by *Chlorella* in the dark following preillumination. In: Studies on Microalgae and Photosynthetic Bacteria. *Plant Cell Physiol.,* (Special Issue) 493-504.

Bell, L.N. & N.P. Shuvalova (1971): Blue radiant energy storage uncoupled from oxygen evolution. *Photosynthetica,* 5: 113-123.

Benson, A.A. & M. Calvin (1947): The dark reaction in photosynthesis. *Science,* 105: 648-649.

Björkman, O. (1969): The effect of oxygen concentration on photosynthesis in higher plants. *Physiol. Plantarum,* 19: 618-633.

Bowes, G. & W.L. Ogren (1972): Oxygen inhibition and other properties of soybean ribulose-1,5-diphosphate carboxylase. *J. biol. Chem.,* 247: 2171-2176.

Chollet, R. (1973): The effect of oxygen on $^{14}CO_2$ fixation in mesophyll cells isolated from *Digitaria sanguinalis L.* leaves. *Biochem. Biophys. Res. Commun.,* 55: 850-856.

Chollet, R. & W.L. Ogren (1972a): Oxygen inhibits maize bundle sheath photosynthesis. *Biochem. Biophys. Res. Commun.,* 46: 2062-2066.

Chollet, R. & W.L. Ogren (1972b): The Warburg effect in maize bundle sheath photosynthesis. *Biochem. Biophys. Res. Commun.,* 48: 684-688.

Coombs, J., C.W. Baldry & C. Bucke (1973): The C-4 pathway in *Pennisetum purpureum.* I. The allosteric nature of PEP carboxylase. *Planta (Berl.),* 110: 95-107.

Frankiewicz-Jozko, A. (1974): Wzmożone ciemniowe wiazanie $CO_2$ u roślin. (Enhanced dark $CO_2$ fixation by plants.) *Wiad. Bot.,* 18: 111-117.

Forrester, M.L., G. Krotkov & C.D. Nelson (1966): Effect of $O_2$ on photosynthesis, photorespiration in detached leaves. II. Corn and other monocotyledones. *Plant Physiol.,* 41: 428-431.

Gaffron, H., E.W. Fager & J.L. Rosenberg (1950): Intermediates in photosynthesis. *Fed. Proc.,* 9: 535-542.

Gaffron, H. & E.W. Fager (1951): The kinetics and chemistry of photosynthesis. *Annu. Rev. Plant. Physiol.,* 2: 87-114.

Hatch, M.D. & C.R. Slack (1969): NADP-specific malate dehydrogenase and glycerate kinase in leaves and evidence for their location in chloroplast. *Biochem. Biophys. Res. Commun.,* 34: 589-593.

Hogetsu, D. & S. Miyachi (1970): Effect of oxygen on the light-enhanced dark carbon dioxide fixation in *Chlorella. Plant Physiol.,* 45: 178-182.

Kagawa, T. & M.D. Hatch (1974): Light-dependent metabolism of carbon compounds by mesophyll chloroplasts from plants with the $C_4$ pathway of photosynthesis. *Aust. J. Plant. Physiol.,* 1: 51-64.

Laber, L.J., E. Latzko, C. Levi & M. Gibbs (1971): Light enhanced dark fixation of $CO_2$ by corn and spinach leaves. In Forti, G., Avron, M. & Melandri, A. (ed.): Photosynthesis, Two Centuries after its discovery by Joseph Priestley. Pp 1737-1744. Dr. W. Junk B.V. – Publishers, The Hague.

Lewanty, Z., S. Maleszewski & J. Poskuta (1971): The effect of oxygen concentration on $^{14}C$ incorporation into products of photosynthesis of detached leaves of maize. *Z. Pflanzenphysiol.,* 65: 469-472.

Lowe, J. & C.R. Slack (1971): Inhibition of maize leaf phosphopyruvate carboxylase by oxaloacetate. *Biochim. Biophys. Acta,* 235: 207-209.

Miyachi, S. (1959) Effect of poisons upon the mechanism of photosynthesis as studied by the pre-illumination experiments using carbon-14 as a tracer. *Plant Cell Physiol.,* 1: 1-15.

Miyachi, S. (1960): Effect of some sulfhydryl reagents on light-induced $CO_2$-fixation in *Chlorella. Plant Cell Physiol.,* 1: 117-130.

Miyachi, S. (1962): Effect of preillumination on the incorporation of sulfur in the lipid fraction of *Chlorella* cells. *Plant Cell Physiol.,* 3: 193-196.

Miyachi, S. (1969): Regulation of photosynthetic carbon metabolism by wavelength, light intensity and oxygen. Proc. 11th. Int. Bot, Congr. Seattle, 149.

Miyachi, S., S. Izawa & H. Tamiya (1955): Effect of oxygen on the capacity of carbon dioxide fixation by green algae. *J. Biochem.*, 42: 221-224.

Miyachi, S., S. Hirokawa & H. Tamiya (1957): The 'background' $CO_2$-fixation occuring in green cells and its possible relation to the mechanism of photosynthesis. In Gaffron, H. et al.,(Ed.): Research in Photosynthesis. Pp. 205-212. Intersciense Publs., New York.

Nagy, A.H., A. Bokany, N.G. Doman & A. Faludi-Daniel (1971): Activities of carboxylating enzymes in light treated leaves from normal and carotenoid mutant maize. In Forti, G., Avron, M. & Melandri, A. (Ed.): Photosynthesis, Two Centuries after its discovery by Joseph Priestley. Pp 1861-1868. Dr. W. Junk B.V. – Publishers, The Hague.

Osmond, C.B. (1974): Carbon reduction and photosystem II deficiency in leaves of $C_4$ plants. *Aust. J. Plant Physiol.*, 1: 41-50.

Poskuta, J. (1969): Photosynthesis, respiration and post-illumination fixation of $CO_2$ by corn leaves as influenced by light and oxygen concentration. *Physiol. Plantarum*, 22: 43-54.

Samejima, M. & S. Miyachi (1971): Light-enhanced dark carbon dioxide fixation in maize leaves. In Hatch, M.D., Osmond, C.B. & Slatyer, R.O. (Ed.): Photosynthesis and Photorespiration. Pp 211-217. Wiley-Interscience, New York.

Stamieszkin, I., S. Maleszewski & J. Poskuta (1972): The effect of oxygen concentration during preillumination on enhanced dark $CO_2$-fixation by maize leaves. *Z. Pflanzenphysiol.*, 67: 180-182.

Tamiya, H., S. Miyachi & T. Hirokawa (1957): Some new preillumination experiments with carbon-14. In Gaffron, H. et al., (Ed.): Research in Photosynthesis Pp 212-222. Interscience Publ., New York,

Tamiya, H., S. Miyachi, T. Hirokawa & T. Katoh (1958): Radioisotopes in Scientific Research. IV, 1.

Togasaki, R.K. & C.R. Botos (1971): Enhanced dark $CO_2$ fixation by preilluminated algae; A tool for analysis of photosynthetic mechanisms in vivo. In: Forti, G., Avron, M. & Melandri, A. (Ed.): Photosynthesis, Two Centuries after its discovery by Joseph Priestley. Pp 1759-1772. Dr. W. Junk B.V. – Publishers, The Hague.

Togasaki, R.K. & Gibbs, M. (1963): Preillumination experiments in *Chlorella pyrenoidosa* and *Anacystis nidulans*. In: Studies on Microalgae and Photosynthetic Bacteria. *Plant Cell Physiol.*, (Special Issue), 505-511.

Togasaki, R.K., M. Gibbs (1967): Enhancement dark $CO_2$ fixation by preilluminated *Chlorella pyrenoidosa* and *Anacystis nidulans*. *Plant Physiol.*, 42: 991-996.

Trebst, A.V., H.Y. Tsujimoto & D.I. Arnon (1958): Separation of light and dark phases in the photosynthesis of isolated chloroplasts. *Nature (Lond.)*, 182: 351-355.

Tregunna, E.B., G. Krotkov & C.D. Nelson (1962): Effect of white, red and blue light on the nature of the products of photosynthesis in tobacco leaves. *Can. J. Bot.*, 40: 317-326.

Usuda, H., M. Samejima & S. Miyachi (1973): Distribution of radioactivity in carbon atoms of malic acid formed during light-enhanced dark $^{14}CO_2$-fixation in maize leaves. *Plant Cell Physiol.*, 14: 423-426.

Voskresenskaya, N.P. (1972): Blue light and carbon metabolism. *Annu. Rev. Plant Physiol.*, 23: 219-234.

# STUDIES ON THE CONTENT OF P 700 AND CYTOCHROMES IN *SINAPIS ALBA* DURING GROWTH UNDER TWO DIFFERENT LIGHT INTENSITIES *

H. GRAHL & A. WILD **

*Fachbereich Biologie der Universität, Didaktik, 6-Frankfurt am Main – 90, Sophienst. 1-3, B.R.D.*
** *Institute for General Botany, University of Mainz, B.R.D.*

*Abstract*

*Sinapis alba* was grown under two extremely different light intensities. We obtained two different modifications which hold several physiological and biochemical characteristics comparable to the relations in light- and shade plants. In the strong light the maximum rate of $CO_2$-uptake increases within five days to a three-fold rate. At the same time the content of several redox systems of photosynthetic electron transport increases while the content of P 700 remains equivalent in both types of plants. In the high-light plants we find a ratio P 700: Cytochrome f of 1 but a ratio of 3 in the low-light plants. These ratios could be probably essential features of various sun- and rainforest species. The present results cannot indicate the specific role of P 700 in photosynthetic electron transport especially in the low-light plants.

## Introduction

In previous publications we indicated that plants of *Sinapis alba* can be modified into two different phenotypes when the plants are grown under two extremely different light intensities (Grahl & Wild, 1972, 1973). The rates of $CO_2$-uptake, the morphological structure, and the ratios of pigments in both phenotypes are in agreement with the relations in light- and shade plants according to the traditional literature.

Studies on the relation of chlorophyll to lipophylic plastid quinones, ferre-doxin and cytochromes indicated that more redox systems of photosynthetic electron transport exist in the plants which are grown under strong light conditions than in those grown under weak ones (Grahl & Wild, 1972, 1973; Wild et al., 1973). According to these results a correlation between the maximum rate of $CO_2$-uptake and the content of redox systems was evident. Further investigations of Wild et al. (1973) led to the somewhat unexpected result that the content of P 700 is the same in both the light- and the shade phenotypes of *Sinapis alba*. Physiological investigations on P 700 and cytochromes are barely mentioned in the literature yet, and thus the present study will support additional data on these components in *Sinapis alba* concerning a longer growth period under two different light intensities.

* This work was supported by a research grant from the Deutsche Forschungsgemeinschaft.

*Abbreviations:* Chl = chlorophyll a + chlorophyll b; Cyt b 559 HP = cytochrome b 559 high potential; Cyt b 559 LP = cytochrome b 559 low potential; Cyt b 563 = cytochrome b 563; Cyt f = cytochrome f; Fd = ferredoxin; PQ = plastoquinone; $PQH_2$ = plastohydroquinone.

# Methods

Details about the cultivation of the plants under different light intensities have already been reported by Grahl & Wild (1972). Chloroplasts were isolated as described by Wild et al. (1973). The content of P 700 was determined by the oxidized-minus-reduced difference spectrum according to the method of Marsho & Kok (1971). The content of the cytochromes was also determined by difference spectra according to the method of Bendal et al. (1971) and only the evaluation of the cytochrome b-peaks was slightly modified. The following millimolar extinction coefficients were used:

| redox system | $\varepsilon mM^{-1} \times cm^2$ | reference |
|---|---|---|
| P 700 | 64 | Hiyama & Ke (1972) |
| cytochrome f | 17,7 | Bendal et al. (1971) |
| cytochrome b | 20 | Bendal et al. (1971) |

The content of the chlorophylls was determined in 80% acetone according to the method of Ziegler & Egle (1965). The difference spectra of P 700 were recorded in a Shimadzu model UV-200 and those of the cytochromes in a Shimadzu model MPS-50 L spectrophotometer.

# Results and Discussion

Figure 1 shows the maximum rate of photosynthesis during the growth under strong- and weak light conditions as reported earlier by Grahl & Wild (1972). At the beginning of the examined time (marked as 0 days) the very young plants were completely adapted to weak light conditions. When the plants were exposed to strong light the rate of $CO_2$ uptake increased and reached an almost constant level after passing a plain maximum. The plants which remained in the weak light showed a slight decrease of the $CO_2$-uptake. In the diagrams two different scales are used for time determination because of the different growth intensity under the two light conditions.

Table 1 shows the molar ratios of chlorophylls to P 700 and cytochromes in the high- and the low-light plants at the end of the examined growth-period. In the diagrams 2-3 the relative molar contents (molecules of redox system per 1000 molecules chlorophyll) are plotted against the growth-time.

Figure 2 shows that the content of P 700 was the same under both the strong- and the weak light conditions. The content remained constant during the examined growth period. The content of cytochrome f increased after a delay of one day and reached a constant level five days after the plants were exposed to strong light. In the weak light there was a slight decrease of the cytochrome f-content which apparently lasted the whole investigated time.

Figure 3 represents the content of the b-cytochromes. In the low-light plants the content of all three b-types showed a decrease. The decrease was linear for cytochrome b 559 high potential. The cytochromes b 559 low potential and b 563 reached a constant level after approximately ten days. In the strong light the

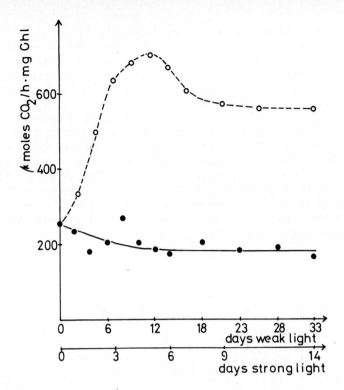

*Fig. 1.* Maximum rate of $CO_2$-uptake in leaves of *Sinapis alba* during growth in weak light
(———) and in strong light (- - - - -).

*Table 1.* Molar ratios of chlorophylls to P 700 and cytochromes in chloroplasts of *Sinapis alba*
during growth in weak light and in strong light.

|  | weak light | strong light |
|---|---|---|
| Chl/P 700 | 250 | 250 |
| Chl/Cyt f | 750 | 280 |
| Chl/Cyt b 559 HP | 2000 | 720 |
| Chl/Cyt b 559 LP | 920 | 250 |
| Chl/Cyt b 563 | 590 | 200 |

content of all three b-types increased and a constant level was reached after approximately six days. There were differences at the beginning of the growth-period
when the plants were brought into the strong light. Cytochrome b 559 high
potential showed a distinct decrease at the first day in the strong light. The initial
content was reached after the following four days. The content of cytochrome b
563 possibly remained constant at the beginning. After the delay of approximately one day we noticed a steep increase. Only the content of cytochrome
b 559 low potential increased almost linear for the first days in the strong light.

The present results indicate that an increase of the $CO_2$-uptake in the strong light is correlated with an increase of cytochromes. Previous studies on the content of other redox systems led us to similar conclusions. We have therefore only little doubt that a higher rate of $CO_2$-uptake in the high-light plants is due to a greater electron-carrying capacity as well as to an increase in the carbon fixation system as found by Björkman & Gauhl (1969).

At the same time we find the same ratio of chlorophylls/P 700 in both types of plants. Therefore the present result cannot indicate the specific role of P 700 in photosynthetic electron transport especially in the low-light plants. Wild et al. (1973) assume that in the low-light plants possibly most of the P 700 is not involved in the non-cyclic electron transport from $H_2O$ to $NADP^+$. It is then likely that in only the high-light plants all P 700 is involved in the non-cyclic electron flow. At the present time we are carrying out studies to prove whether the major part of P 700 is relizable for cyclic phosphorylation in the low-light plants. We can also assume that the ratio P 700/chlorophyll is probably genetic fixed and not

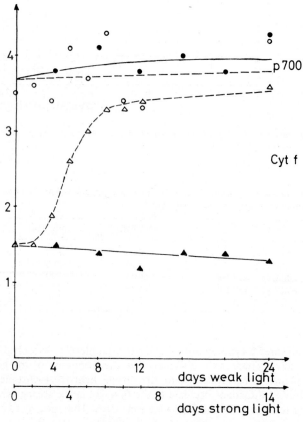

Fig. 2. Relative molar content (molecules per 1000 molecules chlorophyll) of P 700 and cytochrome f in chloroplasts of *Sinapis alba* during growth in weak light (- - - - -). and in strong light (- - - - -).

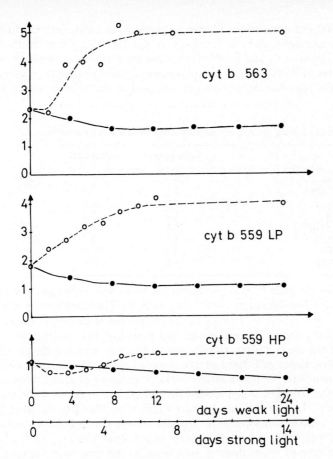

*Fig. 3.* Relative molar content (molecules per 1000 molecules chlorophyll) of b-cytochromes in chloroplasts of *Sinapis alba* during growth in weak light (———) and in strong light (- - - - -).

influenced by light. This hypothesis would mean that the low-light-plants have an abundance of P 700.

If we compare the increase of the maximum rate of $CO_2$-uptake in the strong light with the increase of the various redox systems we find the closest similarity with the appearence of cytochrome f. The ratio chlorophyll/cytochrome f indicates that there are always less molecules of cytochrome f present than of the other redox systems. If we therefore assume one molecule of cytochrome f per one photosynthetic electron transport we can calculate the number of molecules of the other redox systems which are probably associated with one photosynthetic electron transport as it is shown in table 2. We are aware that these data should not be taken as absolute values. This calculation is rather useful for an easy comparison of various redox systems from the aspect of participation in photosynthetic electron transport. The table includes also data of ferredoxin and plastoquinone which were calculated from our previous studies. The data support the view that P 700 and cytochrome f could be possibly the redox systems which are limiting photosynthetic electron transport in the high-light plants. In the low-light plants cytochrome f and ferredoxin are the limiting systems while there are three molecules of P 700 present. The pool-sizes of plastoquinone are comparable in both types of plant but there exists a greater pool of plastohydroquinone in the strong light.

*Table 2.* Molecules of redox systems calculated per one molecule cytochrome f.

|  | weak light | strong light |
|---|---|---|
| Chl | 750 | 280 |
| P 700 | 3,0 | 1,1 |
| Cyt f | 1,0 | 1,0 |
| Fd | 1,2 | 2,1 |
| Cyt b 559 | 1,4 | 1,2 |
| Cyt b 563 | 1,4 | 1,3 |
| PQ | 12,0 | 11,0 |
| $PQH_2$ | 14,0 | 29,0 |

We find that the ratio P 700/cytochrome f is about one in the high-light-plants and three in the low-light-plants. Björkman et al. (1972) found similar ratios in high- and low-light-plants of *Atriplex patula*. Boardman et al. (1972) found that chloroplasts from two rainforest species had three-fold less cytochrome f than chloroplasts from the sun plant *Atriplex patula*. Just now we are in the process of researching the content of P 700 and cytochrome f in several sun- and rainforest species (unpublished). We find a ratio P 700/cytochrome f of about three in the shade plants *Galium odorata* and *Asarum europaeum* but a ratio of one in the sun plant *Zea mays*. Due to the present data which were found in different laboratories with various plant-species it appears reliable that a P 700/cytochrome f-ratio of one is probably an essential feature of sun plants while typical rainforest species have a ratio of about three or even more. At this time we can assume that the ratio P 700/cytochrome f is probably a reliable biochemical indicator for light- and shade plants as it is already well known for pigment ratios.

# REFERENCES

Bendal, D.S., H.E. Davenport & R. Hill (1971): Cytochrome components in chloroplasts of higher plants. In Methods of Enzymology, vol. 23, Pp. 327-343. Academic Press, New York-London.

Björkman, O. & E. Gauhl (1960); Carboxydismutase activity in plants with and without β-carboxylation photosynthesis. *Planta (Berl.)*, 88: 197-203.

Björkman, O., N.K. Boardman, J.M. Anderson, S.W. Thorne, D.J. Goodchild & J.A. Pyliotis (1972): Effect of light intensity during growth of *Atriplex patula* on the capacity of photosynthetic reactions, chloroplast components and structure. In Carnegie Institution Year Book 71, Pp. 115-134. Stanford, California.

Boardman, N.K., J.M. Anderson, S.W. Thorne & O. Björkman (1972): Photochemical reactions of chloroplasts and components of the photosynthetic electron transport chain in two rainforest species. In Carnegie Institution Year Book 71, Pp. 107-114. Stanford, California.

Grahl, H. & A. Wild (1972): Die Variabilität der Photosyntheseeinheit bei Licht- und Schattenpflanzen. Untersuchungen zur Photosynthese von experimentell induzierten Licht- und Schattentypen von *Sinapis alba. Z. Pflanzenphysiol.*, 67: 443-453.

Grahl, H. & A. Wild (1973): Lichtinduzierte Veränderungen im Photosynthese-Apparat von *Sinapis alba. Ber. dtsch.bot.Ges.*, 86: 341-349.

Hiyama, T. & B. Ke (1972): Difference spectra and extinction coefficients of P 700. *Biochim. biophys. Acta,* 267: 160-171.

Marsho, T.V. & B. Kok (1971): Detection and isolation of P 700. In Methods of Enzymology, vol. 23, Pp. 515-522. Academic Press, New York-London.

Wild, A., H. Grahl & H.O. Zickler (1972): Untersuchungen über den Ferredoxingehalt von experimentell induzierten Licht- und Schattentypen von *Sinapis alba. Z. Pflanzenphysiol.,* 68: 283-285.

Wild, A., B. Ke & E.R. Shaw (1973): The effect of light intensity during growth of *Sinapis alba* on the electron-transport components. *Z. Pflanzenphysiol.* 69: 344-350.

Ziegler, R. & K. Egle (1965): Zur quantitativen Analyse der Chloroplastenpigmente. I. Kritische Überprüfung der spektralphotometrischen Chlorophyll-Bestimmung. *Beitr. Biol. Pfl.,* 41: 11-37.

# THE EFFECT OF LIGHT INTENSITY DURING GROWTH OF *SINAPIS ALBA* ON THE ELECTRON-TRANSPORT AND THE NONCYCLIC PHOTOPHOSPHORYLATION

A. WILD, W. RÜHLE & H. GRAHL *

*Institute for General Botany, University of Mainz, B.R.D.*
*Seminary for Didactic of Biology, University of Frankfurt, B.R.D.*

*Abstract*

The photosynthetic rate of *Sinapis alba* can be modified over a wide range by the light intensity during growth. Our present results indicate that there exist regulatory mechanisms in the field of photosynthetic primary reactions. We compared the effect of different light intensities during growth of *Sinapis* plants on the concentrations of soluble proteins, manganese and lipophilic plastid quinones, the electron flow from water to ferricyanide and noncyclic phosphorylation. We further determined the light dependence curves for the uncoupled electron transport with ferricyanide as electron acceptor and methylammonium-chloride as an uncoupler of photophosphorylation.

## Introduction

It is well known that light intensity during growth of plants has a regulatory effect on the maximum rate of photosynthesis. Sun plants have a high capacity for photosynthesis at high light intensities but relatively low rates at low intensities. Shade plants are capable of efficient use of low light intensities for photosynthetic $CO_2$ uptake but are incapable of high photosynthetic rates at high light intensities. The respiratory $CO_2$ release and the light compensation point for $CO_2$ exchange are lower in shade plants than in sun plants.

The adaptability of the photosynthetic apparatus to the light conditions may be determined by the genotype in many plants species and is a result of genetic adaptation to the light climate prevailing in the native habitat.

Björkman & Holmgren (1963) and Gauhl (1969) have found that such genotypically based adapatations exist not only between different species but also between different ecotypes. In other plant species, however, exists a high modification range and such adaptive differences of photosynthesis can be induced in a wide range by the light intensity during growth.

In our preliminary studies we have found that plants of *Sinapis alba* (white mustard) are very suitable for studying the problem of photosynthetic adaptation by plants to different light conditions in their natural environment. It grows under extremely low light conditions comparable to that at a forest floor, as well as under very high light conditions.

*Acknowledgement:* This work was supported by grants from the Deutsche Forschungsgemein-schaft.

# Material and methods

The following growing conditions were used. Seeds of *Sinapis alba* were sown in potting soil and germinated in the laboratory greenhouse under natural daylight. After 4 days the seedling were potted and after another 3 days they were transferred in a climate room under weak light conditions. After growing for a week under low-light conditions, 14 days after sowing, the experiments were started. (In the figures the beginning of the experiments is marked as 0 days). At this time the cotyledons had finished their growth and the primary leaf pair was developing. In parallel series of experiments the plants then were cultivated under two different light conditions, and in this time the experiments were performed. In high light, the plants require about 14 days for flowering, and in extremely low light approximately 33 days. The experiments were finished, when the half of the plants was flowering.

Details about the measurement of $CO_2$-fixation have already been reported by Grahl & Wild (1972). The plastid quinones were separated and quantitative determined according to Lichtenthaler (1968). The chlorophylls were estimated using the extinction factors of Ziegler & Egle (1965). The separation of carotenoids was achieved by thinlayer chromatography according to Hager & Bertenrath (1962). Manganase was determined with a Perkin Elmer atom absorption spectrophotometer. The isolation of chloroplasts and measurement of photochemical activities were achieved according to Strotmann (1970). 0,2% bovine serum albumin and 0,2% pectinase were added to the isolation medium to protect the photochemical activities (Homann & Schmid, 1967).

# Results and Discussion

The photosynthetic rate of *Sinapis alba* can be modified over a wide range by the light intensity during growth (Grahl & Wild, 1973). In Fig. 1 are shown the light saturation curves for the rate of $CO_2$ uptake. The light saturated photosynthetic rate (on the basis of unit leaf area or chlorophyll content) is approximately three times greater for the plants grown at high light than for the plants grown at low light. Similarly, the light intensity required for saturation of the plants grown at the high light level is higher than in the plants grown at the low light level. Plants grown in low light have a lower respiration rate and lower compensation point. When the plants are transferred from low light into high light, they require about five days to attain the maximum photosynthetic rate.

On the other hand, the chlorophyll and carotenoid contents are not very different in the two types of plants (Fig. 2). In high light a greater chlorophyll a:b ratio is found than in low light while the ratio of chlorophylls to carotenoids is greater under weak light conditions.

In the *Sinapis* leaves the content of soluble protein per leaf area also increases with increased light intensity during growth (Rühle, 1974). The protein/chlorophyll ratio becomes about 5 times greater in the high light plants (Fig. 3). It is well known that 'Fraction I' protein constitutes a major portion of soluble leaf protein; Fraction I protein content is correlated with the activity of carboxydismutase (Björkman, 1968; Gauhl, 1969). Compared with soluble protein the

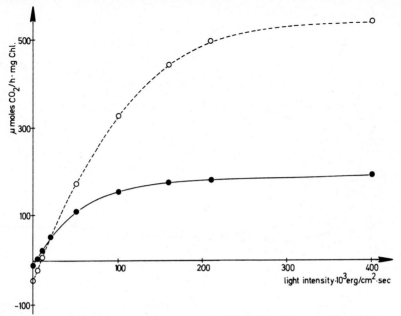

*Fig. 1.* Light saturation curves for photosynthesis of high light (O----O) and low light (●——●) leaves (1200 ppm external $CO_2$, 25°C).

differences in total protein, measured as $N_2$ (Kjeldahl), are not so striking. There-fore the changes are mainly attributed to the soluble part of the proteins.

In the present studies we compared the effect of different light intensities during growth on various photosystem-II components of the photosynthetic light reactions, i.e. the concentration of manganese and plastoquinone as well as the light dependence curves for the Hill reaction and non-cyclic photophosphoryla-tion with K-ferricyanide as electron acceptor.

The content of manganese in chloroplasts from plants grown under different light intensities has been determined. Table 1 clearly shows that the manganese content of *Sinapis* plants depends on the growth conditions. Under weak light conditions we found a manganese-chlorophyll ratio of 1:360 whereas in high light there is a ratio of 1:180. Under natural light conditions of the greenhouse this ratio is 1:290. Compared with that we found in market spinach 1 atom manganese per 140 chlorophyll molecules which agrees well with other published data.

Grahl & Wild (1972) carried out extensive studies on the lipophilic plastid quinones (Fig. 4). The lipoquinone content of the low light plants decreases slight-ly during the experimental period until flowering. Whereas the rate of synthesis of plastohydroquinone and α-tocopherol in high light plants rises sharply during the growth period, the synthesis of plastoquinone and tocopherylquinone rises at a lower rate. At flowering time the plastoquinone concentration is at least twice as much in the high light than in the low light plants.

In a preceding work (Wild et al., 1973) we determined the photochemical activities of subchloroplast fragments fractionated by digitonin according to Boardman (1971). The D-1 and D-10 fragments derived from high light chloro-

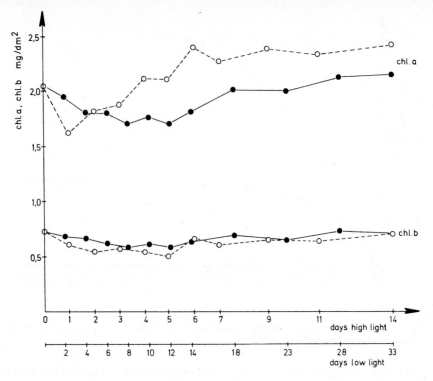

*Fig. 2.* Content of chlorophylls after the change from low to high light conditions (○--○), and the chlorophyll content of plants remaining in the low light (●—●).

*Table 1.* Manganese-chlorophyll ratio of chloroplasts from plants of *Sinapis alba* grown under different light conditions and from market spinach.

| | *Sinapis alba* | | spinach |
|---|---|---|---|
| high light | low light | greening house | |
| 1:180 ± 14 | 1:360 ± 65 | 1:290 ± 69 | 1:140 ± 18 |

plasts show a substantially higher photosystem-II activity than the corresponding fragments from the low light plants. The photosystem-II activity was assayed by DPIP reduction using diphenylcarbazide as the electron donor. In the present studies we have examined the electron transport system from water to ferricyanide in osmotically shocked chloroplasts isolated from high light and low light plants. The basal electron flow rate in the absence of the phosphorylation cofactors, the coupled and uncoupled electron flow as well as the noncyclic phosphorylation were measured (Table 2). We further determined the light dependence curves for the uncoupled electron transport with ferricyanide as electron acceptor and methylammoniumchloride as an uncoupler of photophosphorylation (Fig. 5).

The photochemical activities were measured about 1 hour after the isolation of

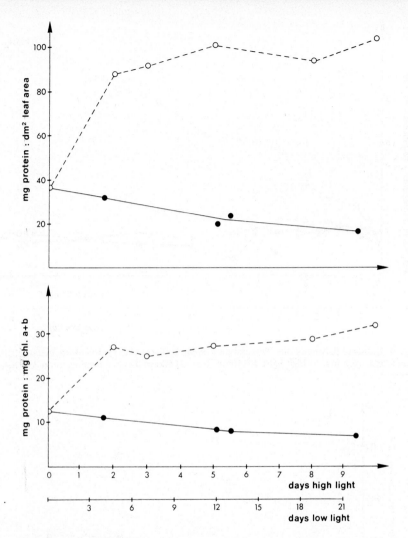

*Fig. 3.* Content of soluble proteins in the primary leaves on the basis of chlorophyll and leaf area of plants transferred from low to high light conditions (○····○) and of plants remaining in the low light (●—●).

*Table 2.* The effect of light intensity during growth of *Sinapis alba* on the electron flow $H_2O \rightarrow K$ ferricyanide and noncyclic phosphorylation ($\mu$ moles/h · mg Chl).

| | electron flow | | | noncyclic phosphor. | P/2e |
|---|---|---|---|---|---|
| | uncoupled | basal | coupled | | |
| low light | $687 \pm 68$ | $169 \pm 48$ | $358 \pm 53$ | $240 \pm 88$ | 1.34 |
| high light | $1150 \pm 134$ | $420 \pm 65$ | $656 \pm 80$ | $367 \pm 91$ | 1.11 |

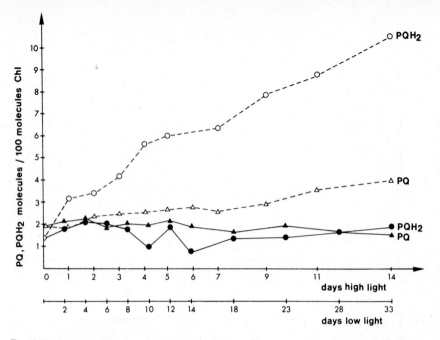

*Fig. 4.* Ratios of plastoquinone (PQ) and plastohydroquinone (PCH$_2$) to chlorophyll of plants transferred from low to high light conditions (- - - - -) and of plants remaining in the low light (———).

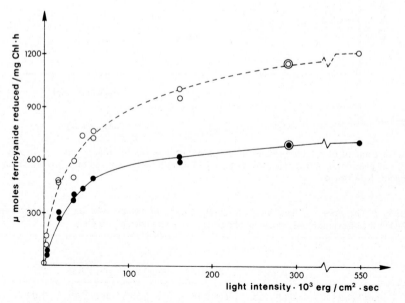

*Fig. 5.* Light saturation curves for the Hill activity with ferricyanide of chloroplasts from high light (O---O) and low light plants (●—●) (uncoupled, 23°C).

chloroplasts. Within the same chloroplast preparation our measurements showed fairly constant results, whereas we found greater variations between different preparations . Electron transport and noncyclic phosphorylation are markedly greater in high light than in low light plants, although we observed a stronger coupling of phosphorylation to electron transport (P/2e) in low light chloroplasts.

Björkman et al. (1972) found in their investigations about light saturation of uncoupled electron flow in *Atriplex* different half maximum saturation between high light and low light chloroplasts. Our experiments with *Sinapis*, however, show half maximum saturation of uncoupled Hill activity at the same light intensity in both types of chloroplasts. The quotient between the measuring points is constant in the linear rise of the curves too.

In our present view in *Sinapis alba* all of the system-II components seem to increase in high light. On the other hand the investigation on the plastoquinone content shows that the increase of this component does not directly correlate with the higher photosynthetic rate. In the following Grahl will report that some redox components of system I behave in a different manner as plastoquinone. More experiments are needed to clarify the changes in the system II of the photosynthetic apparatus of *Sinapis alba* under high and low light conditions.

# REFERENCES

Björkman, O. (1968): Carboxydismutase activity in shade-adapted and sun-adapted species of higher plants. *Physiol. Plant.*, 21: 1-10.

Björkman, O. & P. Holmgren (1963): Adaptability of the photosynthetic apparatus to light intensity in ecotypes from exposed and shaded habitats. *Physiol Plant.*, 16: 889-914.

Björkman, O., N.K. Boardman, J.M. Anderson, S.W. Thorne, D.J. Goodchild & N.A. Pyliotis (1972): Effect of light intensity during growth of *Atriplex patula* on the capacity of photosynthetic reaction, chloroplast components and structure. *Carnegie Institution Year Book*, 71: 115-135.

Boardman, N.K. (1971): Subchloroplast fragments: digitonin method. In San Pietro A. (Ed.): Methods of Enzymology, vol. 23, p. 268-276. Academic Press, New York-London.

Gauhl, E. (1969): Adaptive Differenzierung des Photosynthese-Apparates in Sonnen- und Schatten- Ökotypen von *Solanum dulcamara*. Dissertation, Frankfurt/M.

Grahl, H. & A. Wild (1972): Die Variabilität der Größe der Photosyntheseeinheit bei Licht- und Schattenpflanzen. *Z. Pflanzenphysiol.*, 67: 443-453.

Grahl, H. & A. Wild (1973): Lichtinduzierte Veränderungen im Photosynthese-Apparat von *Sinapis alba*. *Ber. Dtsch. Bot. Ges.*, 86: 341-349.

Hager, A. & T. Bertenrath (1962): Verteilungschromatographische Trennung von Chloro- phyllen und Carotinoiden grüner Pflanzen an Dünnschichten. *Planta (Berl.)*, 58: 564-568.

Homann, P.H. & G.H. Schmid (1967): Photosynthetic reactions of chloroplasts with unusual structures. *Plant. Physiol.*, 42: 1619-1632.

Lichtenthaler, K.H. (1968): Verbreitung und relative Konzentration der lipophilen Plastiden- chinone in grünen Pflanzen. *Planta (Berl.)*, 81: 140-152.

Rühle, W. (1974): Untersuchungen über den Gehalt an löslichem Protein und die Photo- phosphorylierung bei Licht- und Schattentypen von *Sinapis alba*. Diplomarbeit, Frank- furt/M.

Strotmann, H. (1970): Ioneneffekte beim Elektronentransport, $H^+$-Transport und bei der Phosphorylierung isolierter Chloroplasten. *Ber. Dtsch. Bot. Ges.*, 83: 443-446.

Wild, A., B. Ke & E.R. Shaw (1973): The effect of light intensity during growth of *Sinapis alba* on the electron-transport components. *Z. Pflanzenphysiol.*, 69: 344-350.

Ziegler, R. & K. Egle (1965): Zur quantitativen Analyse der Chloroplastenpigmente. I. Kritische Überprüfung der spektralphotometrischen Chlorophyllbestimmung. *Beitr. Biol. Pfl.*, 41: 11-37.

# EFFECT OF WATER STRESS ON THE DECLINE OF LEAF NET PHOTOSYNTHESIS WITH AGE

M.M. LUDLOW [*]

*Division of Tropical Agronomy, CSIRO, Brisbane, Australia*

## Abstract

The normal decline of net photosynthesis with age of young leaves of a $C_4$ grass *(Panicum maximum)* appears to be suspended by leaf water potentials less than ca. -12 bars. This suspension of photosynthetic decline occurred at stress levels down to approximately $-90$ bars, and as long as leaves did not die from the direct effect of stress (as opposed to death associated with normal senescence) their photosynthetic mechanism was not permanently affected. After rewatering, leaves attained rates greater than those of control leaves of the same chronological age but comparable with those of the same physiological age. Suspension of ageing occurred in leaves which had previously experienced water stress and those which had not. Ageing recommenced after rewatering when leaf water potentials rose above ca. $-12$ bars and active net photosynthesis started. Possible explanations for the suspension of leaf ageing are discussed.

## Introduction

It is generally thought that water stress accelerates leaf senescence (Woolhouse, 1967; Addicott, 1969) because it increases the rate of leaf death (Mothes, 1928; Gates, 1964, 1968; Slatyer, 1967), and because the effects of water stress on many metabolic processes (such as protein and nucleic acid synthesis) are similar to those associated with senescence (Brady, 1973; Hsaio, 1973). Although increasing water stress causes progressive leaf death in the $C_4$ grass, *Panicum maximum* var. *trichoglume,* I would like to present evidence which indicates that water stress actually suspends ageing of young leaves as measured by the decline of net photosynthesis.

Measurements commenced when leaves were fully-expanded which corresponds with the beginning of the rapid decline in net photosynthesis (Ludlow & Wilson, 1971). Therefore, the terms ageing and senescence when applied to decline of net photosynthesis of *Panicum* are synonomous (Leopold, 1964; Carr & Pate, 1967; Addicott, 1969). However, other published work referred to has been chosen because it deals specifically with senescence: 'changes caused by factors other than harmful external conditions which are clearly degenerative – ultimately irreversibly so' (Carr & Pate, 1967); as opposed to ageing – changes which occur in time without reference to the natural development of death (Medawar, 1957).

Data presented in this paper are taken, in part, from Ludlow & Ng (1974) and from Ludlow & Ng (in preparation).

[*] I would like to express my gratitude to the Chief, Division of Tropical Agronomy, CSIRO, for allowing me to attend this Symposium and to Roger Davis for writing the computer program.

## Materials and Methods

Plants were grown in pots of soil adequately supplied with mineral nutrients and watered to pF 2 prior to the commencement of water stress treatments. Environmental conditions in the growth rooms were: 500 $\mu E$ m$^{-2}$s$^{-1}$ photosynthetic quantum flux for a 14 hour photoperiod, 30 $\pm$ 1°C day and 22 $\pm$ 1°C night temperature, and 21-23 mb vapour pressure deficit during the day and 1-5 mb at night.

The first experiment consisted of six pairs of plants. The control pair were watered as required to maintain the leaf water potential greater than −5 bars, measured with a pressure bomb. The other five pairs were given three drying cycles separated by a single rewatering back to the control level (Fig. 1). The first cycle commenced when leaf six on main tillers was fully expanded and leaf eight was expanding but had not emerged. In this cycle, leaf water potential decreased to −23 bars in three days, and it was below −12 bars for two days. The second cycle lasted two days and the leaf water potential reached a minimum of only −12 bars. Leaf eight emerged during the first cycle and was fully expanded by the end of the second. In the final cycle the five pairs were allowed to dry out, but at various times one pair of plants was selected at random, rewatered, and subsequently maintained under a water regime similar to the controls (Fig. 1). This experimental design gave five water stress treatments with minimum leaf water potentials from −11 to −85 bars (Table 1).

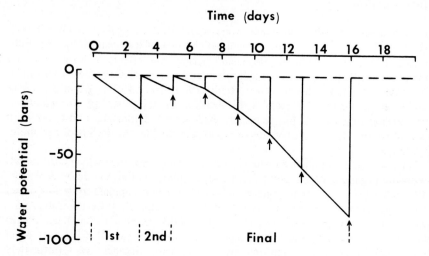

*Fig. 1.* Pattern of water stress imposed on plants in the first experiment. Arrows indicate when all plants were rewatered in the first and second cycles and when pairs of plants were watered in the final cycle.

Plants in the second experiment did not experience water stress above that of the control until leaf eight was fully expanded, when a single drying cycle commenced. This corresponded to the final drying cycle of the first experiment and minimum leaf water potentials ranged from −20 to −92 bars (Table 2).

Table 1. Water stress treatments of plants which had previously experienced stress (first experiment).

| Plant | Minimum leaf water potential (bars) | (a) (bar.days) | (b) (bars) | (c) | (d) (days) | (e) |
|---|---|---|---|---|---|---|
| Control | − 4 | 0 | − | 0 | 0 | 0 |
| B | −11 | 0 | − | 2 | 0 | 2 |
| C | −23 | − 2 | −12 | 2 | 2 | 4 |
| D | −37 | − 13 | −13 | 2 | 4 | 6 |
| E | −57 | − 82 | −12 | 2 | 6 | 8 |
| F | −85 | −170 | −12 | 2 | 9 | 11 |

(a) Integral of the periods of stress experienced above that of the control (From Fig. 3).
(b) Leaf water potential at which net photosynthesis reached zero during the final drying cycle.
(c) (d) and (e) Periods when leaf water potential was less than −12 bars in the first and final drying cycle and their total, respectively.

Table 2. Water stress treatments of plants which had not previously experienced stress (second experiment).

| Plant | Minimum leaf water potential (bars) | (a) (bar.days) | (b) (bars) | (c) (days) |
|---|---|---|---|---|
| Control | − 3 | 0 | 0 | 0 |
| B | −20 | 8 | −12 | 1 |
| C | −34 | 23 | −12 | 2 |
| D | −51 | 57 | −12 | 5 |
| E | −73 | 108 | −13 | 7 |
| F | −92 | 184 | −13 | 9 |

(a) Integral of the periods of stress experienced above that of the control (From Fig. 4).
(b) Leaf water potential at which net photosynthesis reached zero.
(c) Period when leaf water potential was less than −12 bars.

An open gas exchange system was used to measure the leaf net photosynthetic rate of control plants, and of stressed plants during the final drying cycles and after rewatering. The environmental conditions during these measurements were: 1600 $\mu E$ $m^{-2}s^{-1}$ photosynthetic quantum flux, $567 \pm 18$ ng $CO_2$ $cm^{-3}$, $31 \pm 1°C$ leaf temperature and 14-17 mb leaf-air vapour pressure difference. Gas exchange measurements were made on leaf eight in all experiments, except for the three most severe treatments in the first experiment and for the two most severe treatments in the second, where the ninth leaf was used to follow recovery because the eighth leaf was partially killed by water stress. Leaf water potential measurements were made on the eighth leaf of comparable tillers. Preliminary experiments indicated that there was little variation in water potential among leaves of individual plants at each level of stress.

## Results and Discussion

As the leaf water potential decreased the proportion of dead leaves increased until all the leaves were dead at approximately −100 bars (Fig. 2). Leaves died progressively from the oldest to the youngest on a tiller as stress increased, and within a leaf, cells died progressively from the tip to the base.

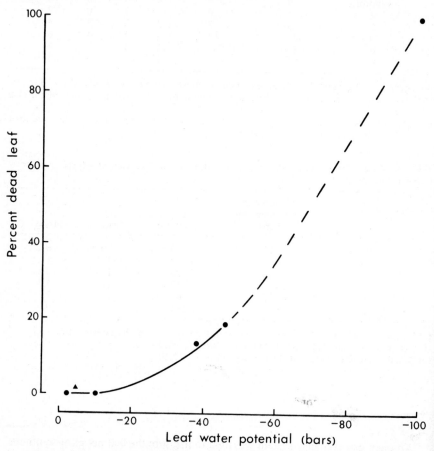

*Fig. 2.* Effect of increasing water stress on the proportion of dead leaf on plants in a continuous protracted drying cycle of about 12 days (●) and on control of comparable age (▲). The value at −100 bars was not measured but obtained by observation.

Net photosynthetic rates and water potentials of the control, and of three of the five young stressed leaves of plants which had experienced stress previously are shown in Figure 3. Details of all stress treatments in the first experiment are given in the Table 1. The net photosynthetic rate of unstressed control leaves declined progressively with time after full expansion. As water was withheld from the remainder of the plants in the final drying cycle both leaf water potential and net photosynthetic rate decreased rapidly. Net photosynthetic rate reached zero when the leaf water potential was approximately −12 bars, and remained at this level until the plants were rewatered. If the leaf did not die as a direct result of water stress, net photosynthesis on rewatering increased to a maximum before declining at a rate similar to that of the control. The maximum value was greater than that of control leaves of the same age, and rather similar among stressed leaves, suggesting that it was independent of the duration and intensity of stress and of age.

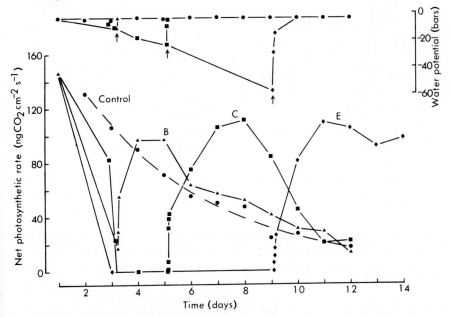

*Fig. 3.* Time trends of leaf water potential and net photosynthetic rate of control leaves, and of stressed leaves (B, C and E, see Table 1) during the final drying cycle and recovery after rewatering in the first experiment. Arrows indicate when plants were rewatered.

Results of the second experiment with plants which had not previously experienced stress were basically similar (Fig. 4, Tabel 2), except the maximum rates on recovery for the two least severe treatments were lower than the remainder.

It could be suggested that the higher net photosynthetic rate of stressed leaves after rewatering compared to controls of the same chronological age was due to a stress-induced enhancement of photosynthetic capacity. This enhancement may be similar to rejuvenation which accompanies partial defoliation (Woolhouse, 1967; Hodgkinson, 1974), increased nitrogen supply (Walkley, 1940), or application of cytokinins (Hall, 1973), and which is characterized by renewed synthesis of chlorophyll, proteins and nucleic acids, and enhanced net photosynthetic rate. In the case of *Panicum* it could have arisen from the partial defoliation caused by leaf death (Fig. 2), together with a burst of cytokinins from the roots (Itai et al., 1968) or nutrients (Gates, 1968) from other parts of the plant after rewatering.

Stress-induced enhancement would seem an unlikely explanation for the following reasons. Firstly, one would expect the maximum net photosynthetic rate

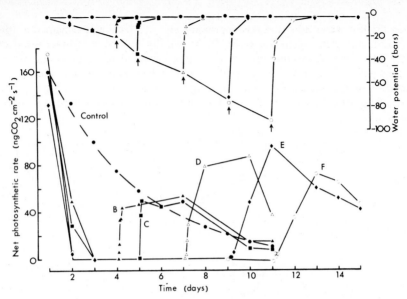

*Fig. 4.* Time trends of leaf water potential and net photosynthetic rate of control leaves, and of stressed leaves (B to F, see Table 2) during the drying cycle and recovery after rewatering in the second experiment.

after rewatering to vary with the degree of stress (and it did not, in general), because defoliation resulting from leaf death increased progressively with stress; secondly, water stress in the shoot reduces rather than increases the cytokinins which are transported to the shoot during recovery (Itai & Vaadia, 1971); thirdly, a burst of growth substances and nutrients might be expected to give a sharper recovery curve with a more rapid decline (as other sinks in the plant competed with leaf 8) rather than a rate of decline similar to that of the control; fourthly, a burst of cytokinins from the roots should lead to a lower stomatal (Pallas & Box, 1970) or intracellular resistance, but stressed plants after recovery had values similar to those of controls of the same physiological age and never less than the minimum of controls; fifthly, the rate of recovery of chlorophyll content (assessed visually to be complete when leaves had regained turgor in about two hours) and net photosynthesis recovered more quickly (2 to 3 days after rewatering) than those of rejuvenating leaves on partially-defoliated plants (Woolhouse, 1967; Hodgkinson, 1974). Therefore another explanation must be found.

Gates (1968) says that, after rewatering, stressed tomato plants exhibited characteristics of a physiologically-younger condition. This suggested that the explanation for the photosynthetic behaviour of stressed *Panicum* leaves might be that they were physiologically younger than their chronological age indicated. This would result if water stress suspended ageing.

To test this hypothesis it is necessary to choose a threshold leaf water potential below which ageing stops and above which it recommences. Net photosynthesis and leaf elongation (Fig. 5) and translocation (T.T. Ng, personal communication) cease at approximately −12 bars and, in the absence of data to the contrary, it seems reasonable to assume that many other physiological processes including

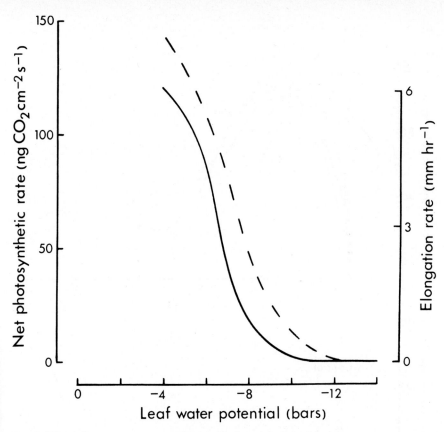

*Fig. 5.* Effect of water stress on net photosynthetic rate (broken line) and leaf elongation rate (solid line).

those associated with ageing would behave similarly.

According to the hypothesis then, stressed leaves should be considered physiologically younger than their chronological age by the number of days during which their leaf water potential was less than −12 bars; for example six days were subtracted from the chronological age of leaf E in Fig. 2. Two more days were subtracted from the chronological age of all leaves of stressed plants in the first experiment to allow for the period during which the leaf water potential was less than −12 bars in the first drying cycle. Thus the leaf E curve then starts on day 1. The data for the other leaves were adjusted in a similar manner (Table 1) and compared with the control (Fig. 6a). Similar adjustments were made to data from the second experiment except that there was no adjustment necessary in addition to that associated with the final cycle (Table 2, Fig. 6b).

The adjusted curves for the stressed leaves coincided reasonably well with that of the control in both experiments suggesting that water stress did indeed suspend ageing. The fact that the maximum value of stressed leaves after recovery coincided reasonably well with the control of comparable physiological age indicates that ageing must have recommenced when leaf water potential was greater than

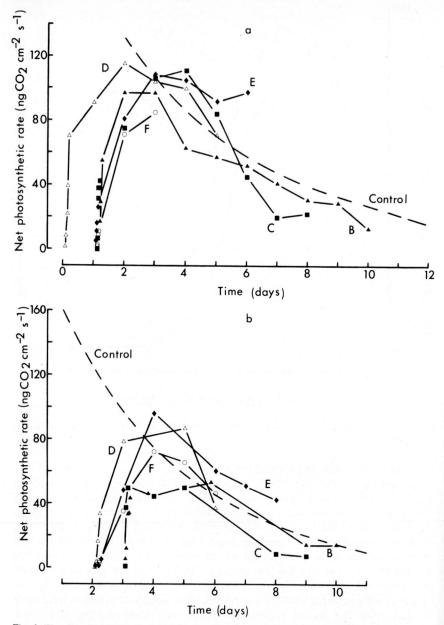

*Fig. 6.* Time trends of net photosynthetic rate of stressed leaves after rewatering adjusted to the same physiological age as control leaves from (a) the first and (b) the second experiment.

ca. −12 bars and net photosynthesis became positive, and that the pattern of net photosynthesis during this period represents the opposing influences of increasing rates associated with recovery and decreasing rates associated with ageing. An error in selecting the threshold value (i.e. $-12$ bars) does not greatly affect the adjustment for physiological age because of the rapid decrease of leaf water potential during the drying cycle and the immediate increase on rewatering. Similar results are obtained using the leaf water potential at wilting ($-9$ bars), which is associated with loss of turgor and the cessation of many processes concerned both directly and indirectly with growth (Slatyer 1967). Similarly, errors in deciding when the threshold value was reached will be small. Some of the variability arose in selecting leaves of comparable age at the commencement of the experiment. The latter is particularly important because of the inherent rapid decline of net photosynthesis during ontogeny. No extra adjustment was made when the ninth leaf was measured instead of the eighth in stressed plants because the error involved is small; the difference in net photosynthetic rate between leaves eight and nine of stressed plants was less than 10 ng $CO_2$ cm$^{-2}$s$^{-1}$ when measured on four occasions at water potentials between $-5$ and $-40$ bars.

The possibility of suspension of the photosynthetic decline by water stress can be tested in another way which does not require any assumptions about the value of the threshold leaf water potential. If the hypothesis is correct and water stress just suspends the photosynthetic decline which recommences when the stress is relieved, then the integrated net photosynthesis performed by stressed and control leaves in their life time should be similar. Because our measurements ceased before leaves died the integrated net photosynthesis was calculated for a period on ten days after leaves were fully-expanded. This period was chosen because one of the control leaves died on the tenth day and the remainder were nearly dead (Figs. 3 and 4). In stressed leaves, net photosynthesis was integrated for the first ten days after full expansion when net photosynthesis was positive; i.e., days when net photosynthesis was zero were ignored. For comparison between experiments the integrated values were expressed as a percentage of the control:

| Treatment | 1st Experiment | 2nd Experiment |
|---|---|---|
| Control | 100 | 100 |
| B | 100 | 96 |
| C | 113 | 77 |
| D | 120 | 114 |
| E | 108 | 117 |
| F | – | 99 |
| Average | 110 | 101 |

Considering the range of treatments (Tables 1 and 2) and the possible errors involved, the integrated net photosynthesis of stressed leaves agree reasonably well with those of the controls.

Senescence and death of leaves are poorly understood (Woolhouse, 1967; Addicott, 1969; Brady, 1973), and it appears that the mechanism of senescence may differ among species (Dyer & Osborne, 1971). Consequently, any explanation of the data presented in this paper must be highly speculative. I believe that there are two opposing processes occuring in *Panicum* leaves during water stress which determine subsequent photosynthetic capacity when the stress is relieved. The fact that a young undamaged leaf dies at approximately $-100$ bars, but recovers completely if rewatered at −90 bars, suggests that there is a threshold below which leaves are killed by water stress and that this effect overrides the mechanism associated with suspension of ageing. Further evidence of the over-riding effect of this premature death is the fact that the chemical composition of killed leaves appears comparable to that of a leaf which is ontogenetically young-er, rather than that of one which has proceeded to death through the normal process of senescence (Wilson & Ng, 1975).

As leaf water potential decreases the progressive death of leaves from the oldest (first-formed) to the youngest (latest-formed) probably occurs because there is a progression of resistance to stress from old to young leaves (Gates, 1964, 1968). A similar explanation may apply to the progressive death from the old cells in the leaf tip to the younger basal cells. Alternatively, if death occurs below a critical leaf water content rather than a critical water potential, later-formed (young) leaves have a higher relative water content per unit leaf water potential than early-formed (old) leaves (Pospisilova, 1973; Ng, Wilson & Ludlow, in prepara-tion), and would take longer to reach the critical water content and die. In other words, progressive death is the result of an increase in drought resistance (Levitt, 1972) from early-formed to later-formed leaves.

A suitable explanation of the mechanism associated with suspension of ageing as measured by photosynthetic decline in young leaves is also difficult to find. Examination of leaf stomatal and intracellular resistances provides no assistance nor does current knowledge of plant growth substances which would predict enhancement rather than suspension of ageing by water stress; cytokinins which delay senescence (Hall, 1973) decrease in water stressed leaves (Itai & Vaadia, 1971), whereas abscisic acid which promotes senescence (Addicott, 1969) rises (Loveys & Kriedemann, 1973; Zabadal, 1974).

Simon (1967) considers that progressive (or sequential) senescence results from continuing competition between the developing leaves of the apex and protein synthesis in mature leaves for amino acids produced by proteolysis in those leaves. Because leaf growth (Fig. 5) and translocation out of leaf eight (T.T. Ng, personal communication) stop at about −12 bars, this competition for and removal of amino acids would cease, and senescence would be suspended. Amino acids re-tained in the leaf would then be available for rapid re-synthesis of proteins when the stress was removed (Lahiri & Singh, 1968). In addition, the metabolism of these leaves must be reversibly suspended by stress to explain the rapid resump-tion of activity after rewatering. Reversible suspension of protein and nucleic acid metabolism by desiccation has been shown in lower plants and seeds of higher plants (Bewley, 1973; Hsiao, 1973). It probably also occurs in leaves of resur-rection plants, which include some $C_4$ grasses (Gaff, 1971) and which are able to withstand air dryness for considerable periods and resume activity when watered (Walter & Kreeb, 1970).

Irrespective of the mechanisms involved in the progressive death and in the suspension of ageing of young leaves, the ecological significance of both phenome-na is clear; progressive leaf death would help to reduce water loss by reducing leaf area as stress increased, and suspension of ageing ensures that surviving leaves will retain their photosynthetic capacity to assist recovery and further growth when the stress is removed.

# REFERENCES

Addicott, F.T. (1969): Ageing, senescence and abscission in plants, phytogerontology. *Hort. Science,* 4: 14-16.

Bewley, J.D. (1973): Polyribosomes conserved during desiccation of the moss *Tortula ruralis* are active. *Plant Physiol.,* 51: 285-288.

Brady, C.J. (1973): Changes acoompanying growth and senescence and effect of physiological stress. In Butler, G.W. & Bailey, R.W. (Eds.): Chemistry and Biochemistry of Herbage. Vol. 2. Pp. 317-50. Academic Press, London.

Carr, D.J. & J.S. Pate (1967): Ageing in the whole plant. In Woolhouse, H.W. (Ed.): Aspects of the Biology of Ageing. Pp. 559-599. Cambridge University Press.

Dyer, T.A. & D.J. Osborne (1971): Leaf nucleic acids. II. Metabolism during senescence and the effect of kinetin. *J. exp. Bot.,* 22: 552-560.

Gaff, D.F. (1971): Desiccation-tolerant flowering plants in southern Africa. *Science, N.Y.,* 174: 1033-4.

Gates, C.T. (1964): The effect of water stress on plant growth. *J. Aust. Inst. agric. Sci.,* 30: 3-22.

Gates, C.T. (1968): Water deficits and growth of herbaceous plants. In Kozlowski, T.T. (Ed.): Water Deficits and Plant Growth, Vol. 2. Pp. 135-90. Academic Press, New York.

Hall, R.H. (1973): Cytokinins as a probe of developmental processes. *Ann. Rev. Plant Physiol.,* 24: 415-444.

Hodgkinson, K.C. (1974): Influence of partial defoliation on photosynthesis and transpiration by lucerne leaves of different ages. *Aust. J. Pl. Physiol.,* 1: 561-578.

Hsiao, T.C. (1973): Plant responses to water stress. *Ann. Rev. Plant Physiol.,* 24: 519-570.

Itai, C., A. Richmond & Y. Vaadia (1968): The role of root cytokinins during water and salinity stress. *Israel J. Bot.,* 17: 187-195.

Itai, C. & Y. Vaadia (1971): Cytokinin activity in water-stressed shoots. *Plant Physiol.,* 47: 87-90.

Lahiri, A.N. & S. Singh (1968): Studies on plant-water relationships. IV. Impact of water deprivation on the nitrogen metabolism of *Pennisetum typhoides. Nat. Inst. Sci. India,* 34-B: 313-322.

Leopold, A.C. (1964): Plant Growth and Development. Pp. 194-204. McGraw-Hill, New York.

Levitt, J. (1972): Responses of Plants to Environmental Stress. Pp. 438-439. Academic Press, New York.

Loveys, B.R. & P.E. Kriedeman (1973): Rapid changes in abscisic acid-like inhibitors following alterations in vine leaf water potential. *Physiol. Plant,* 28: 476-479.

Ludlow, M.M. & G.L. Wilson (1971): Photosynthesis of tropical pasture plants. III. Leaf age. *Aust. J. biol. Sci.,* 24: 1077-1087.

Ludlow, M.M. & T.T. Ng (1974): Water stress suspends leaf ageing. *Plant Science Letters,* 3: 235-40.

Medawar, P.B. (1957): The Uniqueness of the Individual. Basic Books, New York.

Mothes, K. (1928): Die wirkung des wassermangels auf den einweissumsatz in hoheren pflanzen. *Ber. Deut. Bot. Ges.* (Generalversam), 46: 59-67.

Pallas, J.E. & J.E. Box (1970): Explanation for the stomatal response of excised leaves to kinetin. *Nature Lond.,* 227: 87-88.

Pospisilova, J. (1973): Water potential, water saturation deficit and their relationship in leaves of different insertion levels. *Biol. Plant,* 16: 140-3.

Simon, E.W. (1967): Types of leaf senescence. In Woolhouse, H.W. (Ed.): Aspects of the Biology of Ageing. Pp. 215-230. Cambridge University Press.

Slatyer, R.O. (1967): Plant-Water Relations. Pp. 295-299. Academic Press, New York.

Walkley, J. (1960): Protein synthesis in mature and senescent leaves of barley. *New Phytol.,* 39: 362-369.

Walter, H. & K. Kreeb (1970): Die hydration und hydratur des protoplasmas der pflanzen und ihre öko-physiologische bedeutung. *Protoplasmatologia,* 2(6): 1-306.

Wilson, J.R. & T.T. Ng (1975): Influence of water stress on parameters associated with herbage quality of *Panicum maximum* var. *trichoglume. Aust. J. agric. Res.,* 26: 1-10.

133

Woolhouse, H.W. (1967): The nature of senescence in plants. In Woolhouse, H.W. (Ed.): Aspects of the Biology of Ageing. Pp. 179-213. Cambridge University Press,

Zabadal, T.J. (1974): A water potential threshold for the increase of abscisic in leaves. *Plant Physiol.*, 53: 125-127.

# THE EFFECT OF DAYLENGTH ON DAILY CO$_2$ BALANCES OF
## *SINAPIS ALBA L.*

M. MOUSSEAU

*Laboratoire du Phytotron, C.N.R.S., 91190, Gif-sur-Yvette, France.*

*Abstract*

Daily CO$_2$ exchanges have been followed on *Sinapis alba L.*, grown in short or long day conditions. For the CO$_2$ measurements three types of experimental conditions have been used, differing in the level and duration of daylight.

The shape of the night respiration curve is affected by the previous photoperiodic regime, while the shape of daily apparent photosynthesis is not. There is a beneficial effect of lengthening the light period for the same daily energy amount on CO$_2$ balance, specially on the short day grown plant.

In all daylengths, for long and short day grown plants, there is a significant correlation between CO$_2$ gain in the light and loss in the dark. The results suggest that this dependence may be different in short and long day plants.

## Introduction

Studies of various photoperiodic treatments have generally been focused on flowering processes and less attention has been paid to the impact of daylength on vegetative production and its metabolic components. Some work has been done comparing the influence of intensity and duration of light on morphological and growth parameters (Friend, Helson & Fisher, 1962; Hofstra, Ryle & Williams, 1969; Rajan, Betteridge & Blackman, 1971). In particular cases, this kind of study may help to optimise production by agricultural improvements such as lengthening the daylight period, with reduced energy (Hurd, 1973). But generally, the action of the photoperiod is not separated from the well known effect of the level of daily received energy on production.

The hypothesis that changes of metabolic patterns may be related to flower initiation has been stated in many papers. But few facts are presently known about changes in primary metabolism such as respiration and photosynthesis as a function of the photoperiodical treatment, except in crassulacean plants (Gregory, Spear & Thimann, 1954; Queiroz, 1968; Marcelle, 1970). Little is known about the comparative differences of CO$_2$ balance between short and long day flowering plants. Recently, however, it has been shown that there may be a relationship between the photoperiodic responses of plants and the C$_3$ or C$_4$ photosynthetic pathways (Purohit & Tregunna, 1974).

The present study was designed to investigate the effect of daylength on CO$_2$ balances of plants grown in short or long days, using *Sinapis alba,* in which the action of many environmental factors on photosynthesis and respiration is known (Cornic, 1969; Cornic, Mousseau & Monteny, 1970; Cornic, 1974).

## Materials and Methods

*Sinapis alba* L., known to be a long day flowering plant (Bernier & Bronchart, 1964) was used for experimental material. Plants were grown from seeds in vermiculite and nutrient solution in the phytotron of Gif-sur-Yvette. They were divided into two groups which differed only by the length of the light period (Short days: 9 hours daily; Long days: 16 hours daily). Other conditions remained constant; day and night temperature $22°C$, 70% relative humidity, 115 w.m$^{-2}$ irradiance (mixed fluorescent tubes and incandescent lamps, equivalent to 540 $\mu$einsteins m$^{-2}$ sec$^{-1}$, 400-700 nm).

Conventional growth analysis (Watson, 1947) has shown, apart from the known effect of long days on growth i.e. greater net assimilation rate (N.A.R.) and relative growth rate (R.G.R.), that in long or short day conditions, growing plants have an optimal N.A.R. occuring on about the 20th day (Fig. 1). At this age, short day plants show no evidence of flower induction while long day ones show a well formed floral bud. We chose this physiological stage when the growth activity of plant is maximal to measure the daily $CO_2$ balance.

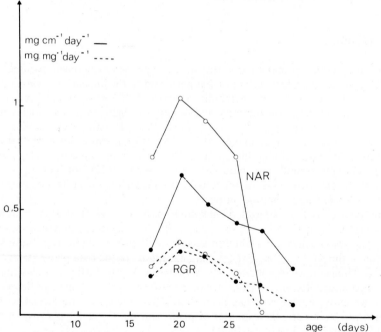

*Fig. 1.* Compared evolution of growth parameters during the life of *Sinapis alba* grown under different photoperiods. ○—○ long day grown plant (LD = 16 hours daily) ●—● Short day grown plant (SD = 9 hours daily).

Carbon dioxide exchange was measured continuously during a daily cycle of 24 hours on entire 20-day old plants with an open system of gas analysis ($CO_2$ analysor Rubis 3 000-COSMA-France) set up inside the conditioned rooms of the phytotron. One daily cycle has been followed on plants for which growing condi-

tions were short or long days (called in the text "short day plant" and "long day plant" or SD and LD plants) under three types of experimental conditions:
—16 hours of total irradiance (LD)
— 9 hours of total irradiance (SD)
—16 hours of reduced irradiance, so that the daily total energy equals the amount of 19 hours total irradiance (LD ≡ SD).

Light flux measured inside the assimilation chamber, at the plant level, was 100 $w.m^{-2}$, or 440 $\mu$einsteins $cm^{-2}.sec^{-1}$ (400-700 nm); in case of reduced level, it was 55 $w.m^{-2}$ or 270 $\mu$einsteins $cm^{-2}.sec^{-1}$.

Air flow was 80 l/h. and inside the assimilation chamber, made of plexiglas, it was mixed with microventilators.

Total leaf area of experimental plants was measured at the end of the experimental with an automatic area-meter (AAM7, Hayashi, Denko C° Ltd, TOKYO).

*Table 1.* Plant growth characteristics after 20 short or long days.

| | Total leaf area mg pl$^{-1}$ | Stem dry weight mg pl$^{-1}$ | Leaf dry weight mg pl$^{-1}$ | Total dry weight mg pl$^{-1}$ | SLA cm$^{-2}$ mg$^{-1}$ |
|---|---|---|---|---|---|
| Short day grown plant | 76 ± 5 | 44 ± 4 | 99 ± 7 | 143 ± 11 | 0.76 |
| Long day grown plant | 207 ± 6 | 165 ± 9 | 342 ± 12 | 508 ± 19 | 0.60 |

Table 1 gives the principal characteristics of plants after 20 days of growth in long or short days. One can see that the percentage of dry matter in the leaves compared to the total dry weight of the plants is nearly the same under the two conditions; leaves weighed twice as much as the stems in each case. Long and short day plants have, however, a slightly different chlorophyll content (higher in LD plants) and a different specific leaf area (S.L.A., higher in SD plants).

The light curve of photosynthesis of an entire plant is shown in figure 2. The saturation level of photosynthesis is higher for the long day plant than for the short day one. Light levels used in the experiments are plotted on the curve: it can be seen that the increase of energy from 55 to 110 $w.m^{-2}$ gives the same percentage increase in photosynthesis in short or long day plants. Moreover in the two cases, measurements of stem photosynthesis after cutting the leaves show that the green stem of *Sinapis alba* does not absorb or evolve $CO_2$; its photosynthetic activity stays at the light compensation point and thus does not contribute to any $CO_2$ exchange of the entire plant.

## Results and Discussion

1. Changes in the daily evolution of net photosynthesis and night respiration with photoperiod:

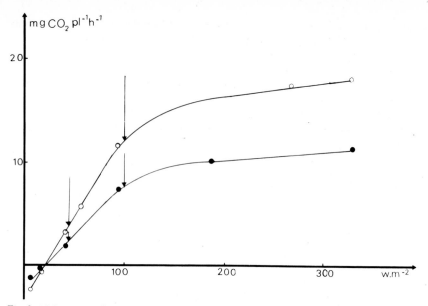

*Fig. 2.* Light curve of apparent photosynthesis of entire plants of *Sinapis alba*, aged 20 days, grown under two different photoperiods. ○ − ○ Long day grown plant; ●—● Short day grown plant. The arrows represent the energy levels used in the measurements of $CO_2$ exchanges.

## a. Photosynthesis

In our experiments, long and short day grown plants were submitted to either a long or a short day cycle; other conditions were kept constant.

−Short day plants
On Figure 3, it can be seen that the short day plant photosynthesis increases continuously from morning to evening under its own growing conditions. For the same type of plant under a long day cycle, the level of photosynthesis reaches a maximum after 9 hours and then remains constant until the light is off. In the two conditions the rate of photosynthesis is higher after the dark period, probably because of the growth of plant area during the night.

−Long day plants
Long day plants show the same feature: the shape of the daily curves of photosynthesis do not differ from those of the short day one. Thus, the photoperiodic treatment does not seem to influence the time course of daily photosynthesis.

## b. Respiration

Under a long day cycle, the night $CO_2$ output follows the same pattern for the two kinds of plants (Figure 4). Respiration is maximum at the beginning of the night, decreases to a minimum after 3 hours, and then goes up again until the end of the short night. This shape does not seem to be influenced by the previous photoperiodical treatment. However, the increase in respiration at the end of the night seems to be more important in short day plants.

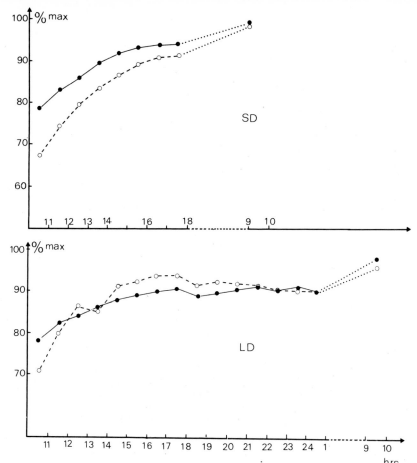

*Fig. 3.* Time course of photosynthesis of long day grown plants (●—●) and short day grown plants (○---○) during a short (9 hours = SD) or a long (16 hours = LD) day. Beginning of the light period: 8h45 – end of the light period: 17h45 or 0h45.

Under a short day cycle, the decrease in respiration goes on during a longer period then a minimum value is obtained. In the case of the long day plant this value remains constant until the next light period.

In the case of the short day plant, a net increase at the end of the night can be observed (Fig. 4A and B). Whatever the length of the preceeding light period is, this increase in respiration at the end of the night is significant, whereas the long day plant adapts the shape of its night output immediately to an opposite photoperiod.

In all cases the shape of the night respiration curve seems linked to the photoperiod preceding the considered night and not to the energy amount received during this photoperiod: Figure 4C shows that if a plant is submitted to a long day of reduced energy so that the total received energy is equivalent to that of a short day, the shape of the respiratory curve remains identical to the one obtained with a long day of full irradiance.

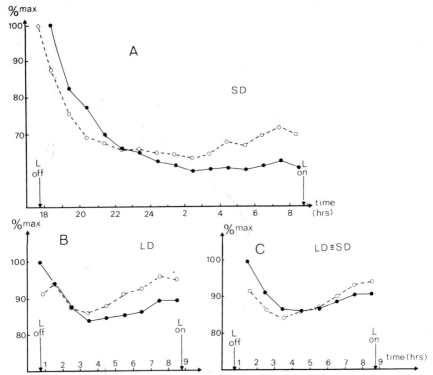

Fig. 4. Time course of night respiration of *Sinapis alba* grown under long days (●—●) or under short days (○---○) following different photoperiods: A–Short day, 9 hours full irradiance (SD); B–Long days, 16 hours full irradiance (LD); C–Long day, 16 hours reduced irradiance (LD ≡ SD). Beginning of the night: 17h45 – End of the night: 8h45 or 0h45.

## 2. $CO_2$ balance and its relation to the photoperiod:

### a. apparent photosynthesis

For long day plants, total apparent photosynthesis during the light period is shown to be related to the daily energy received (Table 2). Thus, $CO_2$ absorption depends at the same time on the level of irradiance and on the duration of the light period. Table 2 shows that the long day plant absorbs an equal amount of $CO_2$ during a short day or a long day during which light was reduced to obtain the same daily amount. From values of Table 3, deduced from Table 2, it can be seen that the net photosynthetic rate is not affected by the length of the photoperiod; a similar result was found by Bate & Canvin on *Populus tremuloïdes* (1971)

For short day plants, the results are slightly different. For the same amount of energy, net photosynthesis seems to be higher when the light is given during a longer period at a reduced level (Table 2: differences between 54 and 45 mg $CO_2$ plant$^{-1}$ day$^{-1}$ are significant at the 0,1 % level). This difference shows that the efficiency of photosynthesis is influenced by the photoperiodical treatment in the

140

*Table 2.* Action of daylength on daily $CO_2$ exchanges of *Sinapis alba* grown under long and short days.

| Plant growth conditions | $CO_2$ exchanges mg $CO_2$ ↗ or ↘ pl⁻¹day⁻¹ | Experimental daylength conditions | | |
|---|---|---|---|---|
| | | LD | LD ≡ SD | SD |
| LONG DAY 16 H | apparent photosynthesis | $164 \pm 19$ | $113 \pm 5$ | $119 \pm 9$ |
| | night respiration | $23 \pm 2$ | $21 \pm 1$ | $39^{**} \pm 3$ |
| | correlation coefficient between photosynthesis and respiration | 0.90 | 0.41[1] | 0.91 |
| | daily net absorption | $141 \pm 17$ | $92 \pm 5$ | $80 \pm 1$ |
| SHORT DAY 9 H | apparent photosynthesis | $74 \pm 7$ | $54^* \pm 6$ | $45^* \pm 3$ |
| | night respiration | $10 \pm 1$ | $10 \pm 1$ | $15^{**} \pm 1$ |
| | correlation coefficient between photosynthesis and respiration | 0.76 | 0.82 | 0.73 |
| | daily net absorption | $64 \pm 7$ | $44 \pm 5$ | $30 \pm 2$ |

[1] this value seems to be accidental – statistical difference $^*P \langle 0,1\%$; $^{**}P \langle 0,05\%$.

case of short day plants. Table 3 shows that the photosynthetic rate itself is affected, which means that the behaviour of the short day plant is different from the long day plant.

## b. Night respiration

$CO_2$ loss by respiration is shown to be 13 to 14% of the gain during the light period in the case of a long day plant in its own photoperiod. This proportion is much higher (32 to 33%) in the case of the short day plant; a fact which probably explain its slower growth and net assimilation rate. As a matter of fact, this hypothesis has been confirmed in Heichel's experiments (1971). But comparing the hourly rate of respiration per plant, one can see that it is always lower in short day plants than in long day plants (Table 3). The presence of the floral bud in the long day plant may account for this difference.

It is interesting to note that only one short day does not disturb the mean respiratory rate per plant of a long day grown plant, whereas only one long day significantly increases the respiratory rate of short day plant (Table 3).

## c. Net daily absorption: $CO_2$ balance

For the long and the short day plant, $CO_2$ balances over a 24 hours period seem to be more or less exactly proportional to the length of the light period: the increase of net $CO_2$ balance between a long and a short day is about 55% in case of the long day plant (Table 2: increase from 80 to 141 mg $CO_2$ plant⁻¹ day⁻¹)

Table 3. Action of daylength on the mean photosynthesis and respiration rates of short and long day grown plants of *Sinapis alba* L.

| | apparent photosynthetic rate mg $CO_2$ $dm^{-2}$ $h^{-1}$ | | | night respiration rate mg $CO_2$ $plant^{-1}$ $h^{-1}$ | | |
|---|---|---|---|---|---|---|
| | LD | LD ≡ SD | SD | LD | LD ≡ SD | SD |
| Long day grown plant | 4.7 | 3.2* | 4.8 | 2.87 | 2.50 | 2.60 |
| Short day grown plant | 5.08* | 3.5* | 5.6* | 1.35 | 1.31 | 1.00* |

and about 45% in the case of the short day plant (increase from 29 to 64 mg $CO_2$ $plant^{-1}$ $day^{-1}$). The results of statistical analysis do not permit an assessment of whether there is any significant difference between these two percentages which happen to correspond to a similar increase in daylength (9h to 16h = 56% increase).

When related to the amount of daily total energy, the $CO_2$ balance does not follow exactly the same rule. Because the losses are greater in short days, the daily balance is slightly better when energy is given over a longer period.

This is particularly true looking at the short day plant behaviour. When the light period is lengthned without any change in total energy, a higher total photosynthesis is added to a decreased night loss. This results in a significant increase in daily net absorption.

On a dry weight basis the results are clearer (Table 4), and differences between experimental conditions appear to be more significant. Night losses represent 14% of day gains in long day conditions for both SD and LD plants. However, these losses increase to 30% when the LD plant is transferred to a short day, whereas they reach 40% when the SD plant is transferred from its own growth conditions to a long day. But despite this higher increase in night respiration, a higher efficiency of photosynthesis per unit dry matter permits a better daily absorption in SD than in LD plants in short day conditions which shows the adaptation of the latter to its own photoperiod.

Lastly, it appears again clearly in Table 4 that there is a beneficial effect of lengthening the light period when giving the same amount of energy. This beneficial effect is again much greater for a short compared to a long day plant.

3. Relation between photosynthesis and respiration:

In all cases, there is a significant correlation between the quantity of $CO_2$ absorbed during the light period and $CO_2$ evolved during the night (Table 2 and 4). Figures 5 and 6 show the most probable slope of the regression curve relating respiration to photosynthesis for LD and SD plants. Statistical analysis of the slopes of the regression curves for different daylengths show that they are not significantly different. However the scattering of experimental values on the figure is such as to suggest that the difference exists in fact, but that our sampling is not

*Table 4.* Daily $CO_2$ exchanges of the unit of dry matter in *Sinapis alba* grown under short or long days.

| Plant growth conditions | $CO_2$ exchanges mg $CO_2$ g$^{-1}$ day$^{-1}$ | Experimental daylength conditions | | |
|---|---|---|---|---|
| | | LD | LD ≡ SD | SD |
| LONG DAY | apparent photosynthesis | $272 \pm 20$ | $201 \pm 19$ | $172 \pm 6$ |
| | night respiration | $39 \pm 3$ | $36 \pm 3$ | $56 \pm 3$ |
| | correlation coefficient between photosynthesis and respiration | 0.79 | 0.85 | 0.72 |
| | daily net $CO_2$ absorption | $233 \pm 18$ | $163 \pm 16$ | $116 \pm 5$ |
| SHORT DAY | apparent photosynthesis | $401 \pm 25$ | $314 \pm 20$ | $237 \pm 19$ |
| | night respiration | $57 \pm 5$ | $63 \pm 5$ | $95 \pm 12$ |
| | correlation coefficient between photosynthesis and respiration | 0.58 | 0.72 | 0.59 |
| | daily net $CO_2$ absorption | $343 \pm 23$ | $250 \pm 19$ | $142 \pm 10$ |

sufficient. It must be noted that the correlation coefficient is always greater for LD plants.

In our experiments, the length of the photoperiod appears to affect the shape of diurnal $CO_2$ exchange. Concerning photosynthesis, the continuous increase observed during the day period does not fit with the findings of Bate & Canvin (1971) who noted that photosynthesis of aspen trees tended to be higher at the start of a photoperiod and to decrease throughout the day. Our result may be due in large part to the extension of leaf area. In effect, the daily mean increase in leaf area, measured on samples of 25 plants between the 20th and the 21th day of growth, was found to be about 30 to 35% of the initial area for all plants in long day conditions, a value which is of the same order of magnitude as the increase of photosynthesis. But in short days, leaf area increase which is about 20-25% for LD plants and only 15-20% for SD plants, cannot account for the 35 to 40% increase in photosynthesis. Another explanation could be a slower opening of stomata in short days, a fact we cannot ascertain because we did not measure transpiration, but which could be different in long and short day plants.

The shape of the $CO_2$ output is linked to the previous photoperiod, but some of its features depend on the photoperiodical growth conditions of plants. Concerning the decline of respiration with time, our results conform with the findings of Heichel (1970) who showed that a constant rate of respiration is reached 2 or 3 hours after darkening after a 12 hours photoperiod and that this base rate is about 50% of the maximal value. Besides this, in our experiments, the speed with which this minimal value is reached varies with the length of the previous photoperiod (Fig. 4). In the case of a long day (Fig. 4B), it is identical for SD and LD plants, but in short days (Fig. 4A), it depends on the photoperiodical growth conditions of the plant considered, and lasts about twice as long as for the LD plant. In long days, the base rate is only 15% of the initial rate. Looking to figure 3B it is

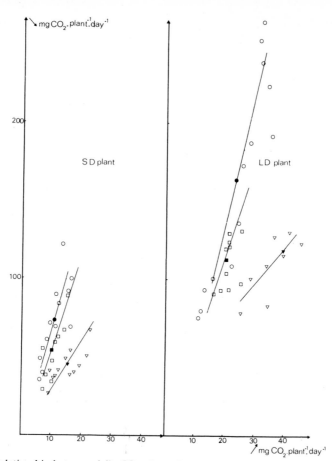

*Fig. 5.* Relationship between daily $CO_2$ absorption and daily $CO_2$ respiration for *Sinapis alba* grown under different photoperiods. LD= Long day, 16 hours light daily; SD = Short day, 9 hours light daily. Experimental conditions: ○—○: 16 hours full irradiance. □—□ 16 hours reduced irradiance; △—△ 9 hours full irradiance. The black points represent the mean value of the distribution. Calculated slopes of the regression line: SD plants ○—○ 0.109; □—□ 0.125; △—△ 0.291. LD plants ○—○ 0.108; □—□ 0.062; △—△ 0.342.

tempting to attribute this smaller decline to the superposition of another decarboxylation process which may begin with the dark period and increase until the end of the night, and which is also reflected on the curves (Fig. 4B) by the increase of $CO_2$ output at the end of the night. But the shape of these curves could as well reflect the variation in stomatal aperture in the dark because it was shown to depend upon the previous photoperiodic regime (Schwabe, 1952). Moreover this behaviour was found only in short day plants (*Chrysanthemum* and *Kalanchoe*).

Our results show clearly that the magnitude of this increase in $CO_2$ output at the end of the night is a phenomenon which seems to be linked to the length of the previous light period and also to the photoperiodical growth conditions of the plants because it appears to be always greater in SD plants.

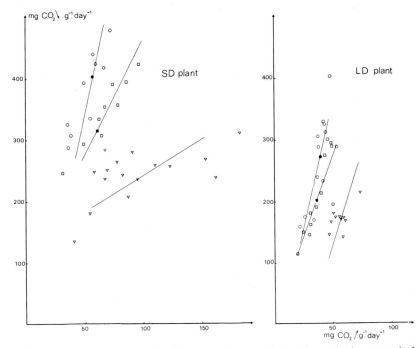

Fig. 6. Relationships between daily $CO_2$ absorption and daily $CO_2$ repiration per unit dry matter of *Sinapis alba* grown under different photoperiod (LD: long days, 16 hours light daily) (SD: short days, 9 hours light daily). Experimental conditions as in Figure 5. Black point: mean value of the distribution. Calculated slopes of the regression line: SD ○—○ 0.117; □—□ 0.185; △—△ 0.366. LD plant ○—○ 0.092 '□—□ 0.062; △—△ 0.151.

Comparison between figure 3B and 3C suggests that this last phenomenon is not related to the amount of energy but to the duration of the light period. Experiments using short days with supplementary light of very low energy would perhaps confirm if phytochrome is involved in the explanation of these results. On the other hand, if there is any link between these patterns of $CO_2$ output and the floral or non floral state, the same measurements made on a short day flowering plant could provide useful comparisons.

The difference of net photosynthetic rate between SD and LD plants (table 3) especially in short day conditions, may be related to differences in photorespiration. It has been shown by Steward et al (1971) that the ratio between $CO_2$ released in the light and $CO_2$ fixed depends on the photoperiod of growth, was lower for leaves of SD plants, and therefore more favorable in terms of the efficiency of $CO_2$ fixation. Whether this lower ratio changes again when a short day plant is transferred to a long day is not known. It is more probable, in this last case, that the decrease in the apparent photosynthetic rate observed is related to the photo oxidation of the chlorophylls reported by Sironval et al. (1961) when a short day plant was transferred to a long day.

Our experiments open again the controversial problem of the relationship between apparent photosynthesis and night respiration. This problem has particularly been studied by 'modelers' of crop productivity in order to know whether respiration is proportional to photosynthesis in the equations of production. When photosynthesis was varied by changing the light level, respiration was found to be proportional to the rate of photosynthesis (McCree & Troughton, 1966; McCree, 1970), and the respiration rate to adapt within 24 hours to another level of light conditions. On the other hand, Heichel (1970) has shown that respiration is in effect related to the previous light level, but not to the $CO_2$ fixation. In their work on aspen tree, Bate & Canvin found no clear relationship between the rate of night respiration of whole plants and their rate of photosynthesis as altered by the photoperiod. It should be self evident that this relation is not simple. Our results would have to contain a much larger amount of experimental data in order to show satisfactorily any influence of an environmental factor such as daylength on this relationship. It has already been found (Penning de Vries, 1973) that the slope of the assimilation-dissimilation relationship does not depend on temperature but varies with nutrition. Our results suggest that this slope is also dependant on the length of the light period, and that this dependence may be different for SD and LD plants.

We must conclude this paper in emphasizing its limits. The results presented here must be considered as preliminary, and nothing can be really stated until we have similar experiments on a short day flowering plant.

# REFERENCES

Bate, G.C. & D.T. Canvin (1971): The effect of some environmental factors on the growth of young aspen trees *(Populus tremuloïdes)* in controlled environments. *Can. J. Bot.,* 49: 1443-1453.
Bernier, G. & R. Bronchart (1964): The steps of floral induction in *Sinapis alba.* L. *Naturwiss.,* 51: 469-470.
Cornic, G. (1969): Evolution de l'activité photosynthétique de feuilles de rang différent chez le *Sinapis alba* en populations de différentes densités. *C.R. Acad. Sci. D.,* 268: 1934-1937.
Cornic, G. (1974): Etude de l'effet de la température sur la photorespiration chez la Moutarde blanche *(Sinapis alba* L.). *Physiol. vég.,* 12: 83-94.
Cornic, G., M. Mousseau & B. Monteny (1970): Importance de la photorespiration dans le bilan photosynthétique au cours de la croissance foliaire. *Oceol. Plant.,* 5: 355-363.
Friend, D.J.C., V.A. Helson & J.F. Fisher (1962): Leaf growth in Marquis wheat as regulated by temperature, light intensity and daylength. *Can. J. Bot.,* 40: 1299-1311.
Gregory, F.G., I. Spear & K.V. Thimann (1954): The interrelation between $CO_2$ metabolism and photoperiodism in *Kalanchoe. Plant Physiol.,* 29: 220-229.
Heichel, G.H. (1970): Prior illumination and the respiration of Maize leaves in the dark. *Plant Physiol.,* 46: 359-362.
Heichel, G.H. (1971): Confirming measurements of respiration and photosynthesis with dry-matter accumulation. *Photosynthetica,* 5: 93-98.
Hofstra, G., G.J.A. Ryle & R.F. Williams (1969): Effects of extending the daylight with low intensity light on growth of wheat and cocksfoot. *Aus. J. Biol. Sci.,* 22: 333-341.
Hurd, R.G. (1973): Long-day effects on growth and flower initiation of tomato plants in low light. *Ann. appl. Biol.,* 73: 221-228.
McCree, K.J. & J.H. Troughton (1966): Prediction of growth rate at different light levels from measured photosynthesis and respiration rates. *Plant Physiol.,* 41: 559-566.
McCree, K.J. (1970): An equation for the rate of respiration of white clover plants grown under controlled conditions. In Productivity of Photosynthetic systems: models and methods. Pp. 221-229. Wageningen. Centre for Agricultural Publishing and Documentation.

Marcelle, R. (1970): Sur l'absence d'une relation simple entre le type d'échange de $CO_2$ et l'induction de la floraison chez *Bryophyllum daigremontianum*. In Bernier (Ed.): Cellular and molecular aspects of floral induction. Pp. 243-251. Longman Publishers; London.

Penning de Vries, F.W.T. (1973): Use of assimilates in higher plants. In Photosynthesis and productivity in different environments. IBP Photosynthesis meeting. Aberystwyth. in press.

Purohit, A.N. & E.B. Tregunna (1974): Carbon dioxide compensation and its association with the photoperiodic response of plants. *Can. J. Bot.*, 52: 1146-1148.

Queiroz, O. (1968): Etude de l'action du photopériodisme et du thermopérodisme sur les réactions de synthèse et de dégradation de l'acide malique chez *Kalanchoe blossfeldiana* 'Tom Thumb'. Thèse Doctorat. Université Paris No CNRS.AO 2253.

Rajan, A.K., B. Betteridge & G.E. Blackman (1971): Changes in the growth of *Salvinia natans* induced by cycles of light and darkness of widely different duration. *Ann. Bot.*, 35: 597-604.

Schwabe, W.W. (1952): Effects of photoperiodic treatment on stomatal movement. *Nature*, 169: 1053-1055.

Sironval, C., W.G. Verly & R. Marcelle (1961): Radioisotopic study of chlorophyll accumulation in soybean leaves, in the conditions of a transfer from one daylength to another ('transfer-effect'). *Physiol. plant.*, 14: 303-309.

Steward, F.C., G.H. Craven, S.P.R. Weerasinghe & R.G.S. Bidwell (1971): Effects of prior environmental conditions on the subsequent uptake and release of carbon dioxide in the light. *Can. J. Bot.*, 49: 1999-2005.

Watson, D.J. (1947): Comparative physiological studies on the growth of field crops. I. Variations in net assimilation rate and leaf area between species and varieties and within and between years. *Ann. Bot.*, 11: 41-76.

# THE INVOLVEMENT OF $CO_2$ UPTAKE IN THE FLOWERING BEHAVIOUR OF TWO VARIETIES OF *ANTIRRHINUM MAJUS*

C.L. HEDLEY & D.M. HARVEY

*John Innes Institute, Colney Lane, Norwich, England.*

*Abstract*

Two varieties of *Antirrhinum majus,* varying in their flowering responses, have been compared for possible variation in components contributing to assimilation rate. Comparable leaves of an early (Pink Ice) and a late (Orchid Rocket) flowering variety demonstrated similar responses, in terms of net $CO_2$ uptake, when leaf temperature, light intensity and external $CO_2$ concentration was varied.

Both varieties produced leaves at similar rates when grown at 15,000 lux and 20°C in 16 hour photoperiods. The first three pairs of leaves of Pink Ice, however, had higher growth rates than corresponding Orchid Rocket leaves and attained a greater maximum leaf area. This increased area enabled these early Pink Ice leaves to fix upto ten percent more $CO_2$ per day than corresponding leaves of Orchid Rocket. The greater $CO_2$ uptake of the early Pink Ice leaves was accompanied by a higher Ribulose-1,5-diphosphate carboxylase activity per leaf.

The data support the suggestion that variation in the photoperiodic responses of *Antirrhinum* can be attributed, at least in part, to differences in assimilation rate.

## Introduction

Commercial varieties of *Antirrhinum* range from types which will produce quality plants at 10°C minimum temperature under the low light, short-day (SD) conditions of midwinter to cultivars which grow optimally under the high temperature, high light intensity and long-days (LD) of summer. Photoperiod has been shown to influence the flowering of many of these varieties during a 'light-sensitive' phase from 40 to 65 days after germination, when the plants have 5 to 10 pairs of leaves, although no absolute dependence on photoperiod was demonstrated (Maginnes & Langhans, 1960, 1961, 1967a, 1967b). Since all of the cultivars tested by Maginnes and Langhans exhibited a quantitative response to LD it is apparent that varieties which require the conditions of late summer for optimal flowering response have requirements for floral evocation which are more exacting than those of varieties which are able to produce quality flowers earlier in the year. The flowering responses of these summer varieties would therefore be more sensitive to environmental parameters other than daylength, such as temperature and light intensity.

Previous studies using two *Antirrhinum* varieties have indicated that the photomorphogenic flowering response varied with light intensity (Hedley, 1974; Hedley & Harvey (in press)). These earlier investigations compared two F1 hybrid cultivars; an early flowering forcing variety — Pink Ice and a summer flowering variety — Orchid Rocket. Pink Ice was found to behave as a typical quantitative long-day plant irrespective of light intensity while Orchid Rocket plants appeared to show

*Acknowledgement:* We would like to thank Mr. A.O. Rowland for his excellent technical assistance.

no response to photoperiod when grown at low light levels. These experiments suggested that at least two major components within the plant must be satisfied prior to floral induction in this species; a primary component which responded to photoperiod and a secondary component which varied with light intensity. The flowering response has also been shown to be modified by temperature and this response also varied with genotype, (Edwards, personal communication).

The most likely effect of light intensity and temperature is on assimilation rate. A sufficient or insufficient assimilation rate could directly affect the photo-morphogenic flowering response by either allowing or inhibiting apical induction respectively, in inductive daylengths. With this hypothesis it is envisaged that the inductive stimulus reaches the apex but apical induction can only occur if the amount of assimilate is beyond a certain limit. Variation within components of assimilation rate will therefore affect the flowering behaviour of genotypes. A large number of components comprising assimilation rate have been recognized and variation within many of these has been reported (see review by Wallace et al., 1972). The current paper reports the results of comparing some of these components in an early and a late flowering variety of *Antirrhinum majus*.

## Materials and Methods

An early (Pink Ice) and a late (Orchid Rocket) flowering F1 hybrid variety of *Antirrhinum majus,* obtained as seed from Suttons Seeds Ltd., Reading, England, were used throughout.

*Growth Conditions.*
Plants were grown under growth-room conditions described in previous communications (Hedley, 1974; Hedley & Harvey (in press)), the light intensity being maintained at 15,000 lux during continuous 16 hour photoperiods. Under these conditions both varieties had similar growth rates and flowering times (Hedley, 1974).

*Measurements of Net $CO_2$ uptake.*
Net $CO_2$ uptake was measured on individual leaves using a multichannel infra-red gas analyser system in which $CO_2$ concentration, light intensity and temperature could be varied independently (Harvey & Hedley 1973; 1974). Using this apparatus both varieties could be analysed simultaneously. The third leaf pair formed from germination were used for measuring responses to temperature, $CO_2$ concentration and light intensity. Changes in net $CO_2$ uptake during development were determined by measuring the $CO_2$ uptake of the first five pairs of leaves at regular intervals.

*Estimate of Ribulose-1,5-diphosphate carboxylase* (E.C.4.1.1.39) was estimated by following the conversion of NADH to NAD in the presence of ribulose-1,5-di-phosphate at 340 nm wavelength (Racker, 1957). The procedure was similar to that described by Downton & Slatyer (1971). The total reaction volume of 3.0 ml included the following: Bicine, $30\mu$ moles; $MgCL_2$ $30\mu$ moles; Dithiothreitol, $0.5\mu$ moles; $NaHCO_3$, $150\mu$ moles; ATP, $10\mu$ moles; NADH, $0.5\mu$ moles; glyceral-dehyde-3-phosphate dehydrogenase, 4 units and 3-phosphoglycerate kinase, 12.5

units. The reaction was started by the addition of ribulose-1,5-diphosphate, 0.2$\mu$ moles. Blanks contained all reagents except the substrate, ribulose-1,5-diphosphate. The pH for optimal activity of ribulose-1,5-diphosphate carboxylase for *Antirrhinum* was 7.8. Enzyme activity was calculated by using the extinction coefficient of NADH (6.22 x $10^6$ $cm^2$/mole).

*Determination of leaf area.*
Leaf areas were determined by applying a constant, specific for each variety, to measurements of width and length.

*Definition of leaf position.*
Individual leaves were numbered from the base of the plant, with the exclusion of the cotyledons.


# Results and Discussion

*Variation in net $CO_2$ exchange (NCE).*
Variation for this complex character has been reported for soybean ([*Glycine max*], Ojima & Kawashima, 1970), dry beans ([*Phaseolus vulgaris*], Izhar & Wallace, 1967) and many other species (Wallace et al., 1972). A comparison of NCE between an early (Pink Ice) and a late (Orchid Rocket) flowering variety of *Antirrhinum* was made using attached leaves of plants grown under the conditions described above. The leaves, which were at similar physiological stages of development, were compared at various temperatures, $CO_2$ concentrations and light intensities.

*Effect of temperature.*
The net $CO_2$ uptake was allowed to reach equilibrium at a range of temperatures and at saturating (222 $Jm^{-2}sec^{-1}$) limiting (41 $Jm^{-2}sec^{-1}$) and intermediate (96 $Jm^{-2}sec^{-1}$) light intensities. The incoming air had a $CO_2$ concentration of 300 $\pm$ 10 ppm and a relative humidity of 60 $\pm$ 2 percent, throughout the experiment.

The two varieties were similar in their responses at each of the temperature and light intensity regimes (Fig. 1). The temperature optimum for leaves of both varieties increased progressively from approximately 15° to 21° with increasing light intensity. The relative $CO_2$ uptake of both varieties was similar at the two lower light levels but there was a tendency at the high light intensity for the net $CO_2$ uptake of the Orchid Rocket leaves to decline more rapidly with increasing temperature.

*Effect of $CO_2$ concentration.*
The effect of varying $CO_2$ concentration on NCE was followed at saturating (222 $Jm^{-2}sec^{-1}$) and limiting (41 $Jm^{-2}sec^{-1}$) light intensities. Leaf temperature was maintained at 20 $\pm$ 1°C throughout and the relative humidity of the incoming air was stabilised at 60 $\pm$ 2 percent. The two varieties behaved similarly at the high and at the low light intensity, but the $CO_2$ saturation concentration for the two light regimes was very different (Fig. 2). At the high light intensity the optimum $CO_2$ concentration was between 500 and 600 ppm while at the low light level the

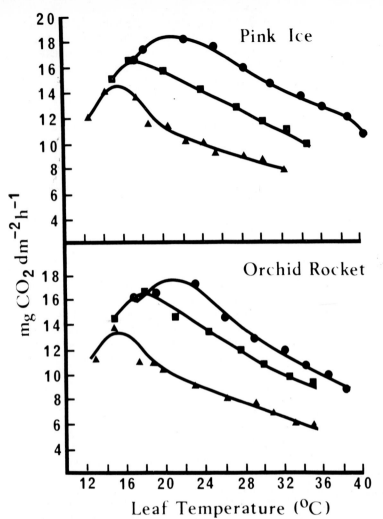

*Fig. 1.* The effect of leaf temperature at three light intensities on the net $CO_2$ uptake of two varieties of *Antirrhinum.* ▲—▲ = 41 $Jm^{-2}sec^{-1}$; ■—■ = 96 $Jm^{-2}sec^{-1}$; ●—● = 222 $Jm^{-2}sec^{-1}$. The incoming $CO_2$ concentration was maintained at $300 \pm 10$ ppm and the relative humidity at $60 \pm 2$ percent, throughout the experiment. Each point is the mean of three determinations.

optimum was about 350 ppm. The $CO_2$ compensation points for both varieties were 80 ppm at saturating light intensity and 50 ppm at the low light level.

*Effect of light intensity.*
Figure 1 illustrated that the temperature optimum for net $CO_2$ uptake varied with light intensity. It was therefore not possible to determine a true curve to illustrate the response of leaves to varying light intensities because it would be necessary to obtain the optimum temperature for each point on the light intensity curve. As a

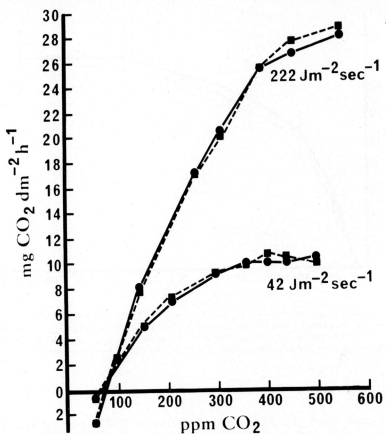

*Fig. 2.* Effect of $CO_2$ concentrations at two light intensities on the net $CO_2$ uptake of Pink Ice (■--■) and Orchid Rocket (●—●). Leaf temperature was maintained at $20 \pm 1°C$ and the relative humidity at $60 \pm 2$ percent, throughout the experiment. Each point is the mean of three determinations.

compromise the leaf temperature was stabilised at $20 \pm 1°C$, the temperature at which the plants had been grown. The $CO_2$ concentration and relative humidity of the incoming air were maintained at $300 \pm 10$ ppm and $60 \pm 2$ percent respectively.

The comparative response to light intensity by the two varieties was repeated several times and the means and standard deviations derived from these experiments are illustrated in Fig. 3. The mean net $CO_2$ uptake of Pink Ice leaves was greater at each point on the curve but the differences were not statistically significant. The light saturation level (ca. 160 Jm$^{-2}$sec$^{-1}$) and the extrapolated light compensation point (ca. 7 Jm$^{-2}$sec$^{-1}$) were similar for both varieties.

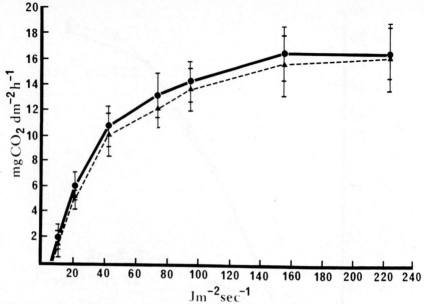

**Fig. 3.** Effect of light intensity on the net $CO_2$ uptake of Pink Ice (●—●) and Orchid Rocket (▲---▲). The leaf temperature was maintained at $20 \pm 1°C$, the incoming $CO_2$ concentration at $300 \pm 10$ ppm and the relative humidity at $60 \pm 2$ percent, throughout the experiment. Means and standard deviations at each point were derived from seven determinations.

## Variation in leaf area

The second major physiological component to be studied was leaf area. This component is in itself a complex character, its immediate components being leaf number and leaf size. The present investigation compared the two varieties for the rate of leaf appearance and the rate of development of individual leaves.

The high light intensity, long-day conditions of the experiment enabled the flowering time of Orchid Rocket ($62.8 \pm 2.09$ days) to approach that observed for Pink Ice ($54.8 \pm 0.63$ days), although the eight day difference was still significant. These apparently favourable conditions also produced a growth-rate for Orchid Rocket, as determined by changes in plant dry weight, which was still significantly less than that observed for Pink Ice (Fig. 4). This difference in growth rate was not reflected in the rate of leaf initiation (Fig. 5a), both varieties producing their first 20 leaves at similar times from germination. Orchid Rocket plants, however, produced on average six more leaves, prior to flowering, than Pink Ice.

The first five pairs of leaves of both varieties had all developed to at least one centimeter in length by day 29 from germination. The combined rate of area increase of these leaves was greater for Pink Ice plants and this was reflected in the final combined area of 144 $cm^2$ for Pink Ice and 132 $cm^2$ for Orchid Rocket (Fig. 5b). These early leaves are of special importance because in the growing conditions used, both varieties become susceptible to induction after the ex-

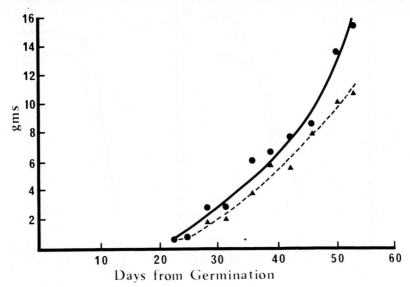

Fig. 4. Effect of constant conditions of 15,000 lux, 20°C and 16 hour photoperiods on the dry weight of Pink Ice (●—●) and Orchid Rocket (▲---▲) plants. Each point is the mean of five plants.

pansion of two to four pairs of leaves (Hedley, 1974).

Comparable leaves of both varieties reached their maximum size after very similar times from initiation (Figs. 5c & d). Varietal differences in the final leaf area must therefore have been due to variation in leaf growth rate. The first three pairs of leaves of Pink Ice had greater areas at full expansion than corresponding leaves of Orchid Rocket, the difference increasing from leaf one (11 percent) to leaf three (17 percent). Leaves four and five of Orchid Rocket, however, had greater areas than corresponding Pink Ice leaves (seven and 22 percent respectively).

The comparative study of net $CO_2$ exchange rates had shown both varieties to be similar. It was apparent, therefore, that any increase in leaf area should result in an increase in the total uptake of $CO_2$. This was illustrated by following the $CO_2$ uptake of the first five pairs of leaves through part of their development and expressing the data as total uptake per leaf (Fig. 6a).

It was apparent that the $CO_2$ uptake for each leaf of both varieties followed similar patterns. In most cases pronounced optima occurred which correlated well with the attainment of maximum leaf expansion. The differences between corresponding leaves of the two varieties reflected the differences in leaf area described previously. Leaves two and three of Pink Ice fixed on average approximately 10 and seven percent more $CO_2$ per day throughout their development than similar leaves of Orchid Rocket. Leaves four and five of Orchid Rocket, however, fixed on average approximately nine and 23 procent more $CO_2$ per day than corresponding Pink Ice leaves.

An estimation of the carboxylation capacity of leaves of the two varieties was determined by following the activity of ribulose-1,5-diphosphate (RuDP) carboxylase (Fig. 6b). Similar patterns were observed for RuDP carboxylase to those

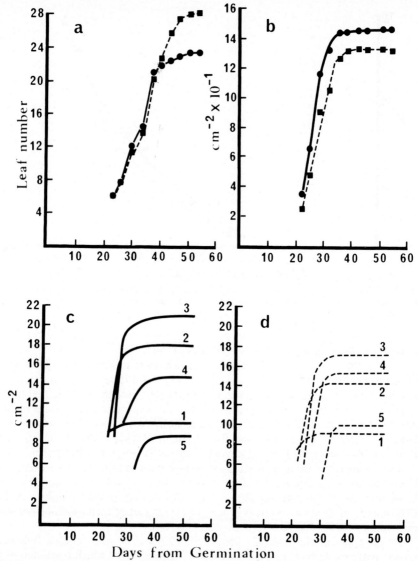

*Fig. 5. a-d.* Effect of constant conditions of 15,000 lux, 20°C and 16 hour photoperiods on leaf number (a); total leaf area of leaves 1-5 (b) and individual leaf growth curves of Pink Ice (c); and Orchid Rocket (d) for leaves 1-5. ●—● = Pink Ice; ■---■ = Orchid Rocket. All curves were derived from determinations on ten plants of each variety.

described for net $CO_2$ uptake. The first three pairs of leaves of Pink Ice had greater activity for this enzyme than corresponding Orchid Rocket leaves, while leaves four and five of Orchid Rocket had higher activities than similar Pink Ice leaves. The peaks of activity corresponded reasonably well with the timing of maximum leaf expansion. It was apparent however that the activity

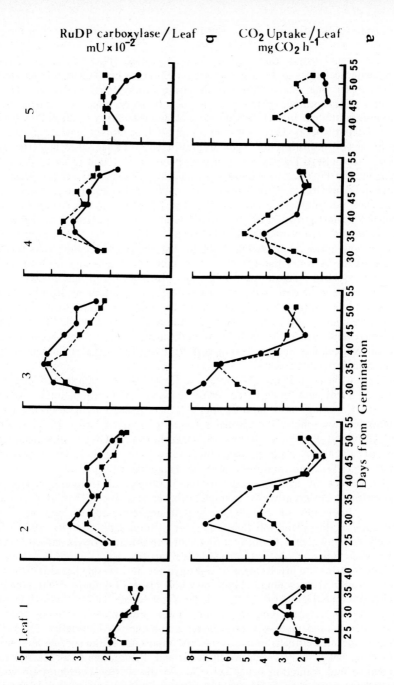

*Fig. 6. a-b*. Developmental patterns for net CO₂ uptake (a) and RuDP carboxylase activity (b) for the first five leaves of Pink Ice (●——●) and Orchid Rocket (■---■). Each point is the mean of duplicated determinations.

declined rapidly after the peaks, and this was especially pronounced in Pink Ice leaves.

These data illustrate that the early flowering variety (Pink Ice) has, by virtue of the greater leaf areas of its early leaves, a potential for net photosynthesis which is much higher than that of the late flowering variety (Orchid Rocket). This increased photosynthetic potential could be crucial in conditions where photosynthesis is limited. In such conditions apical induction may occur in Pink Ice but not in Orchid Rocket. This suggestion is supported by studies in which plants were given light-breaks during the night of an otherwise normal SD regime (Hedley & Harvey (in press)). The leaf areas of such plants were found to be similar to LD controls, the Pink Ice light-break plants having relatively greater leaf areas than corresponding Orchid Rocket plants. The light-breaks, however, did not induce the higher net photosynthetic rates found in LD plants; in this respect they were similar to plants grown in SD. The photosynthetic rates per leaf in the light-break plants were therefore intermediate between the LD and SD plants. Similarly the leaf number prior to flowering, for Pink Ice plants given light-breaks was also intermediate between the LD and SD number. No significant decrease in leaf number was observed for Orchid Rocket plants grown in light-breaks. It can, therefore, be suggested that the greater leaf area induced by light-breaks had increased the photosynthetic potential of the Pink Ice plants to a point where induction could occur earlier than in SD but not as early as in the LD conditions. The relatively smaller difference between the areas of SD and light-break leaves of Orchid Rocket was apparently not sufficient to allow apical induction to occur any earlier in light-breaks.

The involvement of net $CO_2$ uptake in flowering was suggested as long ago as 1913 by Klebs who found that a relatively high level of carbohydrates was necessary for flowering. More recently high $CO_2$ concentrations have been found to inhibit the flowering of the SD plants *Lemna perpusilla* (Posner, 1971), *Pharbitis* and *Xanthium* (Purohit & Treguna, 1974), in short-days. On the other hand *Silene armeria,* a long-day plant, has been induced to flower in short-days when grown under high $CO_2$ concentrations (Purohit & Tregunna, 1974).

The effect of light intensity on the photoperiodic response is also well documented. *Sinapis alba,* for example, behaves as a strict LD plant when grown in light intensities of 22,000 to 28,000 ergs/cm$^2$/sec, whereas at 35,000 ergs/cm$^2$/sec the plant will flower in 7 or 8 hours light (Bernier, 1969). Conversely *Perilla* which is normally a SD plant will initiate flowers under LD conditions when grown at low light intensity (Zeevaart, 1969).

Evidence for the involvement of carbohydrate in flowering has also come from work using *in vitro* techniques, both on whole plants (Takimoto, 1960; Steinberg, 1950) and on explants from roots and stems (see Nitsch, 1967). Cultured explants of *Cichorium intybus* tend to initiate flower buds under normally inductive photoperiods, and vegetative buds under non-inductive daylengths. The sucrose level in the medium was found to be very important. If the sucrose level was at $10^{-2}$M or below only vegetative buds formed, the optimum concentration for vegetative bud formation being $3\times10^{-2}$M. As the sucrose concentration was increased the proportion of flower buds increased until an optimum for flower bud formation was reached at $3\times10^{-1}$M.

Several suggestions have been made to explain the effect of increased assimilation rate on flowering behaviour. Posner (1971) suggests that $CO_2$ and sucrose may inhibit flowering in *Lemna perpusilla* by enhancing glucose-6-P dehydrogenase activity and diverting excessive glucose through the pentose phosphate pathway, resulting in an inadequate production of amino-acid precursors. Kandeler (1968) has suggested lowered levels of ATP as a possible mechanism for the inhibition of flowering of *Lemna gibba*. The suggestion by Purohit & Tregunna (1974) that flowering is controlled solely by a compound(s) 'X' produced during photosynthesis is difficult to reconcile with the well documented photoperiodic experiments using very short light-breaks. (e.g. see review by Evans, 1969).

The observations made on *Antirrhinum* support the idea of a dual requirement for flowering consisting of a primary photoperiodic stimulus and a secondary modifying component linked to assimilation rate. A system may be envisaged similar to that suggested by Cumming (1967) for *Chenopodium rubrum*. Cumming suggested that assimilate accumulated in the light serves as a substrate for phytochrome-Pfr in the dark. From this hypothesis it therefore follows that an increase in any of the components contributing to assimilation rate would either allow induction to occur earlier in the life cycle of the plant, or under poor growth conditions. Although variation in leaf area is suggested to account for differences in the floral induction of Pink Ice and Orchid Rocket, it is not anticipated that variation in this parameter necessarily accounts for variation in the flowering time of other *Antirrhinum* varieties. At present *Antirrhinum* plants are being screened for types which vary for other components of assimilation rate and the flowering behaviour of these types will be compared.

To date all of the experiments on *Antirrhinum* have centered on the interaction of light intensity with daylength; similar investigations using in addition various temperatures and $CO_2$ concentrations are being undertaken with the hope of elucidating further the flowering responses of this species.

# REFERENCES

Bernier, G. (1969): *Sinapis alba L.* In Evans, L.T. (ed): The induction of flowering. Pp. 305–327. MacMillan – Publishers, Melbourne.

Cumming, B.C. (1967): Circadian rhythmic flowering responses in *Chenopodium rubrum*: Effects of glucose and sucrose. *Can. J. Bot.*, 45: 2173–2193.

Downton, J. & R.O. Slatyer (1971): Variation in levels of some leaf enzymes. *Planta* (Berl.), 96: 1–12.

Evans, L.T. (1969): The induction of flowering. MacMillan Publishers Melbourne.

Harvey, D.M. & C.L. Hedley (1973): A system for the study of net carbon dioxide photoassimilation. *Sixty fourth annual report of the John Innes Istitute*. 46–47.

Harvey, D.M. & C.L. Hedley (1974): A multichannel system for the study of net carbon dioxide photoassimilation by leaves. *Laboratory Practice*, 23: 567–568

Hedley, C.L. (1974): Response to light intensity and daylength of two contrasting flower varieties of *Antirrhinum majus* L. *J. hort. Sci.*, 49: 105–112.

Hedley, C.L. & D.M. Harvey (1975): Variation in the photoperiodic control of flowering of two cultivars of *Antirrhinum majus* L. (in press).

Izhar, S. & D.H. Wallace (1967): Studies of the physiological basis for yield differences. III. Genetic variation in photosynthetic efficiency of *Phaseolus vulgaris* L. *Crop Sci.*, 7: 457–460.

Kandeler, R. (1968): Blükinduktion bei Lemnaceen. *Biol. Rundsch.*, 6: 49-57.

Klebs, G. (1913): Uber das Verhaltniss der Aussenwelt zus Entwicklung der Pflanze. *Sitz. Ber.*

*Akad. Wiss. Heidelberg.* Ser. B., No. 5.

Maginnes, E.A. & R.W. Langhans (1960): Daylength and temperature affect initiation and flowering of snapdragons. *N. Y. St. Flower Grs. Bull.,* No. 171: 1-5.

Maginnes, E.A. & R.W. Langhans (1961): The effect of photoperiod and temperature on initiation and flowering of snapdragon (*Antirrhinum majus* – variety Jackpot). *Proc. Amer.Soc. hort. Sci.,* 77: 600-607.

Maginnes, E.A. & R.W. Langhans (1967a): Photoperiod and flowering of snapdragon. *N. Y. St. Flower Grs. Bull.,* No. 260: 1-3.

Maginnes, E.A. & R.W. Langhans (1967b): Flashing light affects the flowering of snapdragons. *N. Y. St. Flower Grs. Bull.,* No. 261: 1-3.

Nitsch, J.P. (1967): Towards a biochemistry of flowering and fruiting: contributions of the *'in vitro"* technique. In Tukey, Sr. H.B. (ed): Proceedings of the XVII International Horticultural Congress Volume III. Pp. 291-308.

Ojima, M. & R. Kawashima (1970): Studies on the seed production of soybean. VIII. The ability of photosynthesis of F3 lines having different photosynthesis in their F2 generations. *Proc. Crop Sci., Soc. Japan.,* 39: 440-445.

Posner, H.B. (1971): Inhibotory effect of carbohydrate on flowering in *Lemna perpusilla.* III. Effects of respiratory intermediates, amino acids, and $CO_2$. Glucose 6-phosphate dehydrogenase activity. *Plant Physiol.,* 48: 361-365.

Purohit, A.N. & E.B. Tregunna, (1974): Effects of carbon dioxide on *Pharbitis, Xanthium,* and *Silene* in short days. *Can. J. Bot.,* 52: 1283-1291.

Racker, E. (1957): The reductive pentose phosphate cycle. I Phosphoribulokinase and ribulose diphosphate carboxylase. *Archs. Biochem. Biophys.,* 69: 300-310.

Steinberg, R.A. (1950): Flowering responses of a variety of *Nicotiana rustica* to organic compounds in aseptic culture. *Amer. Jour. Bot.,* 37: 547-551.

Takimoto, A. (1960): Effect of sucrose on flower initiation of *Pharbitis nil* in aseptic culture. *Plant and Cell Physiol.,* 1: 241-246.

Wallace, D.H., J.L. Osbun & H.M. Munger (1972): Physiological genetics of crop yield. *Advan. Agron.,* 24: 87-146.

Zeevaart, J.A.D. (1969): *Perilla.* In Evans, L.T. (ed): The induction of flowering. Pp. 116-155. MacMillan - Publishers, Melbourne.

# RELATIONS BETWEEN PHOTOSYNTHESIS AND FLOWERING IN *LEMNACEAE*

R. KANDELER, B. HÜGEL & TH. ROTTENBURG

*Botanisches Institut der Hochschule für Bodenkultur, Wien, Austria*

## Abstract

'Feeding' experiments with $CO_2$, sucrose, ADP and ATP as well as experiments with anti-metabolites (arsenate, atebrin, DCMU) lead to the conclusion that floral induction in the long-day plant *Lemna gibba* depends on surplus ATP from photophosphorylation. The short-day plant *Lemna paucicostata,* on the other hand, needs an elevated production of assimilates among other factors for flowering in the long day.

*In vivo,* the effect of sucrose cannot be replaced by glucose, fructose or their combination. In explanted flower primordia, however, all these sugars act alike. We assume, therefore, that products of photosynthesis influence flower formation in an indirect way. Some arguments favour the hypothesis that photosynthesis modifies export of hormones from the leaf.

Fruit growers, gardeners, foresters and ecologists have learned from repeated observations that many plant species may grow in the shade but not come to flowering there (see Lyr et al., 1967; Schuster & Kandeler, 1970). Filzer (1940) investigated this phenomenon in detail at forest edges. He found a threshold level of relative light intensity for each species, below which the plant shows only vegetative growth. With *Thymus serpyllum* this value is 50% of full daylight, with *Brachypodium pinnatum* 11% and with *Asperula odorata* 3%. Several arguments lead to the conclusion that in all these cases the light effect is mediated by photosynthesis. In *Acer mono*  not only the number of flowers increases with increasing light intensity, but also the sugar content in nectar and the amount of dry substance in spring bleeding sap (Gimik, 1962, cited in Lyr et al., 1967). In fruit trees, an interruption of assimilate transport by girdling or strangulation causes both an increase in carbohydrate content and an earlier or stronger flowering in branches above the girdling site (see Kobel, 1954). Moreover, the flower-inhibiting effect of low light intensity can be cancelled by sugar feeding in several plants (see Lang, 1965; Evans, 1969). In the course of investigations with organ cultures, the flower-promoting effect of high sugar concentrations has been amply demonstrated, too (Nitsch, 1967; Deltour, 1967; Nicolas-Prat & Ebrahim-Zadeh, 1968; Van Bragt, 1971). Stem segments of *Plumbago indica* regenerate nothing but vegetative buds in 10 mM sucrose, but only flower buds in 70 mM sucrose (Nitsch, 1968). And finally, the $CO_2$ content of the air can modify the light action on flowering. Increased $CO_2$ content enhances the high-intensity light effect on flower formation. $CO_2$ withdrawal cancels the positive effect of high light intensity (see Lang, 1965; Evans, 1969). In conclusion we may state that a

*Acknowledgement:* Many thanks are due to Dr. H. RICHTER for linguistic revision of the manuscript.

*Abbreviations:* DCMU: 3-(3,4-dichlorophenyl)-1,1-dimethylurea; LD: long day; LDP: long-day plant; SD: short day; SDP: short-day plant.

sufficient amount of photosynthesis seems to be a precondition for generative development.

*Lemnaceae* offer some special advantages for investigating the relations between photosynthesis and flowering. Being water plants, they may be sterilely cultivated in liquid medium. 'Feeding' with organic substances (metabolites and antimetabolites) is easily carried out by simple addition to the medium. The plants show vegetative propagation by formation of daughter fronds. Successive frond generations may be cultivated, until evaluation of experiments, either together in Erlenmeyer flasks or separately in test tubes. It is important, moreover, that the genus *Lemna* comprises both long-day plants (*L. gibba* and L. *minor*) and short-day plants (*L. paucicostata* and *L. perpusilla*). Thus, the two photoperiodic reaction types can be compared as to their demands. (In *Lemna,* an inflorescence, consisting of a spathe, two stamens and one pistil, can be formed in one of the two pockets of each frond. In simplification this inflorescence is often called a flower).

We began our investigations on the significance of photosynthesis for flowering with the LDP *Lemna gibba.* Starting-point for the analysis was an unexpected finding: under certain conditions, both high-intensity light and sucrose addition to the medium inhibit flower formation in this plant. Gassing of cultures with 3,5% $CO_2$ causes a strong inhibition of flowering in the LD, too. In this connection it seems remarkable that under conditions of high light intensity cyclic photophosphorylation is not raised as much as noncyclic photophosphorylation (see Simonis & Urbach, 1973). $CO_2$ gassing causes a consumption of ATP and NADPH for the production of carbohydrates. Assimilation of exogenous sugars is an ATP-consuming process, too (Kandler & Tanner, 1967). The possibility was therefore examined that the flower formation of *L. gibba* depends on surplus ATP from photophosphorylation.

In fact, the flowering of *L. gibba* can be inhibited specifically, that is without simultaneous inhibition of vegetative development, by adding uncouplers to the medium. Arsenate, an uncoupler of photophosphorylation and oxidative phosphorylation, completely stops flowering in a concentration of 500 $\mu$M, whereas frond multiplication even is slightly enhanced. Atebrin, an uncoupler of photophosphorylation, has a very similar effect. In its case, a concentration of 10 $\mu$M suppresses flower formation, diminishing frond multiplication only to a slight degree.

A subsequent series of experiments served to investigate, whether the flower-inhibiting action of sucrose, arsenate and atebrin can be cancelled by addition of ATP and ADP. With concentrations of 100 $\mu$M, this was possible in all cases. In the majority of experiments ADP was more effective than ATP, possibly because of the action of adenylate kinases. Only ATP and ADP were effective in compensating atebrin inhibition, but not so AMP, adenosine or adenine.

Experiments with DCMU provided another argument for the assumption that flower formation in *L. gibba* depends on photophosphorylation. DCMU blocks noncyclic electron transport, but permits cyclic photophosphorylation to proceed. Consequently, assimilation of $CO_2$ is prevented, whereas ATP production continues. With a DCMU concentration of 60 nM, inhibiting vegetative growth moderately, a very strong promotion of flowering was obtained in *L. gibba* under LD. Since diluted medium (1/5 strength) was used in these experiments, controls

*Table 1.* Effect of light, sucrose, $CO_2$, antimetabolites and ADP/ATP on flowering in long day.

| | *Lemna gibba* (long-day plant) | *Lemna paucicostata* (short-day plant) |
|---|---|---|
| High intensity light | inhibiting[a] | promoting[i,k] |
| Sucrose | inhibiting[b,c] or promoting[d] | promoting[i,k,l] |
| $CO_2$ | inhibiting[e] | |
| Arsenate | inhibiting[d,f] | |
| Atebrin | inhibiting[d] | |
| DCMU | promoting[g] | inhibiting[k,l] |
| ADP, ATP | promoting[g,h] | inhibiting[m] |

References: a, unpublished; b, Hillman 1961; c, Kandeler 1968; d, Kandeler 1967; e, Kandeler 1964; f, Oota 1969; g, Kandeler 1969a; h, Kandeler 1969b; i, Esashi & Oda 1964; k, Schuster 1968; 1, Schuster & Kandeler 1970; m, Kandeler 1970.

showed only weak flower formation. Under these conditions, ADP enhances flowering nearly as much as DCMU. Thus, flower formation in the LDP *L. gibba* seems to depend on ATP, which is normally supplied by photophosphorylation.

The previously discussed results are summarized once more in table 1, including references. An additional effect is mentioned: under certain conditions, sucrose may promote flowering in *L. gibba,* especially where another medium is employed (M-medium of Hillman, 1961). Some experimental results obtained with the SDP *L. paucicostata* 6746 and cited here are particularly noteworthy in comparison with the data on *L. gibba.* (Clone 6746, which has been hitherto ascribed to *L. perpusilla,* belongs to *L. paucicostata:* Kandeler & Hügel, in press). It can be seen from table 1, that *L. paucicostata* is very different in its reactions from *L. gibba.* Both high-intensity light and sucrose feeding are factors enhancing, in combination with other factors (copper addition to the medium or senescence of the mother frond), flower formation in this plant under LD conditions. Here, high intensity light and sucrose feeding act in a sense known from many other plants. DCMU prevents flowering under LD in *L. paucicostata.* Sucrose addition cancels this effect, but only with very low concentrations of DCMU. Therefore, the flower-promoting effect of photosynthesis may probably to a certain extent be due to assimilate production. Nevertheless, additional effects of photosynthesis must be taken into account, for instance supply of reduction energy. Surplus photophosphorylation under low light intensity, on the other hand, probably has a flower-inhibiting effect, since ADP addition to the medium inhibits flower formation in *L. paucicostata* under LD very strongly.

These results demonstrate at first sight that photosynthesis intervenes in the processes leading to flowering. For an advanced analysis of the mechanism of photosynthesis action it is of primary importance to clarify, whether the products of photosynthesis affect the actual site of flowering, the apical meristem, or

163

whether they modify the photoperiodic induction processes in the leaf and thus influence flowering only in an indirect way. This question can be attacked if we succeed in cultivating the apical meristem *in vitro,* apart from all leaf organs. There is an obstacle to this approach in *Lemna,* however: flower primordia are not produced by transformation of a vegetative shoot meristem, but as a peculiar outgrowth at the base of a young primordium of leafy character. Therefore, the inducible meristem and the leaf primordium cannot be separated in this case. However, the possibility exists for explantation and *in vitro* development of relatively young primordia of *Lemna* flowers (Kandeler & Hügel, 1974).

We used such organ cultures to investigate the localization of photosynthesis effects. Again and again we have seen during our work, that in *Lemna* all the flower-inducing factors promote at the same time the further development of flower primordia. On the other hand, factors diminishing floral induction often cause the flower primordia formed to dry up. This applies also to the effects of high-intensity light and sucrose. The modification of *in vitro* flower development may therefore serve as an indicator for the effectivity of certain factors during transition to flowering.

In the present connection the action of sugars was of special interest: feeding experiments with intact plants showed that the effect described is restricted to sucrose. LD flower formation in *L. gibba* is strongly inhibited by 30 mM sucrose, but only very slightly so by glucose, fructose or a combination of them in the same concentration. With *L. paucicostata,* moderate SD flowering can be obtained under certain conditions by 30 mM sucrose, but not by glucose, fructose or their combination. These results prompted us to examine, whether such a differential sugar effect might be obtained in explants of flower primordia, too. This is not the case: in both *L. gibba* and *L. paucicostata* all the sugars mentioned show no differences whatsoever in their action on explants. Thus, for example, the inhibitory effect of sucrose present with intact *L. gibba* is lacking with explants.

The results reported suggest the conclusion that the effect of sucrose on flowering is rather indirect. Presumably this holds also for the effects of ATP and ADP: the fact that plasmalemma ATP-ases will most likely prevent transport of these substances in an unaltered state over a longer distance provides a valid argument against their direct action on the apical meristem.

Thus, products of photosynthesis seem to influence the photoperiodic induction processes in the leaf. What can we now say about the transmission of a photoperiodic stimulus from the leaf to the shoot apex? A few years ago, Evans (1969) in cooperation with 20 co-authors summarized the wealth of papers investigating this problem. Evans concludes that the popular belief in a hypothetical flowering hormone is not sufficient to explain the variety of experimental results. He assumes some of the known phytohormones to act as messengers. Our work on *Lemnaceae* together with a survey of the extensive literature have lead us to believe that a specific balance in the concentrations of all the known phytohormones is crucial for the evocation of flowering at the stem apex.

Of special importance are results obtained with organ cultures. Flower primordia of *L. gibba* and *L. paucicostata* develop on a medium containing no hormones other than kinetin. The endogenous hormone production of explants seems to cover the requirements to a sufficient extent. Highly interesting deviations from normal development do, however, occur. In *L. gibba,* pistil development is favored above stamen development; thus, a feminization of the flower results. In *L. paucicostata,* on the other hand, pistil development remains behind the development of stamens; thus, a masculinization occurs.

A normalization of development can be obtained in *L. gibba* by adding gibberellin $A_3$ to the medium. Development in *L. paucicostata* is normalized after addition of CCC (a blocker of gibberellin synthesis) or ethrel (an ethylene-evolving substance) (Hügel & Kandeler, 1974). From these results we conclude that the LDP *L. gibba* produces relatively little gibberellin in the meristem. For normal development, additional gibberellin must be supplied by the leaf or by the agar medium. The meristem of the SDP *L. paucicostata,* on the other hand, produces relatively much gibberellin; therefore, it needs an additional supply of ethylene to keep the hormonal balance for normal development. *In vitro,* normalization may also be brought about by suppressing the endogenous synthesis of gibberellin with CCC.

We think the above-mentioned results to be also of significance for the explanation of floral induction processes. The differing endogenous hormone production in meristems of LDPs and SDPs seems to require a supplementary supply of hormones from leaves to the apical meristems. This supplement has to comprise different sets of hormones in each case in order to obtain the proper flower-inducing hormone equilibrium. In this view, the photoperiodic reaction acts by modifying hormone production in leaves or hormone export from leaves in such a way that hormones needed for the correct equilibrium reach the apical meristem. Many investigations on very different plant species support this hypothesis. In the majority of cases, LD causes an increase in gibberellin and auxin contents; SD, on the other hand, causes an increase in inhibitor (mostly abscisin) content and sometimes in cytokinin content (see Kandeler, in press). For the explanation of contradictory results it must be taken into account that photoperiodic controls may possibly modify hormone transport, too. Furthermore, the possibility of imitating or cancelling daylength effects by means of hormone application is yet another argument supporting the role of phytohormones as mediators of photoperiodic reactions.

Table 2 summarizes the pertinent results obtained with duckweeds. Although the investigations are not completed yet, the following statements can nevertheless be made: 1) Four different phytohormone groups influence flowering of the SDP *L. paucicostata* under LD conditions: gibberellin and indoleacetic acid are inhibiting, low concentrations of abscisic acid and cytokinins are promoting. 2) Low gibberellin concentrations and cytokinins have an opposite effect in LDPs and SDPs. 3) Even in the same plant, different hormone concentrations may have opposite effects (see gibberellin in LDPs, abscisic acid in SDPs). All the results reported here are in agreement with the hypothesis that levels of gibberellin and, possibly, of indoleacetic acid must be moderately raised in LDPs for flowering, whereas the import of abscisic acid and cytokinin has to be throttled simultaneously. SDPs, however, seem to require moderate increases of abscisic acid and, possibly, cytokinin levels in the apex for flowering, accompanied by a diminished import of gibberellin and indoleacetic acid. Further support for this hypothesis was gathered during our investigation of connections between flowering and senescence. In *L. gibba,* flower formation in daughter fronds is inhibited by senescence of the mother frond; in *L. paucicostata,* however, senescence of the

*Table 2.* Effect of hormones and other treatments on flowering in long day

| | *Lemna gibba* (long-day plant) | *Lemna paucicostata* (short-day plant) |
|---|---|---|
| Gibberellin, low concentration | promoting[a] | inhibiting[b] |
| Gibberellin, high concentration | inhibiting[c] | inhibiting[b] |
| Indoleacetic acid, low concentration | | inhibiting[b] |
| Indoleacetic acid, high concentration | inhibiting[a] | inhibiting[b] |
| Abscisic acid, lower concentration | | promoting[d,e] |
| Abscisic acid, higher concentration | | inhibiting[d] |
| Kinetin, Zeatin | inhibiting[a,f] | promoting[b,g] |
| Explants on medium without hormones (except kinetin) | feminization[h] | masculinization[h] |
| Senescence of the mother frond | inhibiting[i] | promoting[i] |
| Lithium | inhibiting[k] | promoting[k] |
| Acetylcholine | inhibiting[l] | promoting[l] |

References: a, Oota 1965; b, Gupta & Maheshwari 1970; c, Cleland & Briggs 1969; d, Kandeler & Hügel 1973; e, Higham & Smith 1969; f, Cleland 1971; g, Gupta & Maheshwari 1969; h, Kandeler & Hügel 1974; i, Kandeler, Hügel & Rottenburg 1974; k, Kandeler 1970; l, Kandeler 1972.

mother frond promotes flowering in daughter fronds. And it has been well known that processes of leaf senescence are linked with decreases in gibberellin and indoleacetic acid production and with an increase in the production of abscisic acid.

The above review of the hormonal regulation of flowering went into some detail in order to obtain a basis for discussing, how photosynthesis may intervene in the processes leading to flowering. We have to ask now for support of the assumption, that photosynthesis products influence production or transport of hormones. A first hint in this direction is given by the fact that both photosynthesis effects and hormonal effects show definite optima: beyond an optimum value both effects decline and finally act in the opposite direction. For *L. gibba* we mentioned flower inhibiting as well as flower promoting effects of sucrose, depending on the type of medium employed. For *L. paucicostata*, Kirkland & Posner (1974) recently reported a pronounced arrest of flower development by continuous high-intensity light. The effect can be partially cancelled by DCMU. At first sight, this result seems to be contradicting the data cited in table 1. However, this apparent contradiction is doubtlessly due to the use of different culture media. This implies that even in one plant the effect of photosynthesis on flowering is not fixed in one direction. Variations in experimental conditions determine the direction of action. The same can be shown to hold for the action of hormones.

An important argument for the influence of photosynthesis on hormone transport is the fact that hormones and assimilates often use the same translocation pathway to the apical meristem (see Crafts & Crisp, 1971). Phloem transport of hormones is doubtlessly determined to some extent by the source-to-sink gradient of assimilate concentration. Pertinent results of King & Zeevaart (1973) are particularly impressive. In the SDP *Perilla,* flowering of individual lateral buds depends on whether the assimilate stream leading to this bud descends from an SD-induced leaf or from a non-induced LD-leaf. Inhibition of flowering by LD leaves increases with increasing light intensity, reaching saturation at the same light level as photosynthesis.

For the time being, the connection between photosynthesis and hormone transport in *Lemnaceae* remains hypothetical. Nevertheless, some results prompt the conclusion that transport conditions are crucial for floral induction in these plants. Lithium ions cause an inhibition of flowering under LD in *L. gibba,* promoting, on the other hand, flowering of *L. paucicostata.* Lithium is known to act as a potassium antagonist in several cases. Potassium, for its part, plays an important role during membrane transport of organic anions (Lüttge, 1973), some phytohormones being organic anions, too. Furthermore, potassium affects sugar transport (Amir & Reinhold, 1971; Ashley & Goodson, 1972; Haeder & Mengel, 1972).

Even more striking are the effects of acetylcholine. In the LD, this substance acts like sucrose and lithium, inhibiting flowering in *L. gibba* and enhancing it in *L. paucicostata.* In many animal nerves, acetylcholine causes a permeability increase of the postsynaptic membrane. In *Lemna,* acetylcholine seems to alter membrane permeability, too. In our experimental cultures this was demonstrated by an acidification of the medium under the influence of acetylcholine. At present we do not yet know, whether the acidification of the medium is due to an excretion of organic acids or to an exchange of hydrogen ions for potassium. In both cases, acidification would be an indication for an effect of acetylcholine on membrane transport processes. The results suggest, in addition, that certain key factors control both flowering and acid metabolism, as shown by Brulfert, Guerrier & Queiroz (1973) for *Kalanchoe blossfeldiana.* An investigation of this problem with duckweeds is a task for the future.

As a conclusion we will summarize in short our ideas on the relationship between photosynthesis and flowering. In order to reach the flower-inducing hormone balance, the endogenous hormone set of the apical meristem must be completed by the import of hormones, the type of hormones needed depending on the plant. Hormones imported come from the leaves, where the photoperiodic reaction provides for the necessary changes in hormone metabolism and hormone export. Photosynthesis modifies hormone export from the leaf through both ATP supply from cyclic photophosphorylation and production of assimilates like sucrose.

# REFERENCES

Amir, S. & L. Reinhold (1971): Interaction between K-deficiency and light in [14]C-sucrose translocation in bean plants. *Physiol. Plantarum,* 24: 226-231.

Ashley, D.A. & R.D. Goodson (1972): Effects of time and plant K status on [14]C-labelled photosynthate movement in cotton. *Crop Sci.,* 12: 686-690.

Bragt, J. van (1971): Effects of temperature, sugars and GA$_3$ on *in vitro* growth and flowering of terminal buds of tulip. *Meded. Landbouwhogeschool Gent,* 36: 479-483.

Brulfert, J., D. Guerrier & O. Queiroz (1973): Photoperiodism and enzyme activity: balance between inhibition and induction of the crassulacean acid metabolism. *Plant Physiol.,* 51: 220-222.

Cleland, C.F. (1971): Influence of cytokinins on flowering and growth in the long-day plant *Lemna gibba* G3. *Plant Physiol.,* 47 (Suppl.): 13.

Cleland, C.F. & W.S. Briggs (1969): Gibberellin and CCC effects on flowering and growth in the long-day plant *Lemna gibba* G3. *Plant Physiol.,* 44: 503-507.

Crafts, A.S. & C.E. Crisp (1971): Phloem Transport in Plants. Freeman & Co., San Francisco.

Deltour, R. (1967): Action du saccharose sur la croissance et la mise à fleurs de plantes issues d' apex de *Sinapis alba* L. cultivés *in vitro. C.R. Acad. Sci.,* Paris, 264: 2765-2767.

Esashi, Y. & Y. Oda (1964): Effects of light intensity and sucrose on the flowering of *Lemna perpusilla. Plant Cell Physiol.,* 5: 513-516.

Evans, L.T. (1969): The Induction of Flowering. MacMillan of Australia, South Melbourne.

Filzer, P. (1940): Lichtökologische Untersuchungen an Rasengesellschaften. *Beihefte Bot. Centralbl.,* 60 (Abt.B): 229-248.

Gupta, S. & S.C. Maheshwari (1969): Induction of flowering by cytokinis in a short-day plant, *Lemna pauciostata. Plant Cell Physiol.,* 10: 231-233.

Gupta, S. & S.C. Maheshwari (1970): Growth and flowering of *Lamna paucicostata* II. Role of growth regulators. *Plant Cell. Physiol.,* 11: 97-106.

Header, H.E. & K. Mengel (1972): Translocation and respiration of assimilates in tomato plants as influenced by K nutrition. *Z. Pflanzenernähr.,* 131: 139-148.

Higham, B.M. & H. Smith (1969): The induction of flowering by abscisic acid in *Lemna perpusilla* 6746. *Life Sciences* 8/II: 1061-1065.

Hillman, W.S. (1961): Experimental control of flowering in *Lemna.* III. A relationship between medium composition and the opposite photoperiodic responses of *L. perpusilla* 6746 and *L. gibba* G3. *Amer. J. Bot.,* 48: 413-419.

Hügel, B. & R. Kandeler (1974): Hormonbedarf *in vitro* bei den Blütenanlagen einer Kurztagpflanze (*Lemna paucicostata* 6746) und einer Langtagpflanze (*Lemna gibba* G1). Abstracts Botaniker Tagung, 1974, Würzburg.

Kandeler, R (1964): Wirkungen des Kohlendioxyds auf die Blütenbildung von *Lemna gibba. Naturwiss.,* 51: 561-562.

Kandeler, R. (1967): The role of photophosphorylation in flower initiation of the long-day plant *Lemna gibba.* European Photobiol. Symp. 1967, Hvar, Yugoslavia, Book of Abstracts. P. 45.

Kandeler, R. (1968): Blühinduktion bei Lemnaceen. *Biol. Rundsch.,* 6: 49-57.

Kandeler, R. (1969a): Förderung der Blütenbildung von *Lemna gibba* durch DCMU und ADP. *Z.Pflanzenphysiol.,* 61: 20-28.

Kandeler, R. (1969b): Hemmung der Blütenbildung von *Lemna gibba* durch Ammonium. *Planta,* (Berl.), 84: 279-291.

Kandeler, R. (1970): Die Wirkung von Lithium und ADP auf die Phytochromsteuerung der Blütenbildung. *Planta,* (Berl.), 90: 203-207.

Kandeler, R. (1972): Die Wirkung von Acetylcholin auf die photoperiodische Steuerung der Blütenbildung bei Lemnaceen, *Z.Pflanzenphysiol.,* 67: 86-92.

Kandeler, R. (in press): Photomorphogenese: Die Rolle der Hormone. *Ber. Dtsch. Bot. Ges.*

Kandeler, R. & B. Hügel (1973): Blütenbildung bei *Lemna paucicostata* 6746 durch kombinierte Anwendung von Abscisinsäure und CCC. *Plant Cell Physiol.,* 14: 515-520.

Kandeler, R. & B. Hügel (1974): Development *in vitro* of flower primordia of *Lemnaceae.* Abstracts 3rd Intern. Congr. Plant Tissue and Cell Culture, 1974, Univ. Leicester, Paper No. 160.

Kandeler, R. & B. Hügel (in press): Wiederentdeckung der echten *Lemna perpusilla* TORR. und Vergleich mit *L. paucicostata* HEGELM. *Plant System. Evolution*, Vienna.

Kandeler, R., B. Hügel & Th. Rottenburg (1974): Gegensätzliche Wirkung der Sprossalterung auf die Blütenbildung bei *Lemna paucicostata* und *Lemna gibba*. *Biochem. Physiol. Pflanzen*, 165: 331–336.

Kandler, O. & W. Tanner (1967): Die Photoassimilation von Glucose als Indikator für die Lichtphosphorylierung *in vivo*. *Ber. Dtsch. Bot. Ges.*, 79: (48) - (57).

King, R.W. & J.A.D. Zeevaart (1973): Floral stimulus movement in *Perilla* and flower inhibition caused by noninduced leaves. *Plant Physiol.*, 51: 727-738.

Kirkland, L. & H.B. Posner (1974): The role of light in the photoperiodic inhibition of flower development in *Lemna perpusilla* 6746. *Plant Physiol.*, 53 (Suppl.): 3.

Kobel, F. (1954): Lehrbuch des Obstbaus auf physiologischer Grundlage. 2.Aufl. Springer Verlag, Berlin etc.

Lang, A. (1965): Physiology of flower initiation. Encyclopedia of Plant Physiology, Vol. 15, 1. Pp 1380-1536. Springer Verlag, Berlin etc.

Lüttge, U. (1973): Stofftransport der Pflanzen. Springer Verlag, Berlin etc.

Lyr, H., H. Polster & H.-J. Fiedler (1967): Gehölzphysiologie. VEB Fischer Verlag, Jena.

Nicolas-Prat, D. & H. Ebrahim-Zadeh (1968): Appoint glucidique et éclairement dans leurs rapports avec l'organogenèse florale in vitro chez *Nicotiana tabacum* L. Les Cultures du Tissus de Plantes. Pp. 83-91. Ed. Centre Nat. Rech. Scient., Paris.

Nitsch, C. (1968): Induction in vitro de la floraison chez une plante de jours courts: *Plumbago indica* L. *Ann. Sci. natur., Bot.*, Paris, Sèr. 12, 9: 1-91.

Nitsch, J.P. (1967): Towards a biochemistry of flowering and fruiting: Contributions of the 'in vitro' technique. Proc. XVII Intern. Hort. Congress, Vol. III. Pp 291-308.

Oota, Y. (1965): Effects of growth substances on frond and flower production in *Lemna gibba* G3. *Plant Cell Physiol.*, 6: 547-559.

Schuster, M. (1968): Die Bedeutung von Starklicht und Kupfer für die phytochromgesteuerte Morphogenese von *Lemna perpusilla*. Thesis. Univ. Würzburg.

Schuster, M. & R. Kandeler (1970): Die Bedeutung der Photosynthese für die Langtag-Blüte der Kurztagpflanze *Lemna perpusilla* 6746. *Z. Pflanzenphysiol.*, 63: 308-315.

Simonis, W. & W. Urbach (1973): Photophosphorylation in vivo. *Ann Rev. Plant Physiol.*, 24: 89-114.

# DEVELOPMENTAL PATTERNS OF CO$_2$ EXCHANGE, DIFFUSION RESISTANCE AND PROTEIN SYNTHESIS IN LEAVES OF POPULUS X EURAMERICANA [1]

D.I. DICKMANN*, D.H. GJERSTAD & J.C. GORDON**

*Department of Forestry, Michigan State University, East Lansing, MI 48824, USA
**Department of Forestry, Iowa State University, Ames, IA 50010, USA

## Abstract

Patterns of gas exchange and protein synthesis were measured in developing leaves of clonal *Populus* x *euramericana* plants. Net photosynthesis and apparent photorespiration (Warburg effect) were zero in very young leaves, but then increased to maximum levels in recently mature leaves. Both processes declined in old leaves, but the decline in photosynthesis was more rapid. Mitochrondrial (dark) respiration decreased with leaf age and was less in the light than in the dark, except in very old leaves, where it increased sharply in the light. Diffusion resistance and CO$_2$ compensation concentration declined to minimum levels in recently mature leaves; however, resistances increased markedly in older leaves, whereas CO$_2$ compensation remained at a minimum. Soluble protein concentrations and incorporation of $^{14}$C-photosynthate into protein declined throughout leaf development. Protein turnover was slight in expanding leaves, but was substantial after leaves matured. Expanding leaves synthesized predominantly Fraction I protein, but formation of this protein was slight once leaves matured. The significance of these findings in relation to the developmental pattern of net photosynthesis in poplar leaves is discussed.

## Introduction

Because of their rapid growth and amenability to intensive culture, trees of the genus *Populus* have become a prominent part of innovative systems designed to maximize wood fiber production. To provide a foundation of basic information for these practical efforts, a coordinated attempt to understand the developmental physiology of young poplars is under way at several laboratories in the United States. Most of this physiological work has concentrated on the developing leaf zone of native eastern cottonwood (*Populus deltoides*) and various hybrid clones of *Populus* x *euramericana* (*P. deltoides* x *P. nigra*).

Work to date has demonstrated that the key event in cottonwood leaf ontogeny is the attainment of the mature state; i.e., the cessation of leaf expansion. At this point leaf anatomy has essentially stabilized; mesophyll cells are mature and intercellular spaces fully developed; stomatal formation is complete and the leaf vascular system and areoles are fully functional (Isebrands & Larson, 1973). Maturation does not proceed at a uniform rate throughout the leaf, however. The lamina tip matures first, both structurally and functionally, and matura-

1 Journal Paper No. J-8006 of the Iowa Agriculture and Home Economics Experiment Station, Ames, Iowa. Project 1872.

*Abbreviations:* RuDP: ribulose-1,5-diphosphate; LPI: leaf plastochron index; PPO: 2,5-diphenyloxazole; POPOP: p-bis(2-(5-phenyloxazole)benzene; TCA: trichloroacetic acid.

tion then proceeds basipetally, the leaf base and margins maturing last (Isebrands & Larson, 1973; Larson et al., 1972). Maturation of the leaf also is reflected in the stem where secondary vascularization begins in the internode associated with the first mature leaf (Larson & Isebrands, 1974).

A cottonwood leaf that has reached anatomical maturity is functionally mature as well. Net photosynthesis is maximum, $CO_2$ compensation concentration minimum, and dark respiration minimum (Dickmann, 1971a; Dickman & Gjerstad, 1973; Larson & Gordon, 1969). The Hill reaction and activity of RuDP carboxylase first reach maximum levels at this developmental stage also (Dickmann, 1971b). Total nitrogen concentrations reach a stable level in mature leaves after declining during expansion, whereas total chlorophyll and the major leaf peroxidase isoenzyme continue to increase beyond full expansion (Dickmann, 1971b; Gordon, 1971). A mature leaf functions only as an exporter of photo-synthate (Larson & Gordon, 1969), although an expanding leaf may be simul-taneously exporting from mature regions (e.g., the tip) and importing into imma-ture regions (Larson et al., 1972). The destination of transported carbohydrate from any leaf follows fairly restricted channels that can be predicted from a knowledge of plant phyllotaxy (Larson & Dickson, 1973).

Studies reported in this paper were iniiiated to further define the gas-exchange characteristics of expanding, mature and senescing leaves of hybrid cottonwood with respect to net photosynthesis, dark respiration, photorespiration, $CO_2$ com-pensation concentration and resistance to water vapor diffusion. The incorpora-tion of $^{14}C$-photosynthate into protein also was studied to elucidate the develop-mental synthesis of Fraction I protein.

## Material and Methods

*Plant material.*
Ramets of clonal hybrid cottonwood (*Populus* x *euramericana*) were propagated from tip cuttings rooted under mist. Trees were then potted in an artificial soil mix consisting of peat, perlite and vermiculite, and grown in growth chambers maintained at $25°C$ during an 18 hr day and $15°C$ at night. Illumination during the day was approximately 30,000 lux. Plants were grown until 30 to 40 leaves had formed and were fertilized weekly with a commercial 20-20-20 (N-P-K) soluble fertilizer supplemented with Fe-EDTA.

To precisely define and duplicate developmental stages, the plastochron con-cept was employed, wherein each leaf is related to the meristematic stem apex and all other leaves on the plant by LPI (Larson & Isebrands, 1971). In this system, the first unfolding leaf at the apex whose lamina has reached at least 2 cm in length is designated the index leaf and assigned an LPI of 0. The next oldest leaf below the index leaf is LPI 1, the second oldest LPI 2 and so on down the plant. An ontogenetic series is thereby formed which defines at any given time the developmental stage of each leaf and relates leaves at comparable morphological stages (same LPI) on different plants. The first fully mature leaf on 30- to 40-leaf cottonwood plants used in these studies was at LPI 10 to 12.

*Gas exchange measurements.*
While still in the growth chamber, resistance to water vapor diffusion of attached leaves of LPI 4, 6, 10, 15, 20 and the lowest three leaves on the plant (LPI 36 to 43 depending on plant size) was measured by a Li-Cor diffusive resistance meter fitted with a Kanemasu sensor. The same leaves were then enclosed in a water-cooled plastic leaf chamber and connected to a Beckman infrared gas analysis system (Dickmann, 1971a). After appropriate equilibration periods, measurements of net $CO_2$ flux in a closed system and $CO_2$ evolution into $CO_2$-free air were taken in the light and dark in air containing 21% or 2% $O_2$. Light intensity in the leaf chambers was 750 $\mu$Einsteins m$^{-2}$ sec$^{-1}$, 400 to 700 nm. Leaf temperature averaged 25 $\pm$ 2° C. Leaves were then excised and $CO_2$ compensation concentrations determined using the Mylar bag method (Dickmann & Gjerstad, 1973). Six replicate leaves at each LPI were measured.

*Determination of $^{14}C$ incorporation into protein.*
Attached leaves of LPI 4, 8, 19, and 36 on different plants were allowed to photoassimilate 25 $\mu$Ci $^{14}CO_2$ in a closed system for 1 hr at 21% $O_2$ as above. Plants were then returned to the growth chamber and treated leaves were harvested 3 hr and 48 hr after $^{14}C$ treatments were initiated. Thus, each treatment sequence consisted of eight leaves from eight different plants; four LPI's each with two harvest times. Each sequence was replicated three times.

Excised leaves were lyophilized and then ground in a Wiley mill. Extracts of the leaf powder were prepated by a single 15-sec grinding of 100 mg samples in 5 ml of pH 7.4 buffer (0.05 Tris-HCL, 0.1 M in sucrose) in a Duall homogenizer. Homogenates were centrifuged at 105,000 xg for 1 hr. Total $^{14}C$ activity in the crude extracts was determined by adding 0.1 ml samples to 15 ml of scintillation medium (toluene and methyl cellosolve, 5:2, plus 4 g/liter PPO and 50 mg/liter POPOP) and counting in a liquid scintillation spectrometer.

Activity of $^{14}C$ in the soluble protein fraction of the crude extract was determined by adding 1 ml 20% TCA (w/v) to 1 ml of the crude extract, centrifuging at 105,000 xg for 30 min and counting 0.1 ml samples of the resulting supernatant in 15 ml of scintillation medium. Difference in activity in the TCA supernatant and crude extract gave activity in soluble protein. Quantitative protein determinations were run on 1 N NaOH solutions of 20% TCA precipitates of the crude extracts (Lowry et al., 1951).

Acrylamide gel electrophoresis of proteins in the crude extract was performed in a slab gel electrophoresis chamber (Gordon, 1971). After electrophoresis, half the gel from each leaf was stained with amido black for total protein. On the remaining half of the gel for each leaf, the portion between the application slot and the electrophoretic front was divided into 5-mm sections for $^{14}C$ assay. Each section was placed in a scintillation vial with 0.3 ml 30% $H_2O_2$ (v/v). After the gel was depolymerized, 3 ml NCS (Amersham/Searle quaternary ammonium base solubilizer) were added followed by 13 ml of scintillation medium.

173

# Results and Discussion

*Gas change experiments.*

Changes in net photosynthesis and apparent photorespiration with leaf age are presented in Fig. 1. Net photosynthesis (uptake of $CO_2$ in light at 400 ppm $CO_2$ and 21% $O_2$) was zero at LPI 4, reached a maximum in recently mature leaves at LPI 15 and then declined in older leaves to about half the maximum rate. A similar pattern was reported by Dickmann (1971a) for young *Populus deltoides* plants.

The increase in net photosynthesis when $O_2$ tensions were lowered to 2% was used in this study to estimate photorespiration. This $O_2$ enhancement of photosynthesis (Warburg effect) is a well-known phenomenon in $C_3$ plants, but the exact physiological basis for it is unclear. Much evidence implicates glycolate metabolism and photorespiration as being largely responsible for $O_2$ enhancement (Bowes & Ogren, 1972; Zelitch, 1971). An equal possibility exists, however, that lowered $O_2$ tensions decrease the inhibitory effect of $O_2$ on $CO_2$ diffusion and fixation (Ludlow, 1970), although this issue is complicated by Fair et al. (1973), who reported that low $O_2$ levels reduced the activity of RuDP carboxylase. Other recent evidence suggests that the $O_2$-enhancement effect is divided almost equally between an increase in $CO_2$ fixation and an inhibition of photorespiratory $CO_2$ evolution (D'Aoust & Canvin, 1973). $CO_2$ evolution into $CO_2$-free air or the extrapolation method could not be used in this study to estimate photorespiration because of the high mitochondrial respiration rates and low photosynthesis rates of young leaves.

Use of $O_2$ enhancement to estimate photorespiration also presumes that mitochondrial (dark) respiration is unaffected by changes in $O_2$ concentrations. Table 1 shows that throughout the range of leaf ages used in this study, no significant differences in mitochondrial respiration were observed between 2% $O_2$ and 21% $O_2$.

*Table 1.* Influence of $O_2$ on Mitochondrial Respiration in Poplar Leaves of Different Ages.

| Leaf Age | Mitochondrial (Dark) Respiration | |
|---|---|---|
| | 2% $O_2$ | 21% $O_2$ |
| LPI | mg $CO_2$ hr$^{-1}$ dm$^{-2}$ | |
| 4 | 6.4[a] | 6.4 |
| 6 | 4.8 | 5.1 |
| 10 | 2.9 | 2.9 |
| 15 | 2.1 | 1.9 |
| 20 | 1.4 | 1.2 |
| 37 | 0.9 | 0.8 |
| 38 | 1.0 | 1.0 |
| 39 | 1.4 | 1.3 |

[a] All values are averages of six replications. Measurements were made in the dark at 25°C and 400 ppm $CO_2$.

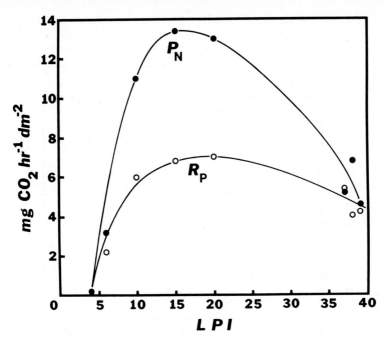

*Fig. 1.* Net photosynthesis ($P_N$) and apparent photorespiration (Rp) in poplar leaves of different ages. $P_N$ = net $CO_2$ uptake at 400 ppm $CO_2$ and 21% $O_2$ . $R_p$ = photosynthetic $CO_2$ uptake at 400 ppm $CO_2$ and 2% $O_2$ – $P_N$ ($O_2$ enhancement). Temperature 25°C; light intensity 750 $\mu$Einsteins m$^{-2}$ sec$^{-1}$, 400 to 700 nm. Points are averages of 6 replications.

Data of Fig. 1 clearly show a pronounced change in apparent photorespiratory activity with leaf development similar to that observed for net photosynthesis. A very immature leaf of LPI 4 showed no photorespiratory activity, but by LPI 6 some photorespiration was evident. Young leaves of citrus and tobacco also show negligible $O_2$ enhancement and low glycolate oxidase activity (Kisaki et al., 1973; Salin & Homann, 1971). By LPI 10 to 20, substantial rates of apparent photorespiration were observed in the present study, amounting to approximately one-third of total photosynthesis. This estimate of photorespiration in mature leaves compares closely with those observed for a number of $C_3$ plants (Samish et al., 1972; Zelitch, 1971), but is considerably higher than the extrapolated values determined by Luukkanen & Kozlowski (1972) for *Populus* leaves. As leaves approached senescence, photorespiratory rates declined slightly (Fig. 1), but at a slower rate than the decline in photosynthesis.

Mitochondrial respiration also operates in the light, but controversy exists over the extent to which it occurs. Jackson & Volk (1970) summarized data supporting an inhibition of mitochrondrial respiration in the light, whereas Zelitch (1971) marshalled evidence pointing to no effect of light on this process. Recent reports have only confounded the issue. Using photosynthetic and tricarboxylic acid inhibitors, Chapman & Graham (1974) concluded that light did not suppress activity of mitochondrial respiration in *Phaseolus aureus*. In contrast, Mangat et al. (1974) analyzed $^{14}CO_2$ and $CO_2$ fluxes of leaves of *Phaseolus vulgaris,* tobacco and corn, and concluded that mitochondrial respiration was about 75% suppressed in the light and was regulated by ATP levels. Data shown in Fig. 2 support the latter report. Rates of mitochrondrial respiration in the light were considerably below those observed in the dark. In very old polar leaves, however, this trend was reversed, and respiration in the light increased sharply and exceeded that in the dark.

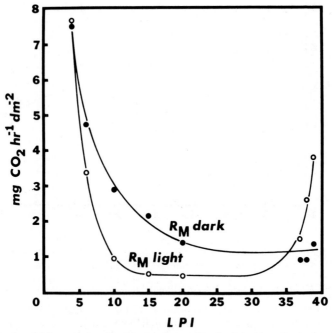

Fig. 2. Mitochondrial respiration ($R_M$) in the dark and in the light in poplar leaves of different ages. $R_M$ in the light = $CO_2$ evolution into $CO_2$-free air at 2% $O_2$. Temperature 25°C; light intensity 750 $\mu$Einsteins m$^{-2}$ sec$^{-1}$, 400 to 700 nm. Points are averages of 6 replications.

The data of Fig. 2 for mitochondrial respiration in the light were derived from measurements of net $CO_2$ evolution into $CO_2$-free air containing 2% $O_2$. In this gas atmosphere, the predominant $CO_2$ flux should be from mitochondrial respiration because low $O_2$ levels inhibit photorespiration, and photosynthesis is suppressed in $CO_2$-free air. Considerable internal recycling of $CO_2$ would still occur under these conditions, however, particularly in older leaves where resistances to diffusion increased markedly (Fig. 3), resulting in underestimates of mitochondrial respiration. Nevertheless, reduction of apparent mitochondrial respiration in the light was sufficient to warrant the conclusion that some suppression of this process occurs in recently mature poplar leaves.

Fig. 3 shows the influence of leaf age on diffusion resistance and $CO_2$ compensation. After declining to a minimum in recently mature leaves, diffusion resistance increased sharply in the oldest leaves. The decrease in resistance as leaves mature can be related to development of intercellular spaces and stomates (Isebrands & Larson, 1973). In older leaves, stomate opening evidently becomes sluggish, causing higher resistances. This conclusion is supported by our observa-

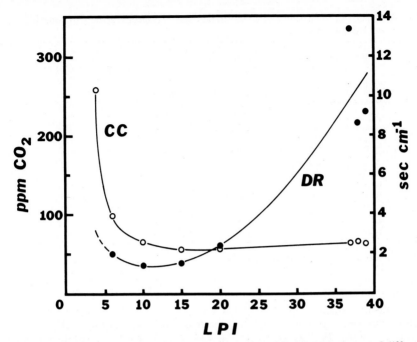

*Fig. 3.* Diffusion resistance (DR) and $CO_2$ compensation (CC) of poplar leaves of different ages. DR was measured with a Li-Cor diffusive resistance meter fitted with a Kanemasu sensor. CC was determined by the Mylar bag method. Temperature 25°C; light intensity 750 $\mu$Einsteins m-2 sec-1, 400 to 700 nm. Points are averages of 6 replications.

tions that steady-state photosynthesis rates after a dark period are achieved much more slowly in old leaves.

$CO_2$ compensation concentrations are thought to be a function of photosynthesis, photorespiration and mesophyll resistances (Luukkanen & Kozlowski, 1972; Samish & Koller, 1968). The present data reflect a more complex relationship. One would expect $CO_2$ compensation to increase gradually in older leaves where photosynthesis is declining faster than photorespiration and where diffusion resistances increase sharply. This did not occur (Fig. 3), although, in most poplar clones that we have measured, $CO_2$ compensation does increase markedly in very old leaves (Dickmann & Gjerstad, 1973). In a developmental sense, then, $CO_2$ compensation concentrations are not accurate indicators of photosynthetic efficiency.

### [14]C-incorporation into protein.

Changes in soluble protein and protein [14]C-specific activity with leaf development are given in Table 2. Protein concentrations declined from LPI 4 through LPI 19, but then increased slightly in the oldest leaves. Incorporation of [14]C-photosynthate into protein, however, declined throughout the ageing series, indicating a general reduction in the level of protein synthesis from photosynthate.

Table 2 also shows that a significant change in total protein turnover (simultaneous synthesis and degradation) occurred during leaf development. Assuming

*Table 2.* Changes in soluble protein concentration and [14]C-specific activity of popular leaves with age and harvest time.

| Leaf Age | Harvest Time | Soluble Protein | [14]C Specific Activity | %Loss in Specific Activity after 48 hr |
|---|---|---|---|---|
| LPI | hr | $\mu$g/mg dry wt | dpm/$\mu$g protein | |
| 4 | 3 | 138[a] | 150 | - |
| 4 | 48 | 122 | 137 | 9 |
| 8 | 3 | 114 | 131 | - |
| 8 | 48 | 112 | 48 | 63 |
| 19 | 3 | 75 | 105 | - |
| 19 | 48 | 78 | 19 | 82 |
| 36 | 3 | 99 | 69 | - |
| 36 | 48 | 83 | 25 | 64 |

[a] All values are averages of three replications.

predominantly open amino acid pools for synthesis, loss in protein specific activity with time after a short [14]C treatment indicates turnover (Huffaker & Peterson, 1974). For leaves treated at LPI 4, only 9% of the protein specific activity was lost after 48 hr; older leaves, however, lost nearly three-fourths of their [14]C in protein after 48 hr. The greater stability of [14]C-protein in young leaves indicates a generally lower turnover rate, or the synthesis of a major stable protein not subject to turnover. Further data presented below support the latter interpretation.

A densely staining band at an $R_F$ of 0.16 to 0.18 was present in electrophoretic gels at all leaf ages. This band was faintest at LPI 4 harvested after 3 hr but darkened considerably after 48 hr. No discernible intensification of band staining occurred in older leaves. This band is Fraction I protein, a prominent constituent of photosynthetic cells and functionally the enzyme RuDP carboxylase (Kawashima & Wildman, 1970). Recent work has demonstrated that this protein also can function as an oxygenase, however, catalyzing the formation of the photorespiratory intermediate phosphoglycolate from RuDP (Andrews et al., 1973; Ogren & Bowes, 1971). The duality in function of this protein may explain in part the close correspondence between photosynthesis and photorespiration during leaf development (Fig. 1).

Patterns of [14]C distribution within electrophoretic gels are shown in Fig. 4. Total activity recovered from the gels declined with leaf age. Also, the total activity recovered after 48 hr was less than that recovered after 3 hr except at LPI 4, where more than twice the activity found at 3 hr was recovered at 48 hr. In addition, the [14]C-labeling profile in the gels showed a distinct pattern with leaf age. At LPI 4 harvested after 3 hr, a fairly uniform profile was found throughout the gel, although a slight peak occurred in the position of the Fraction I band. After 48 hr, however, the Fraction I peak had enlarged dramatically, indicating a rapid synthesis of this protein during the 2-day period following treatment. A

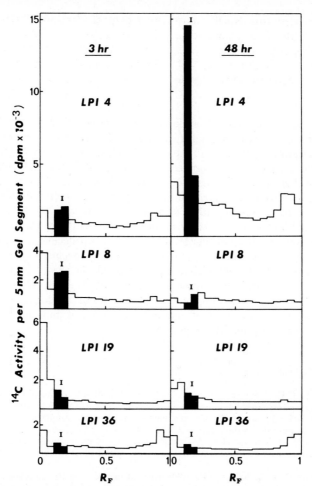

*Fig. 4.* $^{14}C$ profiles for soluble protein in acrylamide electrophoretic gels prepared from leaves of different ages harvested 3 hr or 48 hr after treatment with $^{14}CO_2$. Fraction I protein band is between Rf 0.1 and 0.2 (darkened bars). Figures are averages of 3 replications.

slight peak in $^{14}C$ activity in the Fraction I position also occurred at LPI 8, harvested after 3 hr and 48 hr. These peaks were absent at LPI's 19 and 36.

Woolhouse (1967) concluded that decrease in leaf protein with age was due to 'switching off' of Fraction I protein synthesis. The present study supports this view for *Populus* and shows furthermore that synthesis of this critical protein complex occurs only in expanding leaves. Recent work also has shown that little turnover of RuDP carboxylase (Fraction I protein) occurs under normal environmental conditions, as opposed to other soluble proteins that are rapidly synthesized and degraded (Peterson et al., 1973). The data of Table 2 support this conclusion. Thus, Fraction I protein levels and RuDP carboxylase activity would tend to run down gradually after leaf expansion is completed. Callow (1974) showed that this occurred in cucumber leaves. Because photosynthetic rate is closely correlated with RuDP carboxylase activity (Dickmann, 1971b; Smillie & Fuller, 1959; Steer, 1971) and because the carboxylating step of the Calvin cycle is rate-limiting in photosynthesis at saturating light and normal $CO_2$ levels (Ogren & Bowes, 1971; Neales et al., 1971; Wareing et al., 1968), photosynthesis of ageing poplar leaves would be expected to decline, as shown in Fig. 1.

*Conclusions.*
The photosynthetic pattern of developing poplar leaves (Fig. 1) can be attributed to related morphological, physiological and biochemical events reported in this paper and elsewhere. During leaf expansion increases in photosynthetic $CO_2$ uptake can be related to development of internal leaf structure and stomates (Isebrands & Larson, 1973), decrease in diffusion resistance, chlorophyll synthesis (Dickmann, 1971b), development of physiological and structural integrity of the membrane-bound phosphorylation system of chloroplasts (Dickmann, 1971b; Hernadez-Gil & Schaedle, 1973), Fraction I protein synthesis, increases in RuDP carboxylase activity (Dickmann, 1971b) and a sharp decline in mitochondrial respiration. After expansion is completed, a gradual decline in photosynthesis occurs, which correlates with an increase in diffusion resistance, decay in the synthetic activity of the phosphorylation system of chloroplasts (Hernandez-Gil & Schaedle, 1973), cessation of Fraction I protein synthesis, decline in RuDP carboxylase activity, maintenance of relatively high levels of photorespiration and a sharp increase in mitochondrial respiration.

# REFERENCES

Andrews, T.J., G.H. Lorimer & N.E. Tolbert (1973): Ribulose diphosphate oxygenase. I. Synthesis of phosphoglycolate by Fraction—1 protein of leaves. *Biochemistry*, 12: 11-18.
Bowes, G. & W.L. Ogren (1972): Oxygen inhibition and other properties of soybean ribulose—1, 5-diphosphate carboxylase. *J. biol. Chem.*, 247: 2171-2176.
Callow, J.A. (1974): Ribosomal RNA, Fraction I protein synthesis, and ribulose diphosphate carboxylase activity in developing and senescing leaves of cucumber. *New Phytol.*, 73: 13-20.
Chapman, E.A. & D. Graham (1974): The effect of light on the tricarboxylic acid cycle in green leaves. I. Relative rates of the cycle in the dark and the light. *Plant Physiol.*, 53: 879-885.
D'Aoust, A.L. & D.T. Canvin (1973): Effect of oxygen concentration on the rates of photosynthesis and photorespiration of some higher plants. *Can. J.Bot.*, 51: 457-464.
Dickmann, D.I. (1971a): Photosynthesis and respiration by developing leaves of cottonwood (*Populus deltoides* Bartr). *Bot. Gaz.*, 132: 253-259.
Dickmann, D.I. (1971b): Chlorophyll, ribulose-1, 5-diphosphate carboxylase and Hill reaction activity in developing leaves of *Populus deltoides. Plant Physiol.*, 48: 143-145.
Dickmann, D.I. & D.H. Gjerstad, (1973): Application to woody plants of a rapid method for determining leaf $CO_2$ compensation concentrations. *Can. J. forest Res.*, 3: 237-242.
Fair, P., J. Tew & C.F. Cresswell (1973): Enzyme activities associated with carbon dioxide exchange in illuminated leaves of *Hordeum vulgare* L. II. Effects of external concentrations of carbon dioxide and oxygen. *Ann. Bot.*, 37: 1035-1039.
Gordon, J.C. (1971): Changes in total nitrogen, soluble protein and peroxidases in the expanding leaf zone of eastern cottonwood. *Plant Physiol.*, 47: 595-599.
Hernandez-Gil, R. & M. Schaedle (1973): Functional and structural changes in senescing *Populus deltoides* (Bartr.) chloroplasts. *Plant Physiol.* 51: 245-249.
Huffacker, R.C. & L.W. Peterson (1974): Protein turnover in plants and possible means of its regulation. *Annu. Rev. plant Physiol.*, 25: 363-392.

Isebrands, J.G. & P.R. Larson (1973): Anatomical changes during leaf ontogeny in *Populus deltoides. Amer. J. Bot.*, 60: 199-208.

Jackson, W.A. & R.J. Volk (1970): Photorespiration. *Annu. Rev. plant Physiol.*, 21: 385-432.

Kawashima, N. & S.G. Wildman (1970): Fraction I protein. *Annu. Rev. plant Physiol.*, 21: 325-358.

Kisaki, T., S. Hirabayashi & N. Yano (1973): Effect of age of tobacco leaves on photosynthesis and photorespiration. *Plant cell Physiol.*, 14: 505-514.

Larson, P.R. & R.E. Dickson (1973): Distribution of imported $^{14}C$ in developing leaves of eastern cottonwood according to phyllotaxy. *Planta (Berl.)*, 111: 95-112.

Larson, P.R., R.E. Dickson & J.C. Gordon (1969): Leaf development, photosynthesis and $^{14}C$ distribution in *Populus deltoides* seedlings. *Amer. J. Bot..*, 56: 1058-1066.

Larson, P.R. & J.G. Isebrands (1971): The plastochron index as applied to developmental studies of cottonwood. *Can. J. forest Res.*, 1: 1-11.

Larson, P.R. & J.G. Isebrands (1974): Anatomy of the primary-secondary transition zone in stems of *Populus deltoides. Wood Sci., & Technol.*, 8: 11-26.

Larson, P.R., J.G. Isebrands & R.E. Dickson (1972): Fixation patterns of $^{14}C$ within developing leaves of eastern cottonwood. *Planta (Berl.)*, 107: 301:314.

Lowry, O.H., N.J. Rosebrough, A.L. Farr & R.J. Randall (1951): Protein measurement with the Folin phenol reagent. *J. biol. Chem.*, 193: 265-275.

Ludlow, M.M. (1970): Effect of oxygen concentration on leaf photosynthesis and resistances to carbon dioxide diffusion. *Planta (Berl.)*, 91: 285-290.

Luukkanen, O. & T.T. Kozlowski (1972): Gas exchange in six *Populus* clones. *Sivae Genet.*, 21: 220-229.

Mangat, B.S., W.B. Levin & R.G.S. Bidwell (1974): The extent of dark respiration in illuminated leaves and its control by ATP levels. *Can J. Bot.* 52: 673-681.

Neales, T.F., K.J. Treharne & P.F. Wareing (1971): A relationship between net photosynthesis, diffusive resistance and carboxylating enzyme activity in bean leaves. In Hatch, M.D., Osmond, C.B. & Slatyer, R.O. (Ed.): Photosynthesis & photorespiration. pp. 89-96. Wiley-Interscience, New York.

Ogren, W.L. & G. Bowes (1971): Ribulose diphosphate carboxylase regulates soybean photorespiration. *Nat. new Biol.*,230: 159-160.

Peterson, L.W., G.E. Kleinkopf & R.C. Huffacker (1973): Evidence for lack of turnover of ribulose-1, 5-diphosphate carboxylase in barley leaves. *Plant physiol.*, 51: 1042-1045.

Salin, M.L. & P.H. Homann (1971): Changes of photorespiratory activity with leaf age. *Plant Physiol.*, 43: 193-196.

Samish, Y. & D. Koller (1968): Estimation of photorespiration of green plants and of their mesophyll resistance to $CO_2$ uptake. *Ann. Bot.*, 32: 687-694.

Samish, Y., J.E. Pallas, Jr., G.M. Dornhoff & R.M. Shibles (1972): A reevaluation of soybean leaf photorespiration. *Plant Physiol.*, 50: 28-30

Smillie, R.M. & R.C. Fuller (1959): Ribulose-1, 5-diphosphate carboxylase activity in relation to photosynthesis by intact leaves and isolated chloroplats. *Plant Physiol.*, 34: 651-656.

Steer, B.T. (1971): The dynamics of leaf growth and photosynthetic capacity in *Capsicum frutescens* L. *Ann. Bot.* 35: 1003-1015.

Wareing, P.F., M.M. Khalifa & K.J. Treharne (1968): Rate-limiting processes in photosynthesis at saturating light intensities. *Nature* 220: 453-457.

Woolhouse, H.W. (1967): The nature of senescence in plants. *Symp. Soc. exp. Biol.*, 21: 179-213.

Zelitch, I (1971): Photosynthesis, photorespiration and plant productivity. Academic Press, New York.

# CARBON DIOXIDE EXCHANGE OF YOUNG TOBACCO LEAVES IN LIGHT AND DARKNESS

P.H. HOMANN

*Department of Biological Science, Florida State University, Tallahassee, Fla. 32306, U.S.A.*

*Abstract*

Differences between the $CO_2$ exchange patterns of young and old tobacco leaves can be explained by differences in stomatal resistances. No evidence was obtained that any fundamental changes occur in the activities of the photosynthetic and respiratory systems during the maturation of young tobacco leaves.

## Introduction

Previous studies in my laboratory (Salin & Homann 1971, 1973) have shown that the gas exchange characteristics of leaves from tobacco and citrus plants change during their development. For example, young, not yet unfolded leaves did not photoassimilate more $^{14}CO_2$ when the oxygen tension in the gas phase was reduced from 21 to 2%, and they had a very small $CO_2$ burst when the light intensity was lowered. We concluded that such leaves have an overall smaller photorespiratory gas exchange than older leaves because of lower activities of photorespiratory enzymes and higher resistances to gas diffusion due to a very dense packing of the mesophyll cells in young leaf tissue.

In more recent experiments the differences between the gas exchange of young and old tobacco leaves were found not to be as pronounced as reported earlier. Any observed differences could be fully explained by limitations of net photosynthetic $CO_2$ fixation in young leaves by high stomatal resistances, low chlorophyll concentrations, and high rates of respiratory $CO_2$ production.

## Materials and Methods

The tobacco (*Nicotiana tobacum*) plants were grown from March to July under natural light periods in a greenhouse as described earlier (Salin & Homann, 1973). Significant changes in the qualitative differences between the gas exchange of young and old leaves were not observed when the plants were grown instead outside in full sunlight or under partially shaded conditions, or in a water cooled greenhouse with day temperatures not exceeding about 28° C. 'Thin' young leaves were picked from the flower stalks. They were unfolded and dark green. 'Thick'

*Acknowledgements:* These studies were supported by grant GB 39897 from the National Science Foundation. The technical assistance of Mr. Donald Grunzweig during part of this investigation is gratefully acknowledged. The author also thanks Dr. Marvin L. Salin for helpful advice.

*Abbreviations:* IRGA, infrared gas analyzer; JWB and NB, tobacco varieties John Williams Broadleaf and North Carolina, resp.

young leaves, on the other hand, were cut from the growing tip of the vegetative shoot. They looked yellowgreen, were very pubescent, partially folded, and had veins which protruded strongly from the lower leaf surface.

The $CO_2$ exchange was measured with a Beckman Model 215A IRGA as described in our previous publication. However, the preilluminated leaves or leaf sections were analyzed while standing upright in a water jacketed glass vial with their base in 2 mm $H_2O$. During determinations of the $CO_2$ compensation points, the rapid (4 1/min) circulation of air through the small leaf chamber (35 ml) and the necessity of $H_2O$ vapor removal by $CaCl_2$ at the entrance to the IRGA cell often resulted in wilting of the leaf. Hence, the leaf chamber as well as the analyzer were in by-passes to the main air circuit with flowrates in them adjusted to 250 ml/min. $^{14}CO_2$, liberated from $Na_2^{14}CO_3$ into air, was made available to the leaves in the same gas circuit. The temperature was always 27° C. Recordings were discarded when the saturation rate of photosynthesis declined by more than 10% during the experiment.

Glycolate oxidase activity was analyzed as described earlier after extraction, and in a medium, according to Tolbert et al. (1969). Gas mixtures and pure gases were purchased from AIRCO or Matheson Comp. Other mixtures were prepared in the laboratory from such factory analyzed gases. With these it was found that the response of the IRGA in the range from 0 to 200 ppm $CO_2$ deviated significantly from a straight line. This had not been detected in our earlier investigations due to a lack of a proper series of calibration gases, and explains some of the differences between the values for the $CO_2$ compensation points reported now, and those published previously (Salin & Homann, 1973).

## Results and Discussion

The data in Tables 1 and 2 describe the leaves used in this present study. The 'young' leaves of our earlier investigation (Salin & Homann, 1971, 1973) were intermediate between the 'thin' and 'thick' young leaves described here in respect to weight and chlorophyll content. The thin young leaves resembled the older ones in all important aspects, while the thick ones were distinctly different in that they contained relatively little chlorophyll, had a much higher specific fresh and

*Table 1.* Some characteristics of tobacco leaves.

| variety | leaf type | width (cm) | $\mu$g chl./cm$^2$ | mg fresh wt./cm$^2$ | mg dry wt./cm$^2$ |
|---------|-----------|------------|--------------------|---------------------|-------------------|
| NC | old | 21 | 48 | 17.5 | 2.3 |
| NC | young, thin | 1.3 | 38 | 19.7 | 2.7 |
| NC | young, thick | 2 | 13 | 41 | 3.9 |
| JWB | old | 18 | 52 | 18.4 | 2.4 |
| JWB | young, thick | 2 | | | 4.5 |
| | tip half: | | 19 | 32 | |
| | base half: | | 10 | 42 | |

All determinations were made on a representative leaf sample without midribs.

184

*Table 2.* Some gas exchange characteristics of tobacco leaves (varieties JWB and NC 95)

| Leaf type | $CO_2$ uptake net photosynthesis | $CO_2$ evolution respiration | Warburg Effect* | Dimming burst** ($\mu$moles $CO_2$/mg chl.) |
|-----------|--------------------------------|------------------------------|-----------------|---------------------------------------------|
|           | mg $CO_2$/dm$^2$ x hr          |                              |                 |                                             |
| old | 15 – 21 | 0.8 – 1.5 | 1.40 – 1.50 | distinct (0.3 – 0.5) |
| young, thin | 10 – 18 | 1.5 – 2.0 | 1.40 – 1.50 | distinct (0.2 – 0.3) |
| young, thick | 1.6 – 2.6 | 2.5 – 3.5 | 1.35 – 1.65 | variable (0.1 – 0.4; twice 1.2) |
| young, thick; tip half | 2.5 – 3.9 | 2.0 – 3.5 | 1.30 – 1.80 | distinct (0.2 – 0.6) |
| young, thick; base half | 0 – 2.1 | 3.0 – 4.0 | 1.00 – 1.90 | indistinct and broad, if any (0 – 0.3) |

\* net $CO_2$ fixation in 2%$O_2$/net $CO_2$ fixation in 21%$O_2$.
\*\* upon reducing incident light intensity from 200 W/m$^2$ to 40-50 W/m$^2$.

dry weight, and a much lower net rate of photosynthesis. Moreover, their dark respiration was fast, their Warburg Effect variable and often small, and their $CO_2$ burst indistinct, particularly when the gas exchange of only the bottom half of the leaf was measured.

In our earlier studies we had observed consistently that the $CO_2$ burst upon lowering the light intensity ('dimming burst') was much larger in old leaves than in young ones. During this present investigation, the burst of the old leaves was always quite small, and not significantly different from that of most young leaves. This small burst in the old leaves was not reflected by a lower activity of photo-respiratory enzymes. For example, glycolate oxidase activity was even higher than reported earlier, and there was no difference between young and old leaves when the activity was calculated on a chlorophyll basis (120-150 $\mu$moles glycolate oxidized/mg chlorophyll x hr in an air saturated buffer). But on a protein basis, the activity was 2-3 fold higher in extracts from old leaves. However, even if we knew how the enzyme activities ought to be compared, the above data do not permit any conclusions pertaining to the relative magnitude of photorespiratory rates in the leaves. Therefore, no explanation can be given for the small $CO_2$ bursts observed with the old tobacco leaves of this study.

Fig. 1 contains the recording of the $CO_2$ exchange of an old, and a thick young tobacco leaf. Both leaves displayed a large Warburg Effect, but the $CO_2$ burst was quite indistinct in the young leaf. In Fig. 2 the course of the $CO_2$ exchange is recorded separately for the tip and for the base half of a thick young tobacco leaf. While the Warburg Effect of the tip half of this particular leaf remained relatively small, its $CO_2$ burst was indistinguishable from the burst of mature leaves (compare Fig. 1). Obviously there was no quantitative relation between the magnitudes of the burst, and the Warburg Effect.

The Warburg effect of the base half was negligible when calculated on the basis

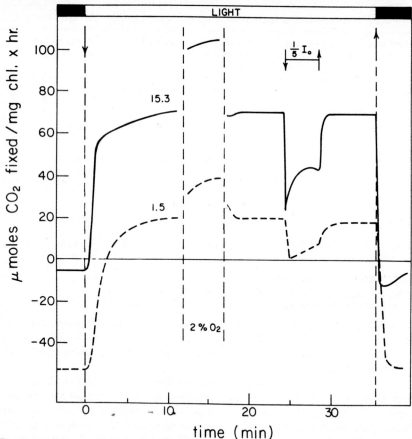

*Fig. 1.* $CO_2$ exchange of detached tobacco leaves, variety NC 95. 0.03 $CO_2$ and 21 or 2% $O_2$ in nitrogen; 27°C; 200 W/m² incandescent light unless indicated otherwise in the figure; further experimental details see Materials and Methods. Numbers in graph indicate net photosynthesis in mg $CO_2$/dm² x hr.———— : section from an old leaf; - - - - - : thick young leaf.

of gross photosynthesis, but perhaps significant when only net photosynthesis is considered. A typical property of thick young leaves, and particularly their base halves, was the slow transition of the $CO_2$ exchange rate from the light to the dark steady state when the light was turned off. The frequently observed broad and indistinct $CO_2$ burst in thick young leaves probably was another expression of the same phenomenon (see Fig. 1).

An evaluation of the significance of small Warburg Effects is made more difficult by the observation that $CO_2$ evolution in the dark sometimes was inhibited by as much as 20% when the $O_2$ tension was lowered to 2% (see also Forrester et al., 1966). Hence, if a considerable portion of mitochondrial respiration continued in the light (c.f. Chapman & Graham, 1974), a very small Warburg Effect might very well be due to an inhibition of dark respiratory processes at low $O_2$ concentrations. Indeed, the relatively high $CO_2$ compensation points of thick young leaves at 2% $O_2$ (Table 3) indicated some $CO_2$ production by processes other than photorespiration in such tissue during illumination.

186

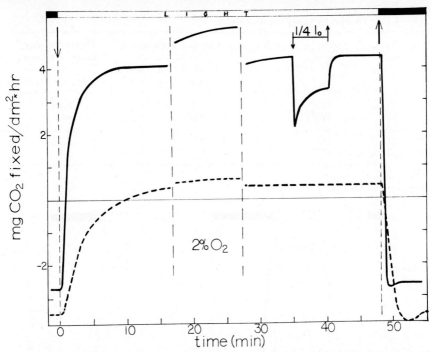

*Fig. 2.* $CO_2$ exchange of sections from a thick young tobacco leaf, variety JWB. Experimental conditions see legend Fig. 1. ——— : tip half of the leaf; ········· : base half of the leaf.

*Table 3.* $CO_2$ Compensation points at $27°C$.

| Plant | leaf type | section | $CO_2$ compensation point ppm $CO_2$ | |
|---|---|---|---|---|
| | | | $21\%O_2$ | $2\%O_2$ |
| Corn, var. Yell.Dent | old | center | 2 | - |
| Tobacco NC 95 | old | center | 34 | 3 |
| | young, thin | whole | 26 | 4 |
| | young, thick | whole | 56 | 14 |
| | young, thick | tip | 42 | 10 |
| | young, thick | base | 115 | 45 |
| Tobacco JWB | old | center | 27 | 2 |
| | young, seedling | whole | 36 | 5 |
| Tobacco Su/su | old | center | 32 | 3 |
| | young, thick | tip | 47 | 10 |
| | young, thick | base | 70 | 20 |
| Tobacco Xanthi | old | center | 30 | 3 |
| | young | whole | 35 | 3 |

*Table 4.* $CO_2$ fixation by young tobacco leaves.

| % $O_2$ in gas phase | tobacco variety | leaf area $cm^2$ | $CO_2$ fixation measured by IR analysis $\mu$moles $CO_2$ in 10 min | | $^{14}CO_2$ fixation $\mu$moles in 10 min |
|---|---|---|---|---|---|
| | | | gross | net | |
| 21 | JWB | 7 | 0.9 | 0.1 | 0.4 |
| 21 | JWB* | 13 | 1.6 | 0.2 | 0.9 |
| 21 | NC | 9 | 1.0 | 0 | 0.3 |
| 21 | NC | 5 | 0.6 | 0 | 0.13 |
| 2 | NC | 7 | 1.2 | 0.1 | 0.5 |

* two leaves
Only the lower 2/3 of the leaves were used.

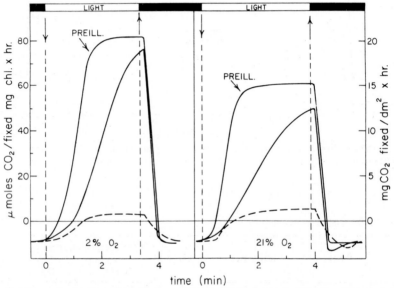

*Fig. 3.* $CO_2$ exchange of sections from an old tobacco leaf, variety NC, after 12 h darkness. Experimental conditions see legend Fig. 1. ——— : lower epidermis peeled off prior to the measurement; „preilluminated" sections had been preilluminated for 20 min with 100 W/cm$^2$ prior to peeling off the epidermis and the measurement to evaluate the influence of the activation of photosynthetic enzymes on the light-on response.

A complete absence of the Warburg Effect in sections incapable of any significant rates of net photosynthesis can be explained in two ways. Either both photosynthetic and respiratory activities are very small in the light, or the gas exchange is severely limited by physical barriers, not by metabolic rates. Table 4 shows that the rate of $^{14}CO_2$ fixation in the light exceeded that of net photosynthetic $CO_2$ exchange, but was much lower than the rate of gross photosynthesis. This finding would be consistent with either possibility. But Fig. 3 suggests that a slow transition of the $CO_2$ exchange rate from state in the light to that in the dark is a result of a sluggish equilibration of the $CO_2$ concentration inside and outside the leaf, and not of a slow resumption of a light inhibited respiratory process. A slow off-response of the $CO_2$ concentration in the gas phase was observed when leaf sections with closed stomates were darkened, and a rapid light-dark transition could be re-established by simply peeling off the epidermis. Similarly, removal of the epidermis from a succulent leaf hastened the off-response, permitted a net photosynthetic $CO_2$ exchange, and a large Warburg Effect (Fig. 4).

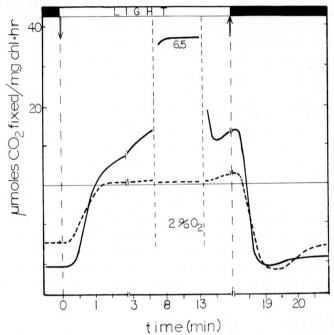

*Fig. 4.* $CO_2$ exchange of a detached leaf from *Kalanchoe fedtschenkoi marginata*. Experimental conditions see legend Fig. 1. ----- : intact leaf with only margins cut off; ——— : same section, but without epidermis.

The data presented in this contribution confirm earlier observations (Salin & Homann, 1971, 1973) that the photosynthetic $CO_2$ exchange a very young tobacco leaves may differ from that of older ones. Differences regarding photosynthetic and respiratory rates between young and mature leaves have been reported also for another tobacco variety (Kisaki et al., 1973) and for cottonwood (Dickmann, 1971, and this symposium). No evidence was obtained during this study which suggested that the relative contribution of photorespiration to the $CO_2$ exchange in the light was less in young tobacco leaves.

From previous experiments we had concluded that the $CO_2$ exchange pattern of young leaves was determined by high resistances to gas diffusion due to very densely packed mesophyll cells in such tissue. Yet, in spite of significant differences between the gas exchange characteristics of the tip and the base half of young leaves (Table 2 and Fig. 2), their density (expressed as weight per leaf area) was quite similar. Light microscopy revealed, however, that the stomates of the bottom half were undeveloped with no or very small openings, while their state of development progressed towards the tip (see also Napp-Zinn, 1973). Hence, the stomatal rather than the mesophyll resistances, augmented by low concentrations of photosynthetic pigments (Table 1), are likely to restrict in certain young leaves net photosynthesis, Warburg Effect, and the expression of a distinct $CO_2$ burst.

As far as the $CO_2$ exchange is concerned, the illuminated base halves of thick young tobacco leaves are a nearly closed system in which the largest portion of the photosynthetically fixed $CO_2$ is respiratory in origin. Since the complement of photosynthetic enzymes in extracts from young leaves was found to be quite normal (Salin & Homann, 1973) in spite of the presence of inactivating contaminants in such preparations, the photosynthetic activity in thick young tobacco leaves should be able to cope with respiratory activities as high as those measured in the dark. The actual respiratory rates in these leaves during illumination are difficult to estimate. Dickmann pointed out in his contribution, and my measurements of the $CO_2$ compensation points in 2% $O_2$ confirm, that photorespiration may not fully account for respiratory activity in the light unless even at low $O_2$ tensions the concentrations of photosynthetically produced $O_2$ in the chloroplasts of young leaves are high enough to permit some photorespiration. If the turnover of $CO_2$ in illuminated thick young leaves were relatively high, a Warburg Effect should occur also when net photosynthesis happens to be negligible in 21% $O_2$ provided the stomates were functional. Forrester et al. (1966) have indeed shown that a Wartburg Effect is not dependent on net photosynthesis. The evidence from this investigations is indirect. According to the model of the Warburg Effect presented by Ogren during this symposium, a 1.9 fold increase of net photosynthesis when the $O_2$ tension is lowered to 2% (table 2) could be explained by assuming that the actual oxygen sensitive rate of $CO_2$ fixation in the chloroplasts at 21% $O_2$ significantly exceeded the net assimilation rate.

## REFERENCES

Chapman, E.A. & D. Graham (1974): The effect of light on the tricarboxylic acid cycle in green leaves. I. *Plant Physiol.*, 53: 879-885.

Dickmann, D.I. (1971): Photosynthesis and respiration by developing leaves of cottonwood (*Populus deltoides* Bartr.). *Bot. Gaz.,* 132: 253-256.

Forrester, M.L., G. Krotkov & C.D. Nelson (1966): Effect of oxygen on photosynthesis, photorespiration, and respiration in detached leaves. *Plant Physiol.,* 41: 422-427.

Kisaki, T., S. Hirabayashi & N. Yano (1973): Effect of the age of tobacco leaves on photosynthesis and photorespiration. *Plant & Cell Physiol.*, 14: 505-514.

Napp-Zinn, K. (1973): Anatomie des Blattes II, Angiospermen A. Encycl. of Plant Anatomy Vol. VIII 2A. Pp 764. Gebr. Bornträger, Berlin & Stuttgart.

Salin, M.L. & P.H. Homann (1971): Changes of photorespiration with leaf age. *Plant Physiol.*, 48: 193-196.

Salin, M.L. & P.H. Homann (1973): Glycolate metabolism in young and old tobacco leaves, and effects of a-hydroxy-2-pyridinemethanesulfonic acid. *Can. J. Bot.,* 51: 1857-1865.

Tolbert, N.E., A. Oeser, R.K. Yamazaki, R.H. Hagemann & T. Kisaki (1969): A survey of plants for leaf peroxisomes. *Plant Physiol.,* 44: 135-147.

# CHARACTERIZATION OF REGULATIVE INTERACTIONS BE-TWEEN THE AUTOTROPHIC AND HETEROTROPHIC SYSTEM IN *PHASEOLUS VULGARIS* AND *TRITICUM AESTIVUM* SEED-LINGS

P. HOFFMANN & Zs. SCHWARZ

*Humboldt-Universität zu Berlin/DDR, Sektion Biologie, Bereich Allgemeine Botanik*

## Abstract

The transition from the heterotrophic to ⸱.e autotrophic phase of development was characterized in seedlings of *Triticum aestivum* L. and *Phaseolus vulgaris* L. in the course of normal development as well as after illumination of etiolated seedlings. The activity of NADP-GADPH increases (induced by light) to the same extent as the efficiency of the photosynthetic apparatus is developed. After initial stimulation, the activity of NAD-GADPH (as well as of the G6PDH/6PGDH system) decrease, an 'enzymatic compensation point' indicating the autotrophic phase of development reached. Experimental delays of the developmental course by NaCl or CCC treatment shift the entry into the autotrophic phase of nutrition, the same ATP contents being ensured by a complex regulative mechanism between photosynthesis and respiration.

## Introduction

In general, a seedling at first develops at the expense of the reserve substances stored in the seed (or in the fruit). In the first phases of this development the preconditions for an efficient photosynthetic apparatus are formed. Already the first photosynthetic products of the young primary leaf are included in the further formation of its own structures (Bradbeer et al., 1974c). Within the hierarchy of the partial systems of a higher plant this principle is followed in a way that the forming leaves are supplied by assimilates of the by now fully differentiated leaves instead of reserves. Only after reaching a total leaf surface due to genotype and environment, reserve or storage substances, respectively, are accumulated. From this aspect alone it follows that the rate, at which the seedling system changes from heterotrophic growth to autotrophic gain of substances, mainly determines the further development of the total plant. Thus the characterization of this transitional phase is both of cognitional and practical relevance.

So far, we have described the transition from the heterotrophic to the autotrophic phase of development via the formation of photosynthetically active chlorophylls (Walter, 1974) as well as the genesis of individual sections of the photosynthetic chain of electron transport (Hieke, not published). Here, this developmental phase will be characterized by means of the synthesis of key enzymes of energy metabolism especially allowing for the regulative processes between heterotrophic and autotrophic metabolism. In doing so, the study of NAD- and

NADP-dependent GAPDH and of the glucose-6-P-dehydrogenase/6-phosphoglu-conate-dehydrogenase system[1] proved to be especially suited.

## Material and Methods

Seedlings of *Phaseolus vulgaris L.* (Cv. Saxanova, harvested in 1969 and 1970) and of *Triticum aestivum L.* (Cv. Carola, harvested in 1971) were used as testing material. In order to follow their normal development, the plants were cultivated at 2.000 lux = 4.960 erg.cm.$^{-2}$s$^{-1}$ permanent light in soil at room temperature. To investigate the development of the photosynthetic apparatus after etiolation, they were at first cultivated in the dark at 24° C, the subsequent illumination was under the same conditions as with permanent-light cultures.

The treatment of seedlings with NaCl (Hoffmann, 1971) or CCC (Hoffmann, 1973) considerably delayed the normal development and the greening process, respectively, and thus made it possible to study the efficiency of these regulative mechanisms. For characterizing the regulative interactions between the systems controlling the energy metabolism the following parameters were determined:
- the chlorophyll content by means of spectralphotometry in acetone extract (Hoffmann & Werner, 1966)
- the gas exchange in the URAS (Hoffmann, 1962) or in the Warburg (Keinzeller, 1965), respectively
- the activities of the NADP-GAPDH exclusively located in the chloroplasts (cp. Latzko & Gibbs, 1968), of NAD-GAPDH and the G6PDH/6PGDH system in the optical test (Schwarz & Schwarz, 1974)
- the adenylates in the enzymatic test (according to Boehringer; Krause, 1974)
- the water-soluble proteins by the microbiuret method (Itzhaki & Gill, 1964).

## Results and Discussion

*Normal development*
There is increasing proof of the fact that regulative interactions between the two systems controlling energy metabolism, respiration and photosynthesis, are responsible for the supply of energy and reduction equivalents (Hoffmann, 1975). In the development of the photosynthetic apparatus, the activity of NAD-GAPDH (glycolyse) reflecting the respiratory rate decreases in bean as well as in wheat seedlings, whereas the NADP-GAPDH of the Calvin cycle increases due to a specific light induction of this enzyme. In the course of this development an 'enzymatic compensation point' is passed which, similar to the compensation point of gas exchange, reflects the entry of the respective organ investigated into the autotrophic phase of nutrition. In the initial development, there is a corresponding correlation of the preponderance of the oxidative pentosephosphate cycle as supplier of pentose and NADPH$_2$. The decrease in the activity of the G6PDH/6PGDH system in the course of development shows that this task is taken over by the reductive pentose-phosphate cycle. The variations repeatedly observed in the activities of this enzyme system in the transitional phase are possibly

[1] G6PDH/6PGDH system.

connected with the special energy demand of tissue at the stage of chloroplast differentiation (cp. also Fig. 2). The $CO_2$ gas exchange behaves in analogy to NAD-GAPDH activity (Fig. 1).

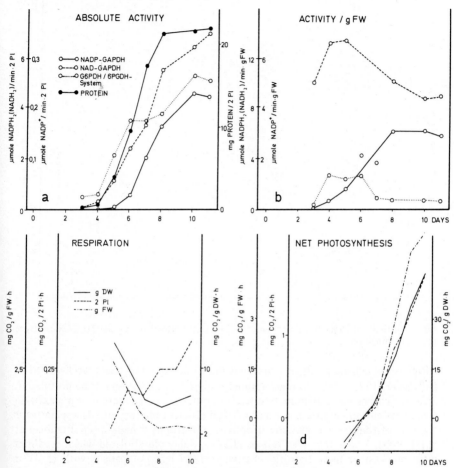

*Fig. 1.* Activity of NADP-GAPDH, NAD-GADH and the G6PDH/6PGDH system and protein content in comparison with gas exchange in the primary leaves of *Phaseolus vulgaris* during normal development (DW - dry weight, FW - fresh weight, PL - primary leaves).

These enzymatic interactions between the heterotrophic and autotrophic metabolic rates can be found accordingly with *Phaseolus* and *Triticum* seedlings and are especially marked during the greening of etiolated seedlings (Fig. 2). Already during their growth in the dark, structural and enzymatic preconditions of the photosynthetic apparatus are developed in the cells. Etioplast structures and enzymes of the Calvin cycle appear – in close correlation with each other – at a certain stage of dark development (Berger & Feierabend, 1967; Bradbeer et al., 1974a). Thus, dependent on the duration of growth in the dark, there are differences between etiolated plants with respect to the degree of development of

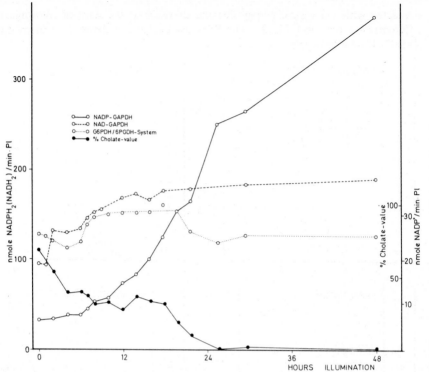

*Fig. 2.* Activity of characteristic enzymes in 11-day-old etiolated leaves of *Triticum aestivum* during illumination.

etioplasts, which are partly due to the degree of depletion of reserves (Obendorf & Huffacker, 1970). Therefore an optimal duration of etiolation can be postulated, after which beginning illumination leads to quickest reorganisation of pre-existing structures and simultaneous formation of an efficient photosynthetic apparatus or the rate of transition from heterotrophy to autotrophy is highest (cp. Bradbeer et al., 1974b). Under the given test conditions, 8-day-old etiolated wheat seedlings were quickest in these transitional processes (17 h in comparison with 25 h for 11-day-old or 28 h for 5-day-old seedlings, respectively, which, besides, showed a distinct lag phase in the kinetics of light induction). The advantage of 8-day-old wheat seedlings is also expressed by the fact that they still supply their energy demand for light-induced syntheses with reserves of caryopses without function-destroying consumption of their own substances in the shoot and that they have a highly economic coefficient of exploitation (while 11-day-old seedlings form 1 mg of new plant substance at a loss of 3,3 mg of dry substance, 8-day-old ones need only 1,9 mg of reserve substances to build up the same mass). Seedlings of highly productive brands, at the same age, need a shorter period to reach the 'enzymatic compensation point' than those of brands not so productive; periods vary between 3 and 26 hours (Schwarz & Schulz, 1974). In general, the $O_2$ compensation point is reached temporally before the $CO_2$ compensation point.

An addition of 2% sodium cholate to the homogenisation medium leads to an increase in the activity of NADP-GAPDH, the separate recording of an in-vivo active and an inactive component becomes possible, beside that of total activity. The percentage of in-vivo inactive enzyme activity (recordable only after addition of cholate in vitro) of total activity continuously decreases during the light-induced formation of the photosynthetic apparatus. Respective observations also under different test conditions suggest as possible the activation of an in-vivo membrane-bound and thus inactive enzyme by sodium cholate (Fig. 2).

*The development under the influence of NaCl*
If etiolated bean or wheat seedlings, respectively, are subjected to a 0,1 (*Phaseolus*) or 0,2 M (*Triticum*) solution of NaCl for 24 h during their greening, a considerable delay in chloroplast differentiation results. This is accompanied by a decreased chlorophyll formation and diminished photosynthetic $CO_2$ assimilation but an increased respiration intensity (Fig. 3). The manifold data on the influence of NaCl on the higher plant have again and again confirmed this opposing behaviour of the photosynthetic and respiratory apparatus, which cannot be attributed to osmotic effects (see refs. in Hoffmann, 1971; Udovenko et al., 1971; Lapina, 1972; Önal, 1973; Heber et al., 1973). In this connection we tested in how far the increase in respiration intensity results from the photosynthesis diminished by NaCl and thus from the shifting between autotrophic an heterotrophic metabolism. This should find its expression in a changed relation between the key enzymes investigated by us. As shown in Fig. 3, the activity of NADP-GAPDH in fact remains lower in comparison with water controls. Here older plants (11 d) show a stronger reaction (67%, diminished ability of compensation, cp. also Önal, 1973) than younger ones (8 d – 21%). As the total activity as well as the activity of the in-vivo active enzyme is decreased, the NaCl effect must at least be partly due to an inhibition of the de novo synthesis of the enzyme.

The absolute activity of NAD-GAPDH, however, is increased by NaCl already in the dark. This is in agreement with the increase in the $O_2$-comsumption already observed by us in comparative investigations of isolated mitochondria as well. The cause of the increased NaCl-induced NAD-GAPDH activity should be hardly a de novo synthesis but rather an activation of the system. Possibly also the allosteric conversion of NAD-GAPDH into its NADP-dependent form may play a role (cp. Pupillo & Piccari, 1973). As the light-induced formation of the effector NADPH is necessary therefore, which is inhibited, however, by NaCl (Santarius, 1969; Heber et al., 1973), the conversion of both enzymes into each other remains prevented and therefore the NAD-GAPDH activity increased.

The extent of the NaCl-induced delay of the 'switching over' from heterotrophy to autotrophy is reflected in the temporal shift of reaching the 'enzymatic compensation point' (with 8-day-old wheat plants from 10 h to 31 h, Fig. 3).

The activity of the G6PDH/6PGDH system is increased by NaCl as expected and thus again confirms the interaction between the oxidative and reductive pentose-phosphate cycle discussed already by Feierabend (1966) and Lendzian & Ziegler (1970). In case of impaired photosynthesis, energy is supplied via an intensification of the respiratory system, the extent of compensation ranging from intensity variations of the individual energy-supplying processes to structural connections and a change of the number of organelles, depending on the degree of the

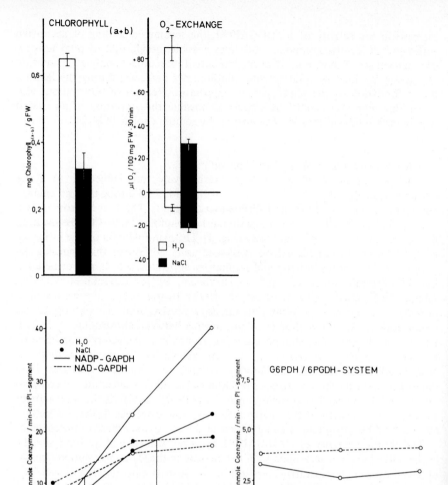

*Fig. 3.* Chlorophyll content, gas exchange and activity of characteristic enzymes in 8-day-old primary leaves of *Triticum aestivum* after NaCI treatment during illumination.

disturbance of the processes (Siew & Klein, 1968, Hecht-Bucholz et al., 1971, Davis & Merret, 1974, Hoffmann, 1975).

Our physiological data agreeing with cytological findings thus confirm once more that the total 'energetic system' of a cell is to be considered as a biochemical and functional unity, the second partial system being stimulated in case of inhibition of the first, due to regulative interactions between both compartments.

In some cases even a direct contact between the mitochondria and the 'consumer', which may be assumed to serve the exchange of respective metabolites.

could be observed, e.g. in the energy supply of regenerating chloroplasts in previously dried *Myrothamnus* leaves (Hoffmann, 1968) or in the energy transmission to the mitotic nucleus (David, 1969; further ref.s in Hoffmann, 1975). To the regulative ability of plastides repeatedly observed by us to develop their enzyme equipment and submicroscopal structure only in connection with the developing or necessary physiological efficiency, the finding corresponds that the chloroplast-specific DNA also renatures more quickly in comparison with the other DNA of the cell (ref.s in Hoffmann, 1975), so that the respiratory system dominates in the total system of the energy metabolism of the plant.

## The development under the influence of CCC

For further testing the efficiency of this regulative principle between autotrophic and heterotrophic energy gain, among others also wheat primary leaves treated with chlorocholinechloride are suited. If 2-cm-long middle segments of 8-day-old etiolated primary leaves are subjected to a $10^{-2}$ M CCC solution for 24 h during greening, they produce at best 50% of the chlorophyll amount of corresponding comparative segments (Hoffmann, 1973). The chloroplast development is delayed (Wellhusen et al., 1973). In the Warburg apparatus, the CCC-treated tissues do not yet reach the compensation point, the respiratory $O_2$ consumption outweighs the photosynthetic $O_2$ production. Respiration is enhanced by the growth regulator.

Enzyme determinations made with respective material yielded the expected data: both absolute and specific activities of in-vivo 'active' NADP-GAPDH were significantly decreased under CCC influence, those of NAD-GAPDH and of the G6PDH/6PGDH system, however, were significantly increased (table 1). Remarkably, there were no differences between control and CCC variant with respect to the 'total activity' of NADP-GAPDH detectable with cholate addition, although CCC decreased the total content of soluble proteins. Therefore the developmental disturbances induced by CCC under the given test conditions should be discussed in agreement with the authors, who explain the effective mechanism by the influence of organelle or cell membranes, respectively (Wellburn et al., 1973).

In the connection interesting here the prospective question for the content of energy-rich compounds in the photosynthetically-inactive tissues arose. When the regulative mechanism between energetic systems characterized by us becomes effective, the same ATP amounts should be detectable also in the weakly greened CCC-treated, due to increased respiration. The data of two test series presented in table 2 confirm this hypothesis. The test segments show even slightly increased adenylate contents, which are mainly due to the increased ATP content. The values of the 'energy charge' are 0,56 and 0,70 and thus correspond to the values typical for wheat seedlings (Krause, 1974). A similar finding was made by Baumann (1973) in the analysis of the amitrol effect on *Avena* seedlings, so that, due to the same reaction in these rather different processes (normal development, greening of etiolated plants, inhibition of normal development or greening, respectively, by NaCl, CCC or amitrol), this regulative principle should have a certain universal character. With respect to its mechanism, adenylates as well as the complex system of key enzymes investigated by us should be of relevance. It is the level of the individual adenylates rather than their amount, which represents the regulative value between the different energy-consuming processes (Lewenstein &

Table. 1. Effect of CCC on the enzymes characterizing the energy metabolism of 8 days old etiolated primary wheat leaves after 24 hr of illumination.

| | NADP-GAPDH −cholate | | NADP-GAPDH +cholate | | NAD-GAPDH | | 6GPDH/6PGDH-system | |
|---|---|---|---|---|---|---|---|---|
| | $H_2O$ | CCC | $H_2O$ | CCC | $H_2O$ | CCC | $H_2O$ | CCC |
| Absolute activities | $61\pm6$ | $37\pm6$ | $86\pm12$ | $75\pm12$ | $75\pm8$ | $104\pm9$ | $11\pm1$ | $15\pm1$ |
| Specific activities | $365\pm34$ | $199\pm21$ | $359\pm50$ | $346\pm81$ | $445\pm51$ | $785\pm76$ | $71\pm10$ | $109\pm14$ |

*Table 2.* Content of adenylates ($\mu M/180$ segments) in segments of *Triticum aestivum*-leaves treated with $10^{-2}M$ CCC (n = 2 x 2)

| n | ATP | ADP | AMP | adenylates | ATP | ADP | AMP | adenylates |
|---|-----|-----|-----|------------|-----|-----|-----|------------|
| | | CONTROL | | | | $10^{-2}M$ CCC | | |
| $a_1$ | 0.062 | 0.047 | 0.041 | 0.150 | 0.083 | 0.045 | 0.024 | 0.152 |
| $a_2$ | 0.054 | 0.045 | 0.040 | 0.139 | 0.087 | 0.033 | 0.033 | 0.153 |
| $b_1$ | 0.072 | 0.031 | 0.039 | 0.142 | 0.083 | 0.047 | 0.025 | 0.155 |
| $b_2$ | 0.056 | 0.047 | 0.051 | 0.153 | 0.093 | 0.033 | 0.025 | 0.151 |
| $\bar{n}$ | 0.061 | 0.042 | 0.043 | 0.146 | 0.086 | 0.040 | 0.027 | 0.153 |

Schneider, 1972; Löppert & Brode, 1973). Besides, it is still unclear if the regulative interactions between NAD- and NADP-GAPDH work via a reversible membrane binding of the enzyme or via a (co-enzyme-dependent) association or dissociation of an (NAD-active, stable) tetramer into (NADP-active) protomers (Schwarz, 1974). ·

# REFERENCES

Baumann, I. (1974): Untersuchungen zum Einfluß von 3-Amino-1,2,4,-triazol (Amitrol) auf Entwicklung und Stoffwechsel von Haferkeimpflanzen unter besonderer Berücksichtigung des Photosyntheseapparates. Diss. A. Mathemat.-Naturwiss. Fak. der Pädagogischen Hochschule „Karl Liebknecht", Potsdam.

Berger, Chr. & J. Feierabend (1967): Plastidenentwicklung und Bildung von Photosyntheseenzymen in etiolierten Roggenkeimlingen. *Physiol. vèg.*, 5: 109-122.

Bradbeer, J.W., H.M.M. Ireland, J.W. Smith, J. Rest & H.J.W. Edge (1974a): Plastid development in primary leaves of *Phaseolus vulgaris*. VII. Development during growth in continuous darkness. *New Phytol.*, 73: 263-270.

Bradbeer, J.W., A.D. Gyldenholm, H.M.M. Ireland, J.W. Smith, J. Rest & H.J.W. Edge (1974b): Plastid development in primary leaves of *Phaseolus vulgaris*. VIII. The effects of the transfer of dark-grown plants to continuous illumination. *New Phytol.*, 73: 271-279.

Bradbeer, J.W., A.O. Gyldenholm, J.W. Smith, J. Rest & H.J.W. Edge (1974c): Plastid development in primary leaves of *Phaseolus vulgaris*. IX. The effects of short light treatments on plastid development. *New Phytol.*, 73: 281-290.

David, H. (1969): Struktur der Mitochondrien. In Bielka, H. (Ed.): Molekulare Biologie der Zelle. VEB G. Fischer Verlag, Jena.

Davis, B & M.J. Merret (1974): The effect of light on the synthesis of mitochondrial enzymes in division-synchronized *Euglena* cultures. *Plant Physiol.*, 53: 575-580.

Feierabend, J. (1966): Enzymbildung in Roggenkeimlingen während der Umstellung von heterotrophem auf autotrophes Wachstum. *Planta Berl..*, 71: 326-355.

Heber, U., L. Tyankova & K. Santarius (1973): Effects of freezing on biological membranes in vivo and in vitro. *Biochim. Biophys. Acta*, 291: 23-37.

Hecht-Buchholz, Chr., R. Pflüger & H. Marschner (1971): Einfluß von Natriumchlorid auf Mitochondrienzahl und Atmung von Maiswurzelspitzen. *Z. Pflanzenphysiol.*, 65: 410-417.

Hoffmann, P. (1962): Untersuchungen zur Photosynthese und Atmung von Laubblättern verschiedenen Alters. *Flora*, 152: 622-654.

Hoffmann, P. (1968): Pigmentgehalt und Gaswechsel von Myrothamnus-Blättern nach Austrocknung und Wiederaufsättigung. *Photosynthetica*, 2: 245-252.

Hoffmann, P. (1971) Wechselbeziehungen zwischen Chloroplasten und Mitochondrien (russ.). Biokimija i biofizika fotosinteza. Pp 94-98. Akad. nauk SSSR, Irkutsk.

Hoffmann, P. (1973): Vergleichende pigmentphysiologische Untersuchungen an CCC-behandelten Weizenkeimpflanzen. *Photosynthetica.*, 7: 213-225.

Hoffmann, P. (1975): Photosynthese. Akademie-Verlag, Berlin. (in press)

Hoffman, P & D. Werner (1966): Zur spektralphotometrischen Chlorophyllbestimmung unter besonderer Berücksichtigung verschiedener Gerätetypen. *Jenaer Rundschau.*, 11: 300-303.

Itzhaki, F.R. & D.M. Gill (1964): A microbiuret method for estimating proteins. *Analyt. Biochem.*, 9: 401-411.

Kleinzeller, A. (1965): Manometrische Methoden und ihre Anwendung in Biologie und Biochemie. VEB G. Fischer Verlag, Jena.

Krause, Chr. (1974): Die Dynamik des Adenylatgehaltes in Keimpflanzen von *Triticum aestivum* L. unter besonderer Berücksichtigung des Energiestoffwechsels. Diss. A Humbold-Universität zu Berlin, DDR.

Lapina, L.P. & S.A. Bikmukhametova (1972): Einfluß isosmotischer Na$_2$SO$_4$- und NaCl-Konzentrationen auf die Photosynthese und Atmung von Maisblättern (russ.). *Fiziol. Rastenij*, 19: 792-797.

Latzko, E. & M. Gibbs (1968): Distribution and activity of enzymes of the reductive pentose phosphate cycle in spinach leaves and in chloroplasts isolated by different methods. *Z. Pflanzenphysiol.* 59: 184-194.

Lendzian, K. & H. Ziegler (1970): Über die Regulation der Glucose-6-photphat-Dehydrogenase in Spinatchloroplasten durch Licht. *Planta* (Berlin),94: 27-36.

Lewenstein, A. & K. Schneider (1972): The level of ATP in Chlorella. *Proceedings of the IInd Internat. Congress on Photosynthesis Res.*, 2: 1371-1378.

Löppert, H.G. & E. Brode (1973): ATP-Gehalt und ATP-Umsatz von Chlorella in Abhängigkeit von Belichtung und Belüftung. *Z. Allg. Mikrobiol.*, 13: 499-506.

Obendorf, R.L. & R.C. Huffacker (1970): Influence of age and illumination on distribution of several Calvin cycle enzymes in greening barley leaves. *Plant Physiol.*, 45: 579-582.

Önal, M. (1974): Die Wirkung der Natriumchloridkonzentration auf den Protein- und Chlorophyllgehalt von *Spergularia salina* und *Suaeda maritima. Rev. de la faculté des Sc. dèUniv. d'Instanbul*, Serie B, 38: 53-65.

Pupillo, P. & G.G. Piccari (1973): The effect of NADP on the subunit structure and activity of spinach chloroplast glyceraldehyde-3-phosphate dehydrogenase. *Arch. Biochem. Biophys.*, 154: 324-331.

Santarius, K.A. (1969): Der Einfluß von Elektrolyten auf Chloroplasten beim Gefrieren und Trocknen. *Planta Berl.*, 89: 23-46.

Schwarz, G. & Zs. Schwarz, (1974): Die Funktionsdauer der Cotyledonen von *Phaseolus vulgaris* L. in Beziehung zu ihrem Energiestoffwechsel und zum sink-source Verhältnis innerhalb der Keimpflanze. *Biol Zentralbl.*, 93: 351-363.

Schwarz, Zs. (1974): Zur Regulation der NADP-abhängigen Glycerinaldehyd-3-phosphat-Dehydrogenase in den Primärblättern von *Phaseolus vulgaris* L. Die Wirkung von Na-Cholat auf die Aktivität des Enzyms in vitro. *Biochem. Physiol. Pflanzen* 166:525-536.

Schwarz, Zs. & H. Schulz (1974): Enzymmolische Untersuchungen zur Charakterisierung des Überganges vom heterotrophen zum autotrophen Stoffwechsel während der Ergrünung etiolierter *Triticum aestivum*-Keimpflanzen. *Wiss. Z. Humboldt-Universität zu Berlin*, 23 (6): 11-17.

Siew, D & S. Klein (1968): The effect of sodium chloride on some metabolic and fine structural changes during the greening of etiolated leaves. *J. Cell. Biol.*, 37: 390-396.

Udovenko, G.V., L.A. Semushina & N.G. Petrocenko (1971): The character and possible explanation of the changed photosynthetizing activity of plants during salinization (russ.). *Fiziol. Rastenij*, 18: 708-715.

Walter, G. (1974): Spektrale Änderungen in vivo während der Chlorophyll-Bildung etiolierter Weizenkeimpflanzen nach experimentell variiertem Anteil der Protochlorophyl(id)-Holochrome P$_{650}$ und P$_{635}$ durch Behandlung mit $\triangle$ -Aminolaevulinsäure. *Photosynthetica*, 8: 40-46.

Wellburn, F.A.M., A.R. Wellburn, J.L.Stoddart & K.J. Treharne (1973): Influence of gibberellic and abscisic acids and the growth retardant, CCC, upon plastid development. *Planta Berl.* 111: 337-346.

# PHOTOSYNTHESIS, PHOTORESPIRATION, RESPIRATION AND GROWTH OF PEA SEEDLINGS TREATED WITH GIBBERELLIC ACID (GA$_3$)

J. POSKUTA, E. PARYS, E. OSTROWSKA & E. WOLKOWA

*Laboratory of Plant Metabolism, Institute of Botany, University of Warsaw, Poland*

*Abstract*

The effect of gibberellic acid (GA$_3$) on the growth and yield of dwarf pea *(Pisum sativum L.* var. 'Bordi'), on the rates of CO$_2$ exchange in the light and in darkness and on the flowering of plants after injection of GA$_3$ to the pods were investigated. It has been shown that GA$_3$ in concentration of 100 ppm applied during seed imbibition strongly stimulated the growth of the seedlings and the yield of plants, enhanced the respiration of seeds, the rates of apparent photosynthesis, photorespiration and respiration of shoots. The injection of GA$_3$ (200 ppm) to the pods containing seeds induced the appearance of new shoots, flower buds and flowers on the fully developed plants. Similar effect was produced after extraction of the seeds from the pods. The appearance of new shoots after GA$_3$ application to the pods induced the transport of photo-assimilates from the seeds to these new organs.

## Introduction

Gibberellins affect all aspects of growth and development of higher plants from germination to flowering, fruit set and senescene. An extensive review of this subject was recently presented by Jones (1973), who emphasized that during the last 5 years plant physiologists concentrated efforts rather to study the physiology of specialized tissues at the expense of the physiology of the whole plant. Such processes as CO$_2$ exchange in light and in darkness and the transport of assimilates would greatly influence plant growth and development if they are in some way controlled by gibberellins or other plant hormones. Recent data indicate that gibberellin influence the photosynthesis of both C-3 and C-4 plants through its action on the ultrastructure of the plastids and on photosynthetic enzymes (Wellburn & Wellburn, 1973; Huber & Sankhla, 1974a, b, c). The interaction between GA$_3$ and the component of plant membranes, phosphatidyl choline, is reported by Wood et al. (1974). The permeability of the membranes play a crucial part in many physiological processes, for instance in the translocation of assimilates. GA$_3$ induced stimulation of translocation of [14]C-sucrose after removing of apical meristeme of soybean plant was reported by Hew et al. (1967). It is well known that gibberellins applied exogenously to genetically dwarf plants almost always produced a stem elongating effect. The action of this hormone on flowering is related to chemical and genetical specifity. Michniewicz & Lang (1962) have demonstrated that only GA$_7$ induced flowering among 9 gibberellins tested. Wittwer & Bukovac (1962) have showed a better effect of GA$_1$, GA$_3$, and GA$_7$ on the flowering of Great Lakes lettuce. Wellensiek (1969, 1973) has observed that the induction of flowering by GA$_3$ depends on the genotype.

The work presented below deals with the effect of gibberellic acid ($GA_3$) on the following processes of dwarf pea: germination and respiration of seeds, growth and flowering, distribution of photoassimilates among plant organs after administration of $GA_3$ to the pods, the rates of $CO_2$ exchange in light and in darkness.

## Material and Methods

Germination and respiration of seeds were examined as follows: seeds of dwarf pea, *Pisum sativum L.* cv. 'Bordi', were placed in Petri dishes on disks of filter paper. Each dish contained 20 seeds. 10 ml of either distilled water or $GA_3$ (10 or 100 ppm) were added and the germination of seeds from 10 dishes was followed during the consecutive 5 days at $22°C$. Respiration of seeds was determined by means of a Clark oxygen electrode introduced in 60 ml volume test tube filled with the solutions mentioned above. The seeds were placed in the tube carefully isolated from the outside atmosphere. The $O_2$ uptake was monitored during 24 hours at $22°C$.

The growth of plants was examined in either sand soil or water cultures. Since the results were qualitatively similar, the data from sand cultures are presented in this paper. Before sowing, the seeds were soaked during 24 hours in tested solutions an then placed in 5 kg pots of sand. The Knop's nutrients were added and seedlings were watered daily to 70% of sand water capacity. Plants were grown in a chamber under light intensity of 10.000 lux from fluorescent tubes at a day temperature of $28°C$ during 16 hours and a night temperature of $22°C$. Under these conditions, flowering was observed after 21-22 days; after 44 days, pods with fully developed seeds were obtained. During the vegetation period growth was measured and fresh and dry weights were determined by periodical harvestings.

The rates of apparent photosynthesis (APS), photorespiration (PR), respiration (DR) were measured by means of an infra red $CO_2$ analyzer (Beckman 215 A) arranged in a closed system (volume: 996 ml; air flow rate: 3 l per minute). Attached shoots were placed in a plexiglass photosynthesis chamber illuminated with saturating light ($2.5 \times 10^5$ ergs cm$^{-2}$ sec$^{-1}$). The light source was four 500 watt photofloodlamps filtered through a 15 cm thick screen of running water to reduce infra red irradiation. Measurements were made during 4 consecutive light-dark cycles and the data presented here are typical for 3 separate experiments. The $GA_3$ effect on flowering was studied as follows: after completion of vegetative growth of plants still green, different treatments were applied to the plants:

a. 0,5 ml $GA_3$ (200 ppm) or distilled water injected to fully developed pods;

b. 0,5 ml $GA_3$ (200 ppm) or distilled water also injected after extraction of seeds from the pods;

c. excision of the pods.

The $GA_3$ effect on the distribution of radioactivity among the plant organs was examined as follows: after injection of 0,5 ml of a 200 ppm $GA_3$ solution to the pods, plants stayed for the next 3 days until new shoots were formed on the plants. Then old and new shoots were enclosed in a plexiglass photosynthesis chamber and connected to a closed system (volume: 1270 ml). The initial $CO_2$

concentration was 500 ppm. Plants were adapted to these conditions during 20 minutes at light  intensity of $2.5 \times 10^5$ ergs $cm^{-2}$ $sec^{-1}$; $^{14}CO_2$ (250 $\mu$Ci) was then introduced in the system. After 20 minutes of photosynthesis in $^{14}CO_2$, plants stayed in dim light for 24 hours. Then shoots were detached and divided in the following parts: leaves, stems, pods, seeds and new shoots. Total radioactivity of the ethanol soluble and insoluble material was determined using the G-M thin window counter.

## Results and Discussion

The germination of seeds was little affected by both concentrations of $GA_3$: the percentage of germinating seeds were 87, 92 and 91 on water, $GA_3$-10 ppm and $GA_3$-100 ppm respectively. The $O_2$ uptake during imbibition was enhanced with both $GA_3$ concentrations compared to control (Figure 1). Moreover a much longer time was required by seeds imbibed in water to obtain the steady state of respiration.

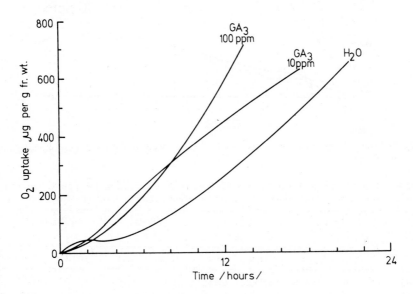

*Fig. 1.* Respiration of pea seeds during imbibition in either water or $GA_3$ solutions.

Figure 2 presents the elongating growth of seedlings obtained from seeds pretreated with $GA_3$, 10 or 100 ppm. It is seen that only 100 ppm $GA_3$ produced a clear elongating effect. The large increase in both fresh and dry weights of plant organs is visible only after pretreatment of seeds with $GA_3$ in concentration of 100 ppm (Table 1). Figure 3 illustrates the effect of $GA_3$ on fresh weight of plants harvested after various growth periods. It is visible that up to 2 weeks of vegetation, no effect of both $GA_3$ concentrations can be observed. After 3 weeks, however, $GA_3$ in concentration of 100 ppm enhanced the production of fresh

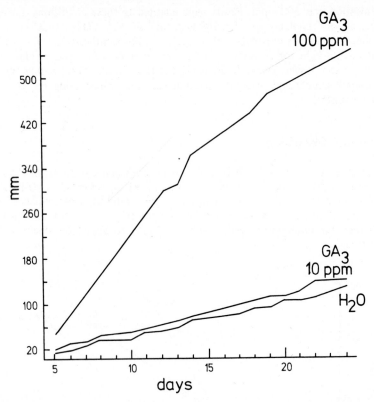

**Fig. 2.** Effect of the imbibition of pea seeds in either water or GA₃ solutions on elongating growth of seedings.

*Table 1.* Effect of the pretreatment of seeds with either water, 10 or 100 ppm GA₃ on the production of fresh and dry weights of pea plants (average from 27 seedlings 37 days old).

| | Leaves g | | Stems g | | Pods g | | Roots g | |
|---|---|---|---|---|---|---|---|---|
| | fresh | dry | fresh | dry | fresh | dry | fresh | dry |
| H₂O | 1,18 | 0,211 | 0,152 | 0,152 | 1,30 | 0,191 | 1,210 | 0,095 |
| GA₃ 10 ppm | 1,52 | 0,254 | 1,170 | 0,190 | 1,72 | 0,220 | 1,650 | 0,136 |
| GA₃ 100 ppm | 4,00 | 0,636 | 3,65 | 0,699 | 6,00 | 0,869 | 3,00 | 0,420 |

weight and this tendency was even stronger during the next 3 weeks of plant growth. It should be noted here that a similar picture was observed for shoot and root growth.

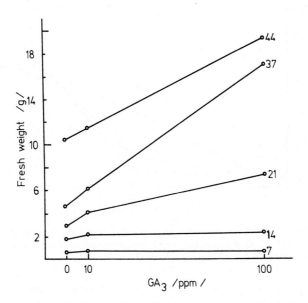

*Fig. 3.* Effect of the imbibition of pea seeds in either water of GA$_3$ solutions on fresh weight of seedings. (Numbers = days to harvesting of plants).

Since GA$_3$ in concentration of 100 ppm produced such large stimulatory effect on growth and yield of plants, it was of interest to examine the rates of CO$_2$ exchanges in light and in darkness. The results of experiments with 21 days old plants are presented on Figure 4. It is clear that only GA$_3$ in the concentration of 100 ppm stimulated these processes. Therefore the GA$_3$ stimulated growth and yield can be related to the increase in the rates of apparent photosynthesis, photorespiration and respiration of plants.

On table 2 are presented the results of 3 separate experiments on the effect of either injection of GA$_3$ to the pods of fully developed plants, excision of the pods, injection of GA$_3$ to the pods after extraction of the seeds or solely after extraction of the seeds from the pods. As is seen from these data, all treatments induced the appearance of flowers and the formation of new pods. It should be noted here that new shoots have appeared on both dwarf and normal plants. Such effect of GA$_3$ on the induction of new shoots and the formation of flowers and pods predict changes in the distribution of assimilates between plant organs after GA$_3$ application to the pods. The results of preliminary experiments are shown on figures 5, 6 and 7. The application of GA$_3$ to the pods of dwarf plants did not induced the appearance of new shoots after 3 days; as a consequence, no GA$_3$ effect can be seen on the distribution of radioactivity among plants organs. In tall plants however the GA$_3$ application already induced in 3 days the appearance of new shoots; in this case, almost all radioactivity was no more found in the seeds but in the new shoots. Similar results were obtained with dwarf plants 6 days after GA$_3$ application with the appearance of new shoots also on these plants.

The pea var. 'Bordi' is a long-day and an early maturing cultivar which can

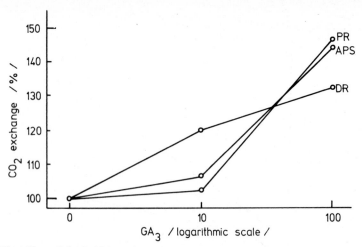

*Fig. 4.* The effect of the imbition of pea seeds in either water or $GA_3$ solutions on apparent photosynthesis (APS), photorespiration (PR), and respiration (DR). Control values: APS: 221.7; PR: 64.3; DR: 34.1 $\mu$g $CO_2$ $g^{-1}$ fr.wt. $min^{-1}$.

flower in continuous light and is little sensitive to photoperiod. In our growing conditions, reproductive growth was completed within about 6 weeks. The $GA_3$ treatment of the seeds modified vegetative growth but had no effect on the time required for flowering and development of pods and seeds. The enlongating and vegetative growth and the yield of plants however were strongly stimulated by imbibition of the seeds in 100 ppm $GA_3$. These effects on plant growth were correlated with the stimulation of the rates of apparent photosynthesis, photorespiration and respiration of the shoots. Our results therefore confirmed and extended observations of many authors on gibberellin stimulation of the fixation of $CO_2$ by plants (Coulombe & Paquin, 1959; Bystrzejewska et al., 1971; Lester et al., 1972). $GA_3$ injection to the pods induced the appearance of new shoots and produced a crucial change in the distribution of photoassimilates among the plant organs. It has been shown that practically all radioactivity was transported from the seeds to the new shoots. The results indicate that new shoots on an old plant are powerfull sinks for photoassimilates. The best source of assimilates utilized by new shoots is located in the seeds. Since such treatments of plants with $GA_3$ induced also the appearance of new flowers and pods, it is reasonable to assume that the seeds are the site not only of food reserves for the development of new shoots but also of some inhibitors responsible for ending vegetative growth of plants. Barendse et al. (1968) have studied the fate of gibberillin $GA_1$ injected to excised pods of pea. They observed that part of the radioactivity of labelled gibberellin was converted into biologically active compounds. At present we have no evidence that there was a direct action of $GA_3$ injected to the pods on the processes of flowering induction and movement of assimilates or that new active compounds responsible for these effects were produced. The transition from vegetative growth to formation of flower buds, flowers and fruits involves undoubtedly crucial changes in the whole metabolism of cells and particulary meristem cells. In our experiments the induction of flowering was obtained by

Fig. 5

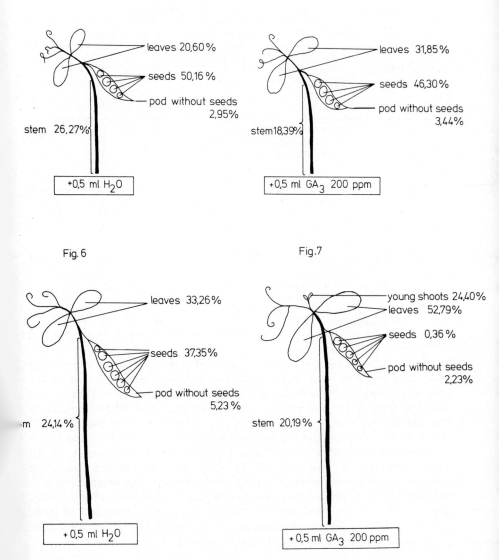

leaves 20,60%

seeds 50,16%

pod without seeds
2,95%

stem 26,27%

+0,5 ml H₂0

leaves 31,85%

seeds 46,30%

pod without seeds
3,44%

stem18,39%

+0,5 ml GA₃ 200 ppm

Fig. 6

Fig.7

leaves 33,26%

seeds 37,35%

pod without seeds
5,23%

m 24,14%

+0,5 ml H₂0

young shoots 24,40%
leaves 52,79%

seeds 0,36%

pod without seeds
2,23%

stem 20,19%

+0,5 ml GA₃ 200 ppm

*Fig. 5,6,7.* The distribution of radioactivity among plants organs after injection to the pods of water or GA₃ 200 ppm. Fig. 5: dwarf plants. Figs. 6 and 7: tall plants.

Table 2. Effect of either: injection of 200 ppm GA₃ to the pod (P + GA₃), excision of the pods (EP), extraction of the seeds from the pods (ES) or extraction of the seeds from the pods plus injection of 200 ppm GA₃ (ES + GA₃) on the appearence of flower buds, flowers and new pods on the fully developed pea plant. The plants were obtained from seeds pretreated with water (control), 10 ppm or 100 ppm GA₃.

| Variant | Plant | Per plant | | |
|---------|-------|-----------|--------|------|
| | | Flower buds | Flowers | Pods |
| P+GA$_3$ | Control | 3 | 2 | 2 |
| | GA$_3$ 10 ppm | 4 | 4 | 1 |
| | GA$_3$ 100 ppm | 3 | 3 | 3 |
| Days to response | | 20 | 24 | 35 |
| EP | Control | 4 | 2 | 2 |
| | GA$_3$ 10 ppm | 4 | 4 | 3 |
| | GA$_3$ 100 ppm | 3 | 3 | 3 |
| Days to response | | 12 | 16 | 31 |
| ES | Control | 3 | 3 | 0 |
| | GA$_3$ 10 ppm | 4 | 3 | 1 |
| | GA$_3$ 100 ppm | 4 | 4 | 2 |
| Days to response | | 12 | 20 | 31 |
| ES+GA$_3$ | Control | 3 | 2 | 2 |
| | GA$_3$ 10 ppm | 4 | 4 | 1 |
| | GA$_3$ 100 ppm | 3 | 3 | 3 |
| Days to response | | 14 | 20 | 31 |

injection of GA$_3$ to the pods. Similar effect was obtained after extraction of the seeds from the pods. Presumably seeds are the site of factors preventing the continuous vegetative growth and responsible for the senescence of plants. The action of these factors can be overcome by gibberellic acid. It is of importance that the induction of flowering by GA$_3$ was obtained on both dwarf and tall plants. In the literature there is no agreement that stem elongation and flower formation are identical processes (Cleland & Zeevaart, 1970; Wellensiek, 1973). The experiments of Wellensiek (1973) with *Silene armeria* had shown the existence of the juvenile phase after GA$_3$ shock treatment. Our observations with pea showed also an appearance of the juvenile phase eg. formation of new shoots, flowers and pods on either dwarf or tall plants after GA$_3$ treatment of pods. These data seem to support the suggestion of Wellensiek (1973) that stem elongation and flower formation are different processes.

## REFERENCES

Barendse, G.W.M., H. Kende & A. Lang (1968): Fate of radioactive gibberellin A$_1$ in maturing and germinating seeds of peas and Japanese Morning Glory. *Plant Physiol.,* 43: 815-822.
Bystrzejewska, G., S. Maleszewski & J. Poskuta (1971): Photosynthesis, [14]C-photosynthetic

208

products and transpiration of bean leaves as influenced by gibberellic acid. *Bull. Acad. Pol. Sci.*, Ser. B., 19: 533-536.

Cleland, F.C. & A.D. Zeevaart (1970): Gibberellins in relation to flowering and stem elongation in the long day *Silene armeria. Plant Physiol.*, 46: 392-400.

Coulombe, L.J. & R. Paquin (1959): Effet de l'acide gibberellique sur le metabolisme des plantes. *Can. J. Bot*, 37: 897-902.

Hew, C.S., C.D. Nelson & G. Krotkov (1967): Hormonal control of translocation of photosynthetically assimilated [14]C in young soybean plants. *Amer. J. Bot.*, 54: 252-256.

Huber, W. & N. Sankhla (1974a): Activity of malate dehydrogenase in leaves of abscisic and gibberellic acid treated seedlings of *Pennisetum typhoides. Z. Pflanzenphysiol.*, 71: 86-89.

Huber, W. & N. Sankhla (1974b): Effect of gibberellic acid on the activities of photosynthetic enzymes and [14]$CO_2$ fixation products in leaves of *Pennisetum typhoides* seedlings. *Z.Pflanzenphysiol.*, 71: 275-280.

Huber, W & N. Sankhla (1974c): Untersuchungen uber den einflus von Abscisin und Gibberellinsäuresbehandlung aüf die Aktivitaten einiger Enzyme des Kohlenhydratstoffwechsels in Blattern von *Pennisetum\typhoides* Keimlingen. *Planta (Berl.)*, 116: 55.64.

Jones, R.L. (1973): Gibberellins: their physiological Role. *Ann. Rev. Plant Physiol.*, 24: 571-598.

Lester, D.C., O.G. Carter, F.M. Kelleher & D.R. Laing (1972): The effect of gibberellic acid on apparent photosynthesis and dark respiration of simulated swards of *Pennisetum Clandestinum.* Hochst. *Aust. J. Agric. Res.*, 23: 205-213.

Michniewicz, M. & A. Lang (1962): Effect of nine gibberellins on stem elongation and flowers formation in cold-requiring and photoperiodic plants grown under non-inductive conditions. *Planta (Berl.)*, 58: 549-563.

Wellensiek, S.J. (1969): The rates of floral deblocking in *Silene armeria* L. *Z. pflanzenphysiol.*, 61: 462-471.

Wellensiek, S.J. (1973): Gibberellic acid, flower formation and stem elongation in *Silene armeria. Neth. J. Agric. Sci.*, 21: 245-255.

Wellbrun, F.A.M. & A.R. Wellburn (1973): Influence of gibberellic and abscisic acids and the growth retardant CCC upon plastid development. *Planta (Berl.)*, 111: 337-346.

Wittwer, S.H. & M.B. Bukovac (1962): Exogenous plant growth substances affecting floral initiation and fruit set. Proc. Plant Sci. Symp. on Fruit Set. 65-93. Campbell Soup Co., Camden, N.J.

Wood, A., G.L. Paleg & T.M. Spotwood (1974): Hormone-phospholipid interaction: a possible hormonal mechanism of action in the control of membrane permeability. *Austr. J. Plant Physiol.*, 1: 167-169.

# THE EFFECTS OF CCC AND GA ON SOME BIOCHEMICAL AND PHOTOCHEMICAL ACTIVITIES OF PRIMARY LEAVES OF BEAN PLANTS.

G. OBEN & R. MARCELLE

*Laboratory of Plant Physiology, Research Station of Gorsem, B-3800 Sint-Truiden, Belgium.*

## Abstract

The growth inhibition induced by CCC is parallelled by a decrease of apparent photosynthesis (APS), whilst an increase of it is coupled with the growth promotion caused by GA. Early work could not explain the APS changes by effects neither on stomatal regulation, nor on the activities of the carboxylating enzymes ribulose-1,5-diphosphate carboxylase and phosphoenol pyruvate carboxylase.

In this paper we report the results obtained by measuring some key enzymes of the amino acid biosynthesis (GDH, GOT and GPT). These activities were found to parallel the modifications of protein and chlorophyll content, induced by both growth regulators.

In isolated chloroplasts the activities of both photosystems were investigated as well as cyclic photophosphorylation. No significant differences could be found between treated and untreated leaves.

It is suggested in the discussion that the effects of CCC and GA on APS might be searched in the relation between chloroplasts and the rest of the cell as well as at the level of photosynthate transport.

## Introduction

T he growth regulators CCC and GA not only affect plant growth, but they also influence the leaf contents in chlorophylls, carotenoids, proteins, free amino acids, soluble sugars and starch (Marcelle et al., 1974). These chemicals also induce changes in photosynthetic activity: CCC, at growth inhibiting concentrations, decreased apparent photosynthesis (APS), whilst GA increased it (Marcelle & Oben, 1973); apparent light respiration (ALR) was also modified in the same direction as APS by both regulators. From this it can be concluded that not only apparent photosynthesis, but also true photosynthesis (= APS + ALR) was diminished by CCC and enhanced by GA.

Our transpiration measurements did not reveal any significant difference be-

*Acknowledgements:* We wish to thank the I.W.O.N.L. for financial support. Moreover, Oben G., being a lecturer at the 'Economische Hogeschool Limburg' is indebted to this institute, and also to the 'Vereniging voor Wetenschappelijk Onderwijs Limburg' (V.W.O.L.) for many facilities.

*Abbreviations:* CCC: (2-chloroethyl)-trimethylammoniumchloride; GA: gibberellic acid; GDH: glutamate dehydrogenase (E.C. 1.4.1.2.); GOT: aspartate aminotransferase (E.C. 2.6.1.1.); GPT: alanine aminotransferase (E.C. 2.6.1.2.); DCPIP: 2,6-dichlorophenolindophenol; MV: methylviologen.

tween treated and control plants, in accordance with the results of El Damaty et al. (1965); Plaut et al. (1964) and Imbamba (1973). However, in the case of CCC, it has also been shown that the transpiration could be reduced (Asher, 1963; Mishra & Pradhan, 1971), so that the question of the effects of both chemicals on transpiration is still open.

Changes in the activities of the carboxylating enzymes ribulose-1,5-diphosphate carboxylase and phosphoenol pyruvate carboxylase could not account for the effects of CCC and GA on photosynthetic $CO_2$ fixation (Marcelle et al., 1974).

Looking forward for an explanation, we investigated the effects of both regulators on photochemical activities, and the results will be reported in this paper. As these two products significantly change chlorophyll and protein contents of leaves, we also measured the activities of some leaf enzymes involved in protein synthesis: GDH, GOT and DPT, one of which (GPT) possibly playing also a part in chlorophyll synthesis (Hedley & Stoddart, 1971a).

## Materials and Methods

Dwarf beans (*Phaseolus vulgaris L.* c.v. Limburgse Vroege) were grown in growth chambers as previously described (Marcelle et al., 1974). When the primary leaves were about 6 cm long (8 days after sowing) 10 ml of a $10^{-1}$M CCC solution was given as a soil drench to each plant or 1 ml of a 100 ppm GA solution was sprayed on the primary leaves.

The method of preparation of enzyme extracts has been described earlier (Marcelle et al., 1974). The assays of GDH, GOT and GPT were carried out following the methods of Bergmeyer (1962), with slight modifications. The measurements were performed with a double-beam spectrophotometer Varian-Techtron Model 635 at $25°C$. The reaction mixture contained (in $\mu$moles; final volume of 3 ml):  ·
- for GDH: 300 Tris pH 7.8; 0.8 $NADH_2$; 600 $NH_4Cl$; 30 $\alpha$-oxoglutarate and enzyme extract corresponding to 15 mg fresh material.
- for GOT: 300 Tris pH 7.8; 0.4 $NADH_2$; 150 L-aspartate; 22.5 $\alpha$-oxoglutarate and enzyme extract corresponding to 5 mg fresh material. It was found that the endogenous activity of malate dehydrogenase (MDH) was sufficient.
- for GPT: 300 Tris pH 7.8; 0.4 $NADH_2$; 200 L-alanine; 22.5 $\alpha$-oxoglutarate; 5.5 units of lactate dehydrogenase (LDH) and enzyme extract corresponding to 5 mg fresh material.

Before the extraction of chloroplasts, leaves were washed with cold distilled water and central nerves were discarded. About 4 grams of primary leaf material was mixed in a MSE homogeniser during 2 x 10 sec at top speed in an ice-cold medium (10 ml/g fresh weight) consisting of 0.05 M Tricine-NaOH buffer pH 8.4, 0.4 M sucrose, 0.01 M NaCl, 0.01 M EDTA and 0.2% bovine serum albumine. After grinding, the pH dropped to 8.0. The homogenate was filtered first through 8 layers of cheese-cloth, further through 1 layer of a nylon cloth. The chloroplasts were sedimented at 1200 g for 7 min and resuspended in 10 ml of a cold suspending medium containing 0.05 M Tricine-NaOH buffer pH = 8, 0.4 M sucrose and 0.01 M NaCl. The chloroplast suspension was kept at $4°C$ until use. Chlorophyll estimations were made according to Mackinney (1941). All measurements of the photochemical activities of chloroplasts were performed within 75 min after the beginning of the extraction.

For the photosystem 2 activity photoreduction of DCPIP was recorded at 522 nm with a Zeiss spectrophotometer model PMQII, adapted for illumination. All experiments were performed in cuvettes of 1 cm (volume used 2 ml) at saturating light intensity ($9.10^5$ ergs sec$^{-1}$ cm$^{-2}$) under continuous stirring. In most cases two different linearities were found: the first one, between at least the 5th and about the 30th second, was immediately followed by a second one which was somewhat lower. The activities were calculated for the first period of linearity, using a value for $\epsilon = 8.6 \times 10^3$ Mole$^{-1}$ cm$^{-1}$ (Armstrong, 1964). The reaction mixture contained (in $\mu$moles/ml): 30 Tricine pH = 8; 30 NaCl; 0.09 DCPIP; chloroplasts corresponding to 25 $\mu$g chlorophylls. For the photosystem 1 activity electron transport from DCPIP to MV was followed by measuring the rate of $O_2$ uptake resulting from the spontaneous reaction of reduced MV with molecular $O_2$. All reactions were performed under light saturation ($3.10^5$ ergs sec$^{-1}$ cm$^{-2}$) and the $O_2$ concentration was measured using a Rank Oxygen electrode. The reaction mixture contained (in $\mu$moles/ml): 20 Tricine pH = 8; 20 NaCl; 3 MgCl$_2$; 0.24 DCPIP; 17.5 ascorbate; 0.835 MV; 0.02 DCMU (3-(3,4 dichlorophenyl)-1,1-dimethylurea); 0.835 NaN$_3$ and chloroplasts corresponding to 10 $\mu$g chlorophylls. Calculations were made by assuming that the transport of 2 electrons corresponds to the uptake of 1 molecule $O_2$; all values were corrected for the rate of dark autooxidation of the electrondonor.

Cyclic photophosphorylation reactions were carried out using a Warburg apparatus (Wessels, 1959). The reaction mixture was put into vessels of about 14 ml, and illuminated from below by a Xenon lamp XBF 6000. Light intensity inside the vessels was saturating ($12.10^4$ ergs sec$^{-1}$ cm$^{-2}$) and temperature was maintained at 15° C. Under continuous shaking the reactions were allowed to proceed for 4 min in the light. Phosphate was determined by colorimetry of ammoniumphosphomolybdovanadate. The given activities are the differences between illuminated samples and dark controls. The reaction mixture contained (in $\mu$moles for a total volume of 3 ml): 50 Tris pH 7.8; 10 MgCl$_2$; 1 ADP; 10 phosphate pH 7.8; 125 glucose; 25 U hexokinase; 0.10 phenazine methosulfate (PMS); 10 ascorbate and chloroplasts containing 150 $\mu$g chlorophylls.

## Results and Discussion

In table 1 the activities of the enzymes GDH, GOT and GPT are given. CCC increased all the activities, whilst GA decreased them when expressed on a fresh weight basis. As each of both treatments changed the protein and chlorophyll contents of the leaves in the same direction and in the same proportion as they did with the activities of the enzymes, no differences could be found when these activities were expressed on a protein nor on a chlorophyll basis. These statements completely correspond with those made in connection with the carboxylases (Marcelle et al. 1974), where we also found an increase with CCC and a decrease with GA but only per unit fresh weight.

For GOT and GPT activities analogous effects of CCC and GA were reported when expressed on a fresh weight basis (Treharne et al., 1971; Hedley & Stoddart,

*Table 1.* GDH, GOT and GPT activities in primary leaves of bean plants treated with CCC or GA.

| | mU GDH | | mU GOT | | mU GPT | |
|---|---|---|---|---|---|---|
| | /g fr.wt. | /mg prot. | /g fr.wt. | /mg prot. | /g fr.wt. | /mg prot. |
| Control | 344 | 12.65 | 3400 | 122.5 | 6150 | 222.0 |
| CCC | 404 | 13.05 | 3910 | 125.7 | 6780 | 218.0 |
| | ** | N.S. | ** | N.S. | ** | N.S. |
| Control | 455 | 16.65 | 3890 | 144.0 | 7150 | 260.5 |
| GA | 358 | 16.50 | 3210 | 146.5 | 5590 | 257.0 |
| | ** | N.S. | ** | N.S. | ** | N.S. |

a. The measurements were performed at the 5th day after CCC treatment, and at the 3rd day after GA treatment.
b. Every value is the average of 16 measurements.
c. Significancy level:  ** = at $p = 0.01$ level according to the Wilcoxon matched-pairs signed ranks test (Siegel, 1956).
N.S. = not significant.

1971b). No results per unit protein were given by these authors. Sankhla & Huber (1974) on their turn reported no significant effects of GA on the activity of any of the three enzymes, and this on a protein basis.

In investigating the possible CCC effects on electron transport, we first measured the reduction of DCPIP, using chloroplasts isolated by standard procedure (phosphate buffer, sucrose, final pH 7.5). The values we obtained were not very high (about half the values of table 2), but nevertheless acceptable. They seemed to indicate that differences of APS could be related to effects on electron transport by photosystem 2, a reduced photosystem 2 activity in CCC plants accounting for the reduced APS. But with these chloroplasts no cyclic photophosphorylation with PMS could be detected. The extraction procedure was then modified in order to get chloroplasts able not only of photosystem 1 and 2 activities, but also of cyclic photophosphorylation.

In table 2 it can be seen that CCC and GA tend to increase all photochemical activities, the effect being somewhat more pronounced for CCC, but in no case significant. Addition of ammoniumchloride as uncoupler nearly doubled the activities of both photosystems as well in control as in treated plants. No change in the ratio of the uncoupled and coupled activities was induced by the treatments.

*Table 2.* Photochemical activities and cyclic photophosphorylation of chloroplasts of primary leaves of bean plants, treated with CCC or GA.

| | Photosystem 1 DCPIP → MV (in $\mu$eq./mg chlor. hr) | Photosystem 2 $H_2O$ → DCPIP (in $\mu$eq./mg chlor. hr) | Cyclic phosphorylation (PMS mediated) (in $\mu$mole P/mg chlor. hr) |
|---|---|---|---|
| Control | 484.0 | 231.7 | 353.7 |
| CCC | 514.5 | 246.5 | 369.3 |
| GA | 490.0 | 242.3 | 363.2 |

a. The measurements were performed between the 4th and the 7th day after treatments.
b. Every value is the average of 12 measurements.
c. None of the differences was significant.

It obviously appeared that the differences in APS activity seen on whole attached leaves could not be related to differences in photochemical behaviour of seemingly active chloroplasts isolated at pH 8.0. These results for the photosystem 2 activities are contradictory with those obtained with the poor active chloroplasts we first isolated at pH 7.5 (see before). If we assume, following McCarty & Jagendorf (1965) that the poor activity of bean chloroplasts isolated at pH 7.5 is due to the breakdown of membranes by galactolipid hydrolising enzymes, our contradictory results seem to indicate that a CCC treatment could modify the membrane structure or/and the activity of these lipases, so that the chloroplasts isolated from CCC treated plants became less active than chloroplasts isolated from control plants. Naturally this argument remains valuable even at pH 8.0, and we are not able to ascertain that in the three series of plants (control, CCC and GA treated ones) the activities of the extracted chloroplasts are maintained unaltered or are modified exactly in the same way.

At this moment, it is clear, we have no explanation for the observed changes in photosynthetic $CO_2$ fixation induced by the growth regulators neither at the biochemical nor at the photochemical level. It can always be argued that it is not possible to correlate enzyme or chloroplast activities measured *in vitro* with the actual activities *in vivo*. The conditions are optimized *in vitro* and we do not know how they are *in vivo*.

We can however speculate on the causes of the observed effects; first, the influence of gas diffusion processes is not to be neglected. Gas diffusion could be limited in CCC treated leaves, which are thicker than control ones, whereas it could be enhanced in the thinner leaves of GA treated plants. The observed effects on gas exchanges would then simply arise from alterations of gas conductance in the leaves.

However, two observations have to be considered: the first one is the decrease of oxidative phosphorylation by mitochondria isolated from CCC treated plants (Dalessandro et al., 1972). This decrease could modify the exchanges between mitochondria and the other parts of the cell, for instance chloroplasts.

The second observation is the slight modification of the chloroplast ultrastructure, induced by the treatment with CCC or GA (Marcelle et al., 1974). In chloroplasts of CCC treated plants the total surfaces of thylakoid membranes as well as of granar and intergranar membranes are significantly increased. The GA treatment results in significant inverse effects, except for intergranar membranes. These alterations of the plastid ultrastructure could modify the activity of the chloroplasts, but only to such extent that we hardly can show the effects by means of the techniques we dispose of at this moment. It nevertheless can be noticed that the observed slight increase of the activities of both photosystems in chloroplasts of CCC treated leaves could be related to the modifications of chloroplast ultrastructure; such a correlation however cannot be seen for GA chloroplasts.

In order to gain some insight in the mode of action of both regulators on photosynthesis the problem of the gas diffusion in the leaves should be deepened. Other points of great interest are the relations between chloroplasts and the rest of the cell, just like the control of photosynthesis by the transport of photosynthates, this control naturally playing no part *in vitro*.

# REFERENCES

Armstrong, J. Mc D. (1964): The molar extinction coefficient of 2,6-dichlorophenol indophenol. *Biochim. Biophys. Acta,* 86: 194-197.

Asher, W.C. (1963): Effects of 2-chloroethyltrimethylammonium chloride and 2,4-dichlorobenzyltributyl phosphonium chloride on growth and transpiration of Slash pine. *Nature,* 200: 912.

Bergmeyer, H.U. (1962): Methoden der Enzymatischen Analyse. Verlag Chemie — GMBH — Weinheim/Bergstr.

Dalessandro, G., F. Vita & R. Lavecchia (1972): Inhibition of oxidative phosphorylation on mitochondria isolated from pea seedlings treated with CCC. *Z. Pflanzenphysiol.,* 66: 254-257.

El Damaty, A.H., H. Kühn & H. Linser (1965): Water relations of wheat plants under the influence of (2-chloroethyl)-trimethyl-ammonium chloride (CCC). *Physiol. Plant.,* 18: 650-657.

Hedley, C.L. &J.L. Stoddart, (1971a): Light-stimulation of alanine aminotransferase activity in dark-grown leaves of *Lolium temulentum L.* as related to chlorophyll formation. *Planta (Berl.),* 100: 309-324.

Hedley, C.L. & J.L. Stoddart (1971b): Factors influencing alanine aminotransferase activity in leaves of *Lolium temulentum L.* II. Effects of growth regulators and protein biosynthesis inhibitors. *J. exp. Bot.,* 22: 249-261.

Imbamba, S.K. (1973): Response of Cowpeas to salinity and (2-chloroethyl) trimethyl-ammonium chloride (CCC). *Physiol. Plant.,* 28: 346-349.

Mc Carty, R.E. & A.T. Jagendorf, (1965): Chloroplast damage due to enzymatic hydrolysis of endogenous lipids. *Plant Physiol.,* 40: 725-735.

Mackinney, G. (1941): Absorption of light by chlorophyll solutions. *J. Biol. Chem.,* 140: 315-322.

Marcelle, R., H. Clijsters, G. Oben, R. Bronchart & J.-M. Michel (1974): Effects of CCC and GA on photosynthesis of primary bean leaves. Proceedings of the 8th International Conference of Plant Growth Substances. In press.

Marcelle, R. & G. Oben (1973): Effects of some growth regulators on the $CO_2$ exchange of leaves. *Acta Horticulturae,* 34: 55-58.

Mishra, D. & G.C. Pradhan (1971): Effect of transpiration reducing chemicals on growth, flowering, and stomatal opening of tomato plants. *Plant Physiol.,* 50: 271-274.

Plaut, Z., A.H. Havely & E. Shmueli (1964): The effect of growth-retarding chemicals on growth and transpiration of bean plants grown under various irrigation regimes. *Israel J. agric. Res.,* 14: 153-158.

Sankhla, N. & W. Huber (1974): Enzyme activities in *Pennisetum* seedlings germinated in the presence of abscisic and gibberellic acids. *Phytochemistry,* 13: 543-546.

Siegel, S. (1956): Nonparametric statistics for the behavioral sciences. Mc Graw-Hill Book Comp. Inc., New-York.

Treharne, K.J., J.L. Stoddart & C.L. Hedley (1971): Effects of the growth regulants $GA_3$ and CCC on the formation and activity of photosynthetic enzymes in graminae. In Forti, G., Avron, M. & Melandri, A. (Ed.): Photosynthesis, Two Centuries after its discovery by Joseph Priestly. Pp 2497-2509. Junk, The Hague.

Wessels, J.S.C. (1959): Studies on photosynthetic phosphorylation. III. Relation between photosynthetic phosphorylation and reduction of triphosphopyridine nucleotide by chloroplasts. *Biochim. Biophys. Acta,* 35: 53-64.

# EFFECTS OF CCC ON PHOTOSYNTHESIS IN *EUGLENA*

J.M. MICHEL

*Laboratory of Photobiology, Department of Botany, University of Liège, B-4000 Liège, Belgium.*

## Abstract

Light grown cells of *Euglena gracilis,* cultivated in synchronous cultures, have been used as a tool to study the effect of (2-chloroethyl) trimethylammonium chloride (CCC) on the photosynthetic apparatus. It was found that CCC inhibits the rate of cell division. The biosynthesis of chlorophylls is also affected, the cell content in total chlorophylls is depressed as well as the synthesis of chl. a relative to that of chl. b.

Measurements of light minus dark oxygen exchanges at several intensities by whole cells show an inhibition of the oxygen evolved by the CCC treated cells.

Examination of some photochemical reactions of chloroplasts show that, with respect to the control, chloroplasts extracted from *Euglena* grown in the presence of CCC have a higher Hill-activity but a lower P 700 content.

## Introduction

It is known that in higher plants CCC decreases the rate of leaf photosynthesis and increases the chlorophyll content per gram fresh weight (Birecka & Zebrowski, 1966; Marcelle & Oben, 1973). It is not yet clearly understood if the growth regulator acts upon the photosynthetic rate by modifying the leaf structure in such a way that the stomatal resistance is changed or by modifying the photosynthetic apparatus itself (Marcelle et al., 1974).

Marcelle and coworkers (1974) showed that in CCC-treated beans the rate of photosynthesis of the leaves is lowered and also that the structure of the chloroplasts is modified in such a manner that the total surface of thylakoïds membranes is increased as well as the total surface of grana membranes. They also found that the reduction of DCPIP by chloroplasts extracted from treated leaves was lower than in the control chloroplasts.

To search for a direct action of CCC on the photosynthetic apparatus itself we decided to study the photosynthetic characteristics of *Euglena* grown in the light on a defined medium containing CCC and to compare them with the photosynthetic characters of control cells grown in the absence of CCC.

## Material and Methods

Cell culture: Cells of *Euglena gracilis* Klebs 6 ar. var. bacillaris Pringsheim were

*Abbreviation:* CCC: (2-chloroethyl)-trimethylammonium chloride; DCPIP: 2,6-dichlorophenolindophenol.

*Acknowledgement:* The author thanks The 'Fond National de la Recherche Scientifique, F.N.R.S.', Brussels, Belgium for its financial support.

cultivated on the medium described by Greenblatt & Schiff (1959). They were grown synchronously on a succession of light-dark cycles of 14 hr light and 10 hr darkness. The light source was a battery of four fluorescent tubes giving an incident intensity of 6000 lux at the level of the cultures vessels. The cultures were aerated by bubbling normal air through them. In these culture conditions the cells divide once per 24 hr period.

As the photosynthetic activity and the chlorophyll content of the cells vary during the life cycle (Cook, 1966; Walther & Edmunds, 1972) the cells were systematically harvested at the middle of the light period.

CCC treatment: To study the effect of CCC on the cells, several culture vessels containing the basal medium to which CCC was added at various concentrations ranging from $2,82 \cdot 10^{-4}$ M to $8.10^{-2}$ M and a vessel containing only the basal medium were inoculated with the same volume of a starting culture which was in the exponential phase of its growth curve. The cell population at the start of the experiments was in the range of $10^3$ to $2.10^3$ cells per ml of culture medium. Samples from each culture were periodically withdrawn aseptically for the measurement of the parameters chosen. In the experiments reported here the solutions of CCC in distilled water were sterilized by autoclaving for 20 min at $126°C$; after the autoclaving they were mixed with the basal medium. Recent experiment, not reported here, where the CCC solutions were sterilized by filtration on Millipore filters at room temperature seems to indicate that the CCC has a more pronounced effect on the cells, especially at low concentrations in the range of $10^{-4}$ to $10^{-3}$ M, than with autoclaved solutions. More work is needed to clear that point.

Measurements of the cultures parameters: The cells counts were performed by microscopic examination in a haemacytometer of samples of cultures which cells had been killed by the addition of a few cristals of $NaN_3$. The chlorophylls determinations were made according to Mackinney (1941).

The exchanges of oxygen were measured at $22°C$ with a Clark type electrode connected to a potentiometric recorder on cells preilluminated with low intensity white light of $2.10^4$ ergs.cm$^{-2}$.sec$^{-1}$. The actinic light for photosynthesis was provided by a tungsten lamp equipped with a water filter 20 mm thick, a Balzers type calflex C heat filter ans a Schott RG630 cut-off filter. The intensity of the actinic light was changed by using a Variac to power it. The light intensity measurements were done with a calibrated thermopile, type $E_1$ special (Kipp, Delft, Holland).

Measurements of photochemical activities of the chloroplasts: chloroplasts were extracted from the cells using the method of Katoh & San Pietro (1967).

The Hill activity was measured by following the oxygen evolution, with the aid of the Clark electrode, by freshly prepared chloroplasts suspended in a medium containing: phosphate buffer 0.05 M; pH 7,8; NaCl 0,01 M; sucrose 0.4 M; DCPIP 0,1 mM. The chlorophyll concentration in the cuvette ranged between 15 and 20 microgramm per ml.

Measurement of the $P_{700}$ content: The $P_{700}$ content was determined according to the chemical method of Anderson & coworkers (1971). The difference extinction coefficient used was 64 mequiv$^{-1}$ cm$^{-1}$ at 700 nm (Hiyama & Ke, 1972).

## Results and Discussion

### 1. Effect of the CCC on the cell multiplication

It is a general fact that *Euglena* grown in the light on a synthetic medium containing CCC (Cycocel) divide more slowly than the control culture. We have tried several concentrations of the inhibitor ranging from $2,83 . 10^{-4}$ M to $8,5 \ 10^{-2}$ M. For the higher concentration tested the drug has almost a lethal effect, the cells do not divide anymore, they loose their mobility and acquire a ball-shaped appearance. We mostly worked with a CCC concentration of $2,83 . 10^{-2}$ M which gives a clear cut inhibition of the cell division and of the chlorophylls synthesis but still allows the cells to have a normal aspect and a normal mobility under microscopic examination and to perform a measurable photosynthesis in the light. Concentrations above $5 . 10^{-2}$ appear to be toxic for *Euglena*.

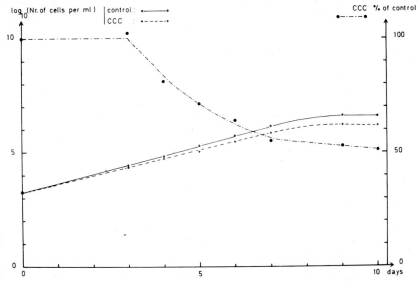

Fig. 1. Growth curves of a control and a culture containing CCC, $2,83 . 10^{-2}$M. Abcissa: days after inoculation. Left ordinate: $\log_{10}$ of the number of cells per ml of culture medium. Right ordinate: cell numbers of the CCC containing culture expressed as percentage of the cell numbers in the control culture.

The effect of CCC on the cell multiplication is illustrated by figure 1. On that figure we have plotted the log in the base ten of the number of cells per ml in two cultures as a function of the age of the cultures. One culture is the control, the other one is grown on the basic medium made $2,83 . 10^{-2}$ M in CCC. The moment of inoculation is chosen as the origin of time, the two vessels are inoculated aseptically with the same number of cells taken from an exponentially growing stock culture. We have also plotted on the figure the number of cells per ml in the CCC treated medium expressed as the percentage of the control culture as a function of time. The inhibitory effect of CCC on the rate of cell divisions is immediately apparent from the curves giving the $\log_{10}$ of cell numbers.

Considering the curve giving the percentage of cells per ml in the CCC treated culture with respect to the control, one sees that after three generations, there is no difference between the treated and the control cultures, this has been checked by applying the Wilcoxon test (Paerson & Hartley, 1972) at the 0,05 significance level.

After four generations the number of cell in the treated culture is significantly different at the 0,05 level of significance. With increasing number of generations, that is time of culture, the percentage difference between treated and control culture increase to reach finally a plateau value between 50 and 60%. This result shows that the inhibitory effect of CCC on the cell division in *Euglena* does not show up from the beginning of the cultivation of the cells on a medium containing the inhibitor since it takes at least four generations to detect a significant effect.

*2. Effect of the CCC on the chlorophyll content of the cultures.*
The Cycocel not only affects the multiplication of *Euglena* but it also has an effect on the biosynthesis of the chlorophylls. This fact is shown by figure 2. We have plotted on that figure the chlorophyll content, expressed as microgramm chlorophyll (a + b) per ml of culture medium, as a function of the time of cultivation for two cultures, a control one and an other one treated with CCC at a concentration of 2,83. $10^{-2}$ M. At the same time, we have plotted the chlorophyll content of the CCC treated culture as the percentage of the control culture at different times after the inocculation of both cultures with the same amount of cells containing the same amount of chlorophyll.

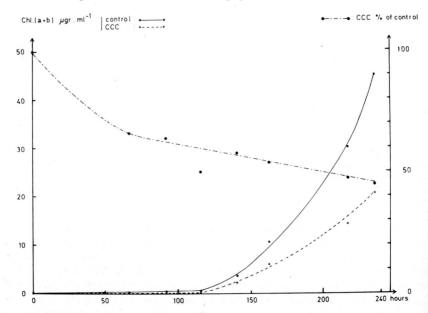

*Fig. 2.* Chlorophyll content of a control and a culture containing CCC, 2,83. $10^{-2}$M as a function of time from inoculation. Left ordinate: chlorophyll expressed as microgramm ch (a+b) per ml of culture medium. Right ordinate: chl (a+b) content in the CCC containing culture expressed as percentage of the control.

The figure shows that the chlorophyll content per ml of culture medium is lower in the presence of CCC than in the control. With increasing age of the culture the inhibition of chlorophyll synthesis in the presence of CCC increases. Interestingly enough is the fact that after 3 generations the chlorophyll content of the treated culture represent only 65% of the control. The percentage reached after 10 days is about 45%, in some cases it can be even less, around 35% of the control. We want to point out that the decrease in chlorophyll content is already important after 3 generations, contrarily to what has been observed about the cell numbers. This means that the effect of CCC on the chlorophyll synthesis does not parallel the effect on the cell division, the inhibition of chlorophyll synthesis occurs more rapidly that the inhibition of cell divisions.

If we express the chlorophyll content as the amount of pigment contained in a given number of cells, rather than in a given volume of culture, we see that the amount of chlorophyll per cell is also reduced in CCC treated cells with respect to the control cells. This is due to the fact that the inhibition of chlorophyll synthesis is higher than the inhibition of cell division. The fact that the chlorophyll synthesis is specifically inhibited in CCC-treated *Euglena* can also be expressed by the calculation of the ratios of the chlorophyll a content to that of chlorophyll b. This ratio is always depressed for the cultures grown on CCC. An example is given in table 1 for cultures grown on different concentrations of CCC. One sees that the chl a/chl b ratio falls from 5,5 in the control (this high ratio is normal for light grown *Euglena gracilis* var. bacillaris) to a value as low as 4,1 in *Euglena* cultivated on a medium containing 2,83. $10^{-2}$ M of CCC. The synthesis of chl a as well as the

*Table 1.* Effect of CCC on the chlorophyll content.

| CCC conc. | chl (a+b) (a) | chl a (a) | chl b (a) | $\dfrac{chl\ a}{chl\ b}$ |
|---|---|---|---|---|
| 0 (control) | 30.42 | 25.78 | 4.64 | 5.55 |
| 2.83. $10^{-4}$M | 25.98 | 21.42 | 4.55 | 4.70 |
| 2.83. $10^{-3}$M | 20.50 | 16.47 | 4.02 | 4.09 |
| 2.83. $10^{-2}$M | 11.17 | 9.21 | 1.96 | 4.69 |
| 4.24. $10^{-2}$M | 7.95 | 6.60 | 1.35 | 4.89 |

(a): chlorophyll content expressed as microgram chlorophyll per ml of culture medium.

synthesis of chl b are both inhibited by CCC, but for any concentration of Cycocel the inhibition of chl a synthesis is always higher that the inhibition of chl b synthesis.

### 3. Photochemical activities and $P_{700}$ content of CCC treated Euglena.

We undertook the study of the photochemical activities of *Euglena* grown on CCC by measuring three parameters. The first one is the photosynthetic oxygen evolution by whole cells. The second one concerns the ability of photosynthetic electrons transport by isolated chloroplasts at the level of photosystem 2, namely the Hill reaction with water as electrons donor and DCPIP as electron acceptor. The third parameter refers to the content of the chloroplasts in $P_{700}$ (reaction center of Photosystem 1). The results reported here concern these measurements made on control cells and on *Euglena* grown on CCC at concentration ranging from 2,83. $10^{-4}$ M to 4,24. $10^{-2}$ M.

The oxygen evolution was measured on whole cells kept in the culture medium. We decided arbitrarily to compute the photosynthetic oxygen evolution from the difference in the slopes of the curves representing the change in $O_2$ concentration with time when the cells are kept in darkness and when they are illuminated. Doing so we admit implicitly that the rate of oxygen consumption measured when the cells are in darkness does not change upon illumination.

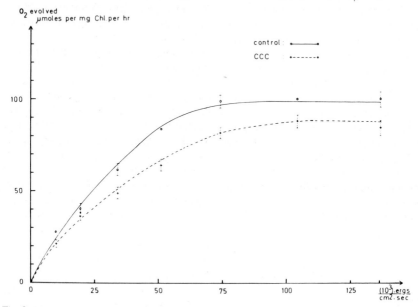

*Fig. 3.* Light saturation curves for oxygen evolution by control *Euglena* and cells grown in the presence of CCC 2,83. $10^{-2}$M.

Figure 3 shows the results obtained when calculating the oxygen evolved, on a chlorophyll basis, as a function of light energy in a typical experiment with a CCC concentration of 2,83. $10^{-2}$ M. The CCC affects the slope of the linear part of the curve and the value reached at saturation, both are decreased with respect to the control. If we consider the oxygen evolution at the plateau we see that the inhibition by CCC amounts to 10 to 15%. The comparison of the figures obtained at light intensities giving an oxygen evolution equal to half the saturated values shows also an inhibition by CCC in the range of 10 to 15%.

If we express the oxygen evolved on a population basis, rather than on a chlorophyll basis, we observe the same general pattern, namely an inhibition by CCC. But as the amount of chlorophyll per cell is lower in the cultures grown on CCC than in the controls, the values of the inhibition of the photosynthetic oxygen evolution are slightly higher than with the data expressed on a chlorophyll basis.

Knowing that CCC has an inhibitory effect on the chlorophyll synthesis and on the photosynthetic oxygen evolution, we tried to localize the site of action of the inhibitor. In this line of reasoning we isolated chloroplasts and we measured their ability to perform a photochemical reaction representative of the photochemistry of the water splitting system to see if the activity of Photosystem 2 is also impaired by CCC. For this purpose we grew *Euglena* on the basic medium and on the same medium enriched with various concentrations of CCC respectively: 2,83. $10^{-4}$ M, 2,83. $10^{-3}$ M, 1,41. $10^{-2}$ M and 4,24. $10^{-2}$ M. After 7 days of growth the cells were harvested and their chloroplasts extracted. The Hill activity was measured with an intensity of a red actinic light of 3.4. $10^4$ ergs cm$^{-2}$ sec $^{-1}$. For that intensity the rate of oxygen evolved by whole cells was still in the linear range of the saturation curve. The results of the measurements are given in Table 2. As can be seen, the Hill activity of the chloroplasts extracted from cells grown on CCC is higher, for all the concentrations tested, than the activity of the control. The increased activity is found with the lowest concentration of CCC used, it is not proportionnal to the CCC concentration. With a CCC concentration of 2,83. $10^{-4}$ M the percentage in the increase of activity is 46% but from 2,83. $10^{-4}$ M to 4,21. $10^{-2}$ M the relative increase is only 5%, which indicate that the CCC effect is obtained with a very low concentration.

*Table 2.* Effect of CCC on the Hill activity and the $P_{700}$ content of chloroplasts.

| CCC conc. | Hill activity (a) | chl/$P_{700}$ (b) |
|---|---|---|
| 0 (control) | 36.9 ± 2,6 | 328 |
| 2,83. $10^{-4}$M | 53.9 ± 2,7 | 1440 |
| 2.83. $10^{-3}$M | 54.0 ± 6,5 | 676 |
| 1.41. $10^{-2}$M | 55.1 ± 4,2 | 1012 |
| 4.24. $10^{-2}$M | 56.5 ± 5,2 | 679 |

(a): Hill activity expressed as microequivalent of electrons transferred per mgr chlorophyll per hr.

(b): chl/$P_{700}$ expressed as the molar ratio of chl (a+b) to $P_{700}$.

The above mentioned results show that the ability of the Photosystem 2 to transfer electrons is enhanced. The problem naturally arose to see what happens to the Photosystem 1 side of the photosynthetic electron transport chain. As a first step we measured the relative content of chlorophyll to the so-called $P_{700}$. This measurement has been made in parallel to the Hill activity measurement on chloroplasts extracted from *Euglena* cells grown on various concentrations of Cycocel. The results are shown on table 2. One sees that the chloroplasts of the control cells contain about 330 molecules of chlorophyll per molecule of $P_{700}$. This figure is in good agreement with the data generally found for higher plants chloroplasts. As compared with the control, the chloroplasts extracted from cells grown on CCC show a marqued increase in the value of the molar ratio of chlorophyll to $P_{700}$. The increase ranges from two to three times, which means that the cells grown on CCC contain between two and three times less $P_{700}$ per chlorophyll than the normal cells. As for the Hill activity it is seen that the lowering of the $P_{700}$ content is not proportionnal to the CCC concentration.

Cycocel is a well-known inhibitor of growth in higher plants, our results show that it has also an inhibitory effect on the growth of a monocellular phytoflagellate. They also show that CCC affects the photosynthetic apparatus itself at several levels from the synthesis of chlorophylls to the functional properties of the chloroplastic electron transport chain as well as of the whole machinery leading to oxygen evolution. The data obtained do not allow us to decide whether the inhibition of cellular growth is dependant of the inhibition of the photosynthetic apparatus or not. As *Euglena* can be cultivated heterotrophically in darkness as well as photoautotrophically or photoheterotrophically; it could be a suitable organism to investigate on the two above-mentioned possibilities. The inhibitory effect of CCC on the chlorophyll synthesis observed in *Euglena* is not in agreement with the observation made on higher plants (Marcelle et al., 1973) where one sees an increase of the chlorophyll content of bean leaves whether the results are expressed on a fresh weight or on a surface basis. This discrepancy has to be explained in our opinion by different specific responses between beans and *Euglena* as it is known, by morphometric studies, that the total surface of thylakoids membranes is higher in CCC treated beans than in the control (Marcelle et al., 1974).

Comparison of the effect of Cycocel on the oxygen evolution by whole cells, on the Hill reaction and on the $P_{700}$ content of the chloroplasts leads to the idea that the inhibition of the whole photosynthesis could result from an unbalance between the photochemical capacity of the two photochemical systems of the chloroplast. As the Hill activity is enhanced in the chloroplasts extracted from cells grown on CCC and the $P_{700}$ content is lowered one can think of a kind of bottleneck existing at the level of Photosystem 1. This tentative interpretation does not preclude however from a possible effect of CCC at the level of the carbon dioxide fixation mechanism. In any case our results show that an effect of CCC on the stomatal resistance as proposed by Birecka & Zebrowski (1966) is not sufficient to account for the observed inhibition of photosynthesis in green leaves.


# REFERENCES

Anderson, J.M., K.C. Woo & N.K. Boardman (1971): Photochemical systems in mesophyll and bundle sheath chloroplasts of $C_4$ plants. *Biochim. Biophys. Acta*, 245: 398-408.

Birecka, H. & Z. Zebrowski (1966): Influence of (2-chloroethyl)-trimethyl ammonium chloride (CCC) on photosynthetic activity and frost resistance of Tomato plants. *Bull. Acad. Polon. Sci.*, 14: 367-373.

Cook, J.R. (1966): Photosynthetic activity during the division cycle in synchronized *Euglena gracilis. Plant Physiol.*, 41: 821-825 .

Greenblatt, C.L. & J.A. Schiff (1959): A pheophytin-like pigment in dark adapted *Euglena gracilis. J. Protozool.*, 6 (1): 23-28.

Hiyama, T. & B. Ke (1972): Difference spectra and extinction coefficients of $P_{700}$. *Biochim. Biophys. Acta*, 267: 160-171.

Katoh, S. & A. San Pietro (1967): The role of C-type cytochrome in the Hill reaction with *Euglena* chloroplasts. *Arch. Biochem. Biophys.*, 118: 488-496.

MacKinney, G. (1941): Absorption of light by chlorophyll solutions. *J. biol. Chem.*, 140: 315-322.

Marcelle, R. & G. Oben (1973): Effects of some growth regulators on the $CO_2$ exchanges of leaves. *Acta Horticulturae*, 34: 55-58.

Marcelle, R., H. Clijsters, G. Oben, R. Bronchart & J.M. Michel (1974): Effects of CCC and GA on photosynthesis of primary bean leaves. Proceedings of the 8th International Congress of Plant Growth Substances (in press).

Paerson, E.S. & H.O. Hartley (1972): Biometrika tables for statisticians, volume II, Pp 46-47 and 227-230. Cambridge University Press.
Walther, W.G. & L.N. Edmunds Jr. (1973): Studies on the control of the rhythm of photosynthetic capacity in synchronized cultures of *Euglena gracilis (Z)*. *Plant Physiol.*, 51: 250-258.

# HORMONAL INFLUENCES ON STOMATAL PHYSIOLOGY AND PHOTOSYNTHESIS

P.E. KRIEDEMANN & B.R. LOVEYS

*C.S.I.R.O. Division of Horticultural Research, Private Mailbag, Merbein, Vic. 3505 and P.O. Box 350, Adelaide, S.A. 5001, Australia*

Abstract

Both environmental and internal effects on gas exchange have been analysed in terms of hormonal physiology. Photosynthetic changes were resolved into stomatal and residual components following concurrent measurements of $CO_2$ assimilation and $H_2O$ evolution by individual leaves under laboratory conditions ($2\% O_2$ gas stream). Both woody perennials and herbaceous annuals provided test material.

Photosynthetic and stomatal changes during moisture stress and subsequent recovery, or after fruit removal or stem girdling, were always associated with changes in the levels of abscisic acid (ABA) and its close relative, phaseic acid (PA). Decreased rates of gas exchange were correlated with increased levels of these compounds and *vice versa*.

While the significance of ABA in stomatal physiology has been amplified in this work, the present data also implicate phaseic acid as a specific inhibitor of photosynthesis *in vivo*. Such inhibition probably contributes to the 'after-effect' of moisture stress, and to other forms of internal control over gas exchange. Phaseic acid extracted from test plants was also tested *in vitro:* photosynthetic activity of both excised leaves (cuvette measurements) and tissue slices (oxygen electrode determinations) was strongly inhibited by buffered solutions of PA at physiological concentrations.

## Introduction

When plants encounter moisture stress, both stomatal physiology and photosynthesis are affected due to the combined effects of hydraulic control and hormonal influences (Liu et al., 1974; Loveys & Kriedemann, 1973). Subsequent recovery after irrigation, again demonstrates internal controls whereby gas exchange is not fully restored despite the improvement in leaf water potential – i.e. the widely documented 'after-effect' of moisture stress.

The massive increase in endogenous levels of abscisic acid (ABA) which occurs during stress (forty fold in wheat for example, Wright & Hiron, 1969) has been viewed as a mechanism enabling the plant to modify water consumption, and was thought to be a factor in this 'after-effect'. Two sets of observations are difficult to reconcile with this view: ABA levels drop precipitously upon rewatering whereas gas exchange shows a more protracted recovery (see for example Wright & Hiron, 1972; Loveys & Kriedemann, 1973) Photosynthetic activity during stress and subsequent recovery can show a non-stomatal component in its inhibition and restoration of stomatal function is not matched by an increase in $CO_2$ fixation (Boyer, 1971; Liu et al., 1974). Since any effect of ABA on photosynthesis is most likely mediated via stomatal effects rather than direct action at sites of $CO_2$

*Acknowledgements:* Grateful acknowledgement is made to Mrs. E. Törökfalvy and Mrs. J.P. Milln for technical assistance.

227

fixation, (Cummins et al., 1971; Kriedemann et al., 1972) we wondered whether some additional factor was involved in regulating photosynthetic recovery during recovery from water stress. Moreover, our observations on test plants of *Vitis vinifera* have revealed stomatal and photosynthetic responses to manipulative treatments, unrelated to water stress, where gas exchange showed a close correlation with endogenous levels of ABA and PA (phaseic acid). These additional observations (Loveys & Kriedemann, 1974) highlighted the possible significance of PA as a specific inhibitor of photosynthesis. Experiments reported here confirm this role.

## Material and Methods

*Experiments on Moisture Stress and Recovery*
Three-year-old rooted cuttings of Concord grapevines (*Vitis labruscana*) were established in 85-litre containers over a period of 2 months on trellis wires ad jacent to a commercial vineyard at Hector in Upper New York State (USA). All lateral shoots were removed and cluster number was reduced to one per shoot to encourage vegetative growth.

A pressure chamber (Scholander et al., 1965) provided data on leaf xylem water potential. Excised leaves were transferred immediately to polyethylene bags and held under shade for no more than 2-3 minutes depending upon sampling frequency and duration of individual measurements. Values for leaf water potential obtained this way were closely related with $\Psi_{leaf}$ (r = 91) derivied from thermocouple psychometry.

Stomatal resistance was measured with a diffusion porometer. The basic design followed Kanemasu et al. (1969) although our instrument was machined out of solid teflon, while leaf and chamber temperatures were measured with thermistors and a bridge circuit (Bingham, 1972).

Photosynthesis measurements were based on $CO_2$ exchange by a portion of single leaves monitored with a Beckman model 315 A IRGA set up for differential measurement. The leaf chamber and gas circuit have been described previously (Kriedemann & Smart, 1971).

Leaf material was sampled for ABA assay immediately after water potential measurements. Brief exposure of leaf material to the high positive pressure inside the Scholander chamber was without effect on its ABA content. Petioles were removed and remaining laminae were weighed and then chopped into cold ($-10°$ C) 80% methanol. These operations were performed as rapidly as possible, and in dim light to minimise any likelihood of ABA isomerisation. Leaf samples were subsequently homogenised and assayed for ABA according to the procedure of Seeley (1971). Levels of PA and ABA released by basic hydrolysis were also examined.

*Manipulative Treatments*

Grapevines (*Vitis vinifera* L. cv. Cabernet Sauvignon) were planted as rooted cuttings and grown in a greenhouse. Inflorescence retention was encouraged by removal of leaves basal to the lowermost inflorescence (Mullins, 1966). Supplementary illumination extended daylength to 16 hours. Vines were well established at the time of experimentation (12 weeks after planting) and had grapes 2-4 mm in diameter. Plants were selected for uniformity of bunch size at the beginning of each experiment. Single shoots were trimmed to 6 fully expanded leaves above the bunch. Lateral buds were excised. In cinctured vines, a collar of tissue, 2 mm wide, was removed from the internode below the fruit-bearing node.

Plants were transferred to growth cabinets 7 days before an experiment was started. Temperature regime was $15/22°C$ (day/night); light intensity was 550 $\mu E$ $m^{-2}$ $sec^{-1}$ ($4.1 \times 10^4$ lux) above the crop.

Stomatal resistance was measured inside the growth cabinet on leaves adjacent to reproductive nodes using a diffusion porometer (Lambda Model Li 60). Photosynthesis of single leaves was measured under controlled conditions using infrared gas analysis for $CO_2$ and $H_2O$ exchange (see Kriedemann, 1971 for details).

ABA and PA were assayed as follows: leaf tissue was removed from the plant and immediately plunged into liquid nitrogen and then homogenised in 80% aquous methanol. The residue was subsequently extracted three times over the following 48 hour period. The bulked methanol was evaporated under reduced pressure at 30°C and the resulting aqueous extract frozen overnight.

After centrifugation to remove suspended material, measured aliquots, representing 5g fresh weight of tissue, were partitioned with ether to yield an ether-soluble acid fraction (Kriedemann et al., 1972). Approximately 1% of the total PA remained in the aqueous phase after ether partition. This could be extracted into ethyl acetate but the final fraction for gas-liquid chromatography (glc) then contained unacceptably high levels of contaminants. Ether was therefore routinely used for the extractions and the small amount of PA remaining in the aqueous phase was neglected. The ether was dried over anhydrous sodium sulphate and separated by thin layer chromatography (tlc) using silica gel $G_{254}$ as support and developing with either benzene: ethyl acetate: acetic acid (50:5:2, v/v) or toluene: ethyl acetate: acetic acid (25:15:2, v/v). Zones opposite marker spots of ABA and PA were removed and eluted with water-saturated ethyl acetate. Although we have never detected significant amounts of 2-trans-ABA in our extracts, care was taken to exclude this zone from the silica removed from the tlc plates. The dried eluates were dissolved in ether containing diazomethane. The methylated extract was dried overnight in a vacuum desiccator and redissolved in methanol (1 ml per 10 g fresh weight of tissue) and subjected to glc. A Packard model 409 instrument fitted with a $^{63}Ni$ electron capture detector was used. Best results were obtained with a 1 m x 3 mm glass column packed with 2.5% OV17 on chromosorb WAW DMCS; 10% methane in argon was used as carrier. Column flow rate was 40 ml $min^{-1}$ and by-pass flow 20 ml $min^{-1}$; column temperature was 205°C, detector temperature 250°C and injection port 260°C. Pulse period for the detector was 100 $\mu s$ and pulse width 1 $\mu s$.

Usually 3 replicates from each aqueous extract were purified and ABA estimated by glc.

Further details on the separation and identification of PA from vine leaf extracts are given by Loveys & Kriedemann (1974).

# Results and Discussion

## After-effects of Moisture Stress

Vines encountering moisture stress showed a steady increase in leaf resistance as $\Psi_{leaf}$ declined and reached maximum values around 20-25 sec cm$^{-1}$ once $\Psi_{leaf}$ had fallen to between -13 and -16 bars. Endogenous levels of ABA showed an associated increase; the extent of this rise depending upon duration of moisture stress. Prolonged periods of low water potential (-12 to -14 bars for 16 days) resulted in ABA increasing from 0.14 (pre-stress) to 2.5 mg kg$^{-1}$ fresh wt.; i.e., an 18-fold increase.

Photosynthetic decline during moisture stress was attributed to the combined effects of higher stomatal resistance and a loss of photochemical efficiency (gauged from light response curves). The vine leaf physiological status prior to rewatering is summarized in Figure 1 (left side).

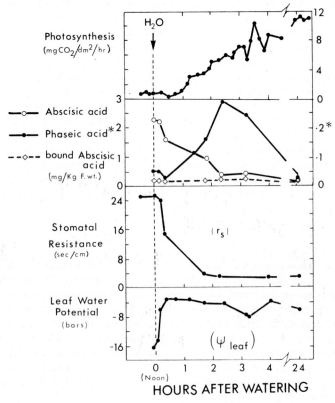

*Fig. 1.* Vine leaf recovery following moisture stress: Time course of gas exchange and hormonal changes on an excised shoot.

Complete restoration of moisture status, stomatal function and photosynthetic capacity normally occurred within 16-18 hours of rewatering, but irrigation water had to percolate throughout the root zone and then overcome resistance to flow within the roots themselves; these limitations would naturally influence the time course of recovery. In a attempt to eliminate these restrictions, we compared recovery on attached or excised portions of the same grapevine; i.e. on a vine consisting of two main trunks and associated shoots, one trunk was severed and then recut under water at the same moment as the remaining portion was irrigated via its roots in the normal way.

Ultimately, both portions of the vine showed full restoration of moisture status, but a disparity between stomatal function and photosynthesis was particularly evident within the first 4 hours of irrigation on excised shoots. Relevant data are shown in Figure 1. Foliage on these severed shoots showed an almost immediate recovery of $\Psi_{leaf}$ and stomatal opening, but the steep drop in stomatal resistance was not matched by a commensurate rise in photosynthesis.

ABA showed a gradual diminution after rewatering, while levels of "bound" ABA remained uniformly low. PA however, showed a substantial increase (10 fold) within 3 hours of rewatering followed by an overnight decline to pre-stress levels:

Stomatal recovery bears some correlation to ABA levels during the first 3 hours but the delayed recovery of photosynthesis is not explained by "residual" ABA. The elevated levels of PA during the recovery period suggest a role for the compound in photosynthetic control, and subsequent experiments substantiated this view.

*Relationship between Stomatal Function and Endogenous ABA and PA.*
Environmental stress certainly leads to changes in levels of endogenous ABA, but in that case, a fall in $\Psi_{leaf}$ probably acts as a stimulus for increased ABA synthesis with attendant effects on stomatal physiology. The present treatments of fruit removal in combination with stem cincturing (treatments 1 to 4 in Figure 2) have no adverse influence on water relations, but do affect the levels of endogenous inhibitors.
Both ABA and PA showed changes within four days of treatment. While ABA levels rose by 42% those of PA showed a massive 480% increase: This correlation between $r_s$ and inhibitor levels, especially PA, was highly significant (see Figure 2).

These responses to fruit removal in combination with stem cincturing (Figure 2) were subsequently confirmed in a repeat experiment. Comparing treatments 1 and 4 for example, $r_s$ rose from 1.41 to 7.14 sec cm$^{-1}$ within 7 days (although effects were apparent within the first 24 hours). ABA and PA levels had risen by 50% and 370% respectively. Relative leaf water content was again measured to ensure that stem cincturing had not influenced $\Psi_{leaf}$ unfavourably. No significant differences were encountered, plants with higher levels of inhibitors actually showed higher water content.

*Relationship between Photosynthesis and Endogenous ABA and PA.*
Any modification of a plant's source/sink relationship may lead to a change in photosynthate level within source leaves, which may in turn modify the rate of $CO_2$ fixation (see Neales & Incoll, 1968, for references). Starch levels in vine leaves were influenced by the manipulative treatments outlined in Figure 2. By the end of this experiment, leaf starch (% dry wt). was 9, 13, 20 and 21% for

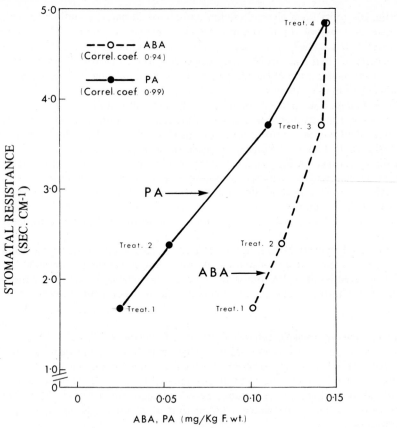

*Fig. 2.* Stomatal resistance as a function of endogenous levels of abscisic acid (ABA) and phaseic acid (PA). The 4 manipulative treatments imposed 4 days previously were as follows: 1) Control (fruit present, stem intact); 2) Fruit removed, stem intact; 3) Fruit present, stem cinctured below fruiting node; 4) Fruit removed, stem cinctured.

treatments 1 through 4 respectively. Higher leaf starch is obviously associated with lower demand for photosynthate, although stomatal adjustment (Figure 2) is not necessarily in parallel. A discontinuity exists between treatments 3 and 4 where $r_s$ increases from 3.69 to 4.82 sec cm$^{-1}$ for a 1% change in starch content. Clearly some other influence is operating and data in Table 1 imply that some internal factor is involved. Fruit removal on intact vines resulted in increases in both stomatal ($r_s$) and internal ($r_r$) resistances and photosynthesis was reduced by 35%. Inhibitor levels, and especially PA were increased by the treatment. PA levels were *trebled* following fruit removal.

*Photosynthetic Response to Exogenous PA*

To confirm the possible significance of PA in the regulation of gas exchange, the photosynthetic effects of this compound were tested directly. Five kg of leaves from mature grapevines (cv. Cabernet Sauvignon) were allowed to wilt and their

*Table 1.* The response of Vine Leaf Photosynthesis and Stomatal Resistance to Fruit Removal plus Associated Changes in Levels of Endogenous Inhibitors.

| Treatment | Photosynthesis[a] $(mgCO_2dm^{-2}hr^{-1})$ | Stomatal Resistance[b] $(sec\ cm^{-1})$ | Residual Resistance $(sec\ cm^{-1})$ | Abscisic acid | Phaseic $(mg\ Kg^{-1}fresh\ wt.)$ |
|---|---|---|---|---|---|
| + Fruit | 11.38 | 2.03 | 16.08 | 0.136 | 0.052 |
| – Fruit | 7.49 | 4.63 | 22.68 | 0.162 | 0.155 |
| LSD 5% | 3.48 | 2.08 | 4.72 | 0.03 | 0.04 |

(a) Measured in an $O_2$ free gas stream.
(b) Calculated from water vapour exchange, measured by infrared gas analysis.

laminae were extracted. ABA and PA were recovered and sufficient PA was then separated (its identity was confirmed by combined glc/mass spectrometry - see Loveys & Kriedemann, 1974, for details), to permit several experiments on excised leaves. Photosynthesis and transpiration were monitored continuously under laboratory conditions and a solution of native PA ($7 \times 10^{-5}$ M in 50 mM phosphate buffer at pH 6.4) was supplied to cut petioles of excised leaves. (Each leaf was previously held in buffer prior to treatment with PA).

Representative data for the time course of photosynthetic response, plus changes in component resistances to gas exchange, are shown in Figure 3. The abrupt decline in photosynthesis within the first 10 to 20 minutes after PA addition clearly stems from a non-stomatal inhibition of $CO_2$ fixation. The subsequent rise in leaf resistance could either result from an elevation in substomatal $CO_2$ concentration following the decline in photosynthesis; or alternatively, PA supplied to the leaf could be accumulating in sufficient quantities to exert a stomatal influence over and above photosynthetic effects.

What appears then to be a selective influence of PA on photosynthesis (Figure 3) has been confirmed in more than 20 similar experiments using excised leaves from a variety of species. The reality of non-stomatal inhibition of photosynthesis by PA was also established on tissue slices ( 6 mm x 400 $\mu$m sections) from vine leaves by monitoring photosynthesis with an oxygen electrode. Preincubation (60 min) with 10 $\mu$M PA in 50 mM PIPES buffer at pH 5.6 virtually eliminated $O_2$ evolution upon addition of $KHCO_3$ (7.5 mM) and illumination (200 $\mu$E m$^{-2}$ sec$^{-1}$), but the effect could be partially reversed by addition of $CaCl_2$ at low concentration (7.5 mM), or by an upward adjustment in pH (to 7.5).

In sharp contrast to ABA effects on gas exchange where the primary effect is on stomatal behaviour, PA exerts a direct effect on the leaf photosynthetic apparatus. This distinction was inferred from time course studies on gas exchange and confirmed with oxygen electrode measurements. An equivalent concentration of ABA (10 $\mu$M) had no immediate effect on photosynthesis in this leaf slice system. Slight inhibition of photosynthesis was observed in the presence of ABA after 6-8 hours of incubation, but metabolic conversion of the exogenousABA into PA would readily account for this observation. PA was also supplied to excised vine leaves whose stomata were being held open in darkness by $CO_2$-free air; no closing reaction was observed.

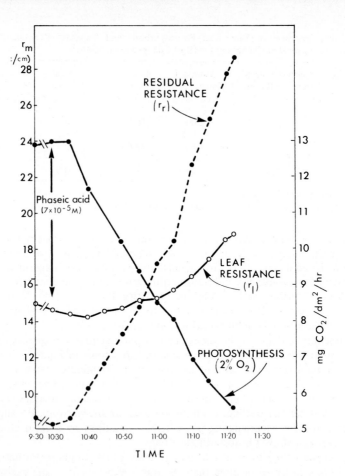

*Fig. 3.* Time course of photosynthetic response (and changes in component resistances) following exogenous supply of phaseic acid to an excised vine leaf.

PA is known to be a product of ABA metabolism in tomatoes (Milborrow, 1970) and the same appears to be true for grapevine (Loveys & Kriedemann, 1974). Label from $2\text{-}^{14}C\text{-ABA}$ supplied to either attached or excised leaves, was incorporated into a fraction that co-chromatographed with a marker spot of PA in a variety of solvent systems. Since PA shows a considerable accumulation after rewatering (Figure 1) or in response to manipulative treatments (Figure 2) we wondered whether an alteration in the metabolism of ABA could account for this increase. Accordingly, $2\text{-}^{14}C\text{-ABA}$ was supplied to leaves showing stomatal responses to manipulative treatments, (treatments 1,2,3 & 4; Figure 2) and its rate of conversion into $^{14}C\text{-PA}$ determined. Activity in the PA fraction showed no increase as a function of treatment, whereas native PA increased some 200% as a result of stem cincturing and fruit removal. Assuming that conversion of exogenously supplied $^{14}C\text{-ABA}$ into PA is a reasonable index of the degradation of native ABA we have to conclude that the abundance of PA in leaves with low rates of photosynthesis and high stomatal resistance is related to metabolic events besides ABA degradation within the lamina.

In photosynthetic terms, the site of action of PA remains obscure. The abbreviated time course of inhibition plus ease of reversibility suggest a photochemical rather than biochemical locus, and some preliminary experiments where PA was tested *in vitro* on crude preparations of RuDP carboxylase, support this view. No reduction in enzyme activity was observed following a simple addition or preincubation with PA (10 $\mu$M at pH 6.5; W.J. Downton, personal communication).

Photosynthetic inhibition following the relief of water stress may well be a consequence of the upsurge in endogenous levels of PA as ABA concentration diminishes (Figure 1), but other forms of internal control (Figure 2 and Table 1) seem more circuitous. Why should fruit removal or stem cincturing, in the absence of any suggestion of moisture stress, lead to an increase in ABA and PA? The present data suggest that fine control over gas exchange can be exerted by the combined action of ABA and PA. These inhibitors would then represent components of a control system modulated by environmental factors and by developing organs and other sites of growth.

# REFERENCES

Bingham, G.E. (1972): Stomatal response in field corn (*Zea mays* L.) and apple (*Malus sylvesterus*), - Ph.D. Thesis, Cornell University, Ithaca, N.Y.

Boyer, J.S. (1971): Non-stomatal inhibition of photosynthesis in sunflower at low leaf water potentials and high light intensities. *Plant Physiol.*, 48: 532-536.

Cummins, W.R., H. Kende & K. Raschke (1971): Specificity and reversibility of the rapid stomatal response to abscisic acid. *Planta* (Berl.), 99: 347-351.

Kanemasu, E.T., G.W. Thurtell & C.B. Tanner (1969): Design, calibration and field use of a stomatal diffusion porometer. *Plant Physiol.*, 44: 881-885.

Kriedemann, P.E. (1971): Photosynthesis and transpiration as a function of gaseous diffusive resistances in orange leaves. *Physiol. Plant.*, 24: 218-225.

Kriedemann, P.E. & R.E. Smart (1971): Effects of irradiance, temperature and leaf water potential on photosynthesis of vine leaves. *Photosynthetica.*, 5: 6-15.

Kriedemann, P.E., B.R. Loveys, G.C. Fuller & A.C. Leopold (1972): Abscisic acid and stomatal regulation. *Plant Physiol.*, 49: 842-847.

Liu, W.T., R. Pool, W. Wenkert & P.E. Kriedemann (1974): Soil-plant water relations in a New York vineyard II. Changes in photosynthesis, stomatal resistance and abscisic acid of *Vitis labruscana* through drought and irrigation cycles. *Physiol. Plant.* (In prep).

Loveys, B.R. & P.E. Kriedemann (1973): Rapid changes in abscisic acid-like inhibitors following alterations in vine leaf water potential. *Physiol. Plant.*, 28: 476-479.

Loveys, B.R. & P.E. Kriedemann (1974): Internal control of stomatal physiology and photosynthesis I. Stomatal regulation and associated changes in endogenous levels of abscisic and phaseic acids. *Aust. J. Plant Physiol.*, 1: 407-415.

Loveys, B.R.; A.C. Leopold & P.E. Kriedemann (1973): Abscisic Acid Metabolism and Stomatal Physiology in *Betula lutea* Following Alteration in Photoperiod. *Ann Bot.*, 38: 85-92.

Milborrow, B.V. (1974): The Metabolism of abscisic acid. *J. Exp. Bot.*, 21: 17-29.

Mullins, M.G. (1966): Test plants for investigations of the physiology of fruiting in *Vitis vinifera* L. *Nature*, 209: 419-420.

Neales, T.F. & L.D. Incoll (1968): The control of leaf photosynthesis rate by the level of assimilate concentration in the leaf: a review of the hypothesis. *The Botanical Review*, 34: 107-125.

Scholander, P.F., H.T. Hammel, E.D. Bradstreet & E.A. Hemmingsen, (1965): Sap pressure in vascular plants. *Science,* 148: 339-346.

Seeley, S.D. (1971): Electron capture gas chromatography of plant hormones with special reference to abscisic acid in apple bud dormancy. Ph.D. Thesis, Cornell University, Ithaca, New York.

Wright, S.T.C. & R.W.P. Hiron (1969): (+) Abscisic acid, the growth inhibitor induced in detached wheat leaves by a period of wilting. *Nature* 224: 719-720.

Wright, S.T.C. & R.W.P. Hiron (1972): The accumulation of abscisic acid in plants during wilting and under other stress conditions. In Carr, D.J. (Ed.): Plant growth substances. Pp 291-299. Springer Verlag, Berlin Heidelberg New York.

# ASPECTS OF [14]C-SUCROSE TRANSLOCATION PROFILES IN *HIBISCUS ESCULENTUS* L. (OKRA)[1]

N.O. ADEDIPE

*Department of Agricultural Biology, University of Ibadan, Ibadan, Nigeria*

*Abstract*

The translocation patterns of [14]C from [14]C-sucrose applied to vegetative okra plants were investigated in relation to duration, leaf position, nitrogen and phosphorus nutrition, and water stress.

When the lowermost, most mature leaf 1 was fed with [14]C-sucrose, distribution was initially at 6 h uniform to the leaves, stem and root; but at 48h, the root accumulated about 60% of the total exported [14]C. The uppermost, expanding leaf 4 distributed a substantial proportion of its minimally exported [14]C initially downward to the root, but progressively upward to the shoot apex. The application of high levels of N and P led to enhanced movement of [14]C out of the fed leaf. While high levels of N stimulated distribution to leaves, high levels of P were inhibitory. A water stress pulse lasting 3 days had no significant effect on the patterns of [14]C translocation but minimally influenced that of [32]P.

It is concluded that in the vegetative okra plant, (a) leaves do not export [14]C assimilates principally in terms of proximity to a given organ, but in response to demands by specific sinks which vary with time; (b) high levels of N and P enhance the movement of [14]C out of the fed leaf, but the effects of N on distribution patterns are generally opposite to those of P; and (c) water stress which does not result in significantly inhibited growth also does not adversely affect [14]C translocation.

## Introduction

The export magnitudes and distribution patterns of [14]C assimilates have been subjects of considerable interest for many years. This has resulted in extensive research with temperate zone cereals like wheat and barley, and many other species, particularly soybean and sugarbeet (Wardlaw, 1968; Zimmerman, 1960). To a lesser degree, some industrial tropical plants, notably sugarcane (Hart et al., 1963) and cotton (Mason & Maskell, 1928) have also been investigated. There are, however, no known reports of tranlocation studies with okra.

The okra plant belongs to the *Malvaceae* family. It produces green fruits which are used as a slimy vegetable or as a relish in many areas of Africa, Asia, West Indies and southern United States. Its essential growth features have been described by McGinty & Barnes (1932). In spite of increasing production the physiology of the okra plant has received little attention. Since the fruit is harvested as early as 6 days following fruit set, there is a delicate balance between the vegetative and the reproductive phases of growth, particularly in relation to the patterns of flowering, fruit set, and fruit maturity. There is therefore the need to define such a relationship in terms of assimilate distribution as it influences dry matter accumulation. This is particularly necessary since it is well recognized that

1) This work was supported by National Research Council of Canada and University of Ibadan Sentae research grants.

the levels of assimilates in leaves determine the rates and magnitudes of photo-synthesis (Neales & Incoll, 1968). Also, in the light of increasing studies of the responses of the okra crop to fertilizers (Ahmed & Tulloch-Reid, 1968; Asif & Greig, 1972), a knowledge of the effects of mineral nutrition and of soil-plant water relations on assimilate transport is needed, since many environmental factors are known to modify translocation patterns in many species (Wardlaw, 1968).

The work reported herein was therefore carried out to obtain information on $^{14}C$ assimilate export and distribution characteristics of leaves of different ages; and to investigate the influence of nitrogen and phosphorus nutrition, as well as water stress.

## Materials and Methods

### Plant Culture

Seeds of okra (*Hibiscus esculentus* L.) cv. 'Perkins Spineless' were sown in flats containing sand. When the primary leaves developed fully, usually 8-10 days, the seedlings were transplanted each into a 15 cm plastic pot, and transferred into growth chambers. They were grown at a day/night temperature regime of $30/25°C$, photoperiod of 12h, light intensity of 16-18 klux, and relative humidity of 80-90%. The plants were watered twice daily and fed with Hoagland solution (Hoagland & Arnon, 1950) on alternate days. In the nitrogen (N) and phosphorus (P) experiments, the nutrient solutions were modified to supply 2, 20, and 200 mg/l, representing low, medium and high levels of N or P.

### Watering Pattern During Soil Water Stress

In the water stress experiment, the plants were grown in a sandy loam soil. The soil characteristics and water contents at the 0.3, 3, and 15 bars of tension have been described in detail (Khatamian et al., 1973).

Watering of plants was controlled by applying known amounts of water following plant removal of water to 15% by weight, using a calibrated resistance — type moisture meter ('Aquaprobe', product of Howard Crane Inc., New York, U.S.A.). Incipient wilting, indicated by drooping of leaf 1, occurred at 12% soil water. The gravimetrically determined soil water contents at wilting and at field capacity were 5 and 30% respectively. The plants were subjected to two treatments, wet and dry, at the 7-leaf stage, 40-43 days after planting. The wet treatment plants were soil-monitored twice daily and were always maintained at field capacity. The dry treatment plants were allowed to remove soil water to 15%, a level just above the incipient wilting point of 12% soil water. The plants removed soil water to this level in about 40 h. About 600 ml of water was required to bring the soil back to field capacity as indicated by the moisture meter calibrations. Two cycles of such a watering pattern, lasting 3 days, were used to subject the plants to water stress. Relative leaf turgidity was determined by the method of Barrs & Weatherley (1962), using 10 leaf discs of 10 mm size punched out of leaf 4.

### $^{14}C$-sucrose Feeding and Radioactivity Counting

Preliminary experiments utilizing a system described earlier (Adedipe & Ormrod,

1974) showed similar translocation patterns of $^{14}C$ from $NaH_2{}^{14}CO_3$-released $^{14}CO_2$ as from $^{14}C$-sucrose. The latter was therefore used to reduce feeding time. When the plants attained pre-determined growth stages, usually the 4-leaf stage when the plants were 30-35 days old, each plant was fed with 0.5 ml solution containing 5 $\mu Ci$ uniformly labelled $^{14}C$-sucrose with a specific activity of 250 mCi/mM. The sucrose was administered as microdroplets to the interveinal areas of the adaxial leaf surface. At predetermined periods the plants were harvested and separated into parts. Each part was homogenized in ethanol. A 1.0 ml aliquot of each homogenate was plated on aluminium planchettes and dried on a warm plate or with a lamp. Radioactivity was determined with a Nuclear Chicago gas flow proportional counter.

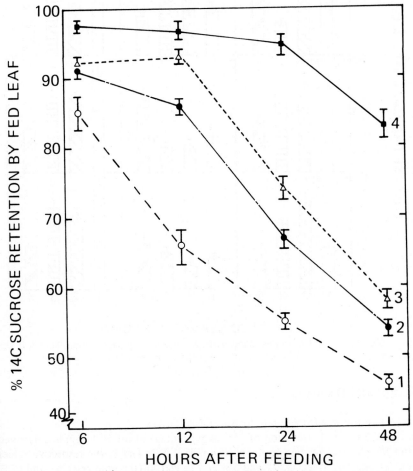

*Fig. 1.* Time course of $^{14}C$ retention by successive leaves (1, 2, 3, 4) of the okra plant. Each leaf was fed with 5 $\mu Ci$ $^{14}C$-sucrose. Each point in the mean of 6 replicates $\pm$ the standard error.

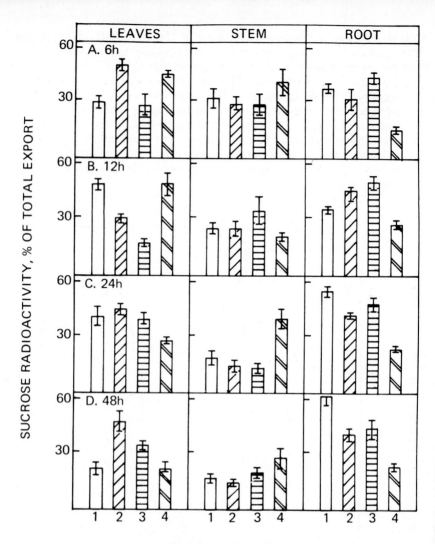

*Fig. 2.* Distribution of [14]C to okra plant parts as influenced by leaf position. Each leaf was fed with 5 $\mu$Ci [14]C sucrose. Each bar is the mean of 6 replicates $\pm$ the standard error.

## Results and Discussion

### [14]C Retention and Distribution by Successive Leaves

In the oldest leaf 1, retention of [14]C as percentage of total [14]C applied declined from 85 6h after feeding to 46 at 48h (Fig. 1.). In leaf 2, the respective values were 91 and 54; in leaf 3, 93 and 58; and in leaf 4, (the youngest), 97 and 83%. While retention declined between 6 and 12h in leaves 1 and 2, there were no changes in leaves 3 and 4. There was no change in retention by leaf 4 even at 24h.

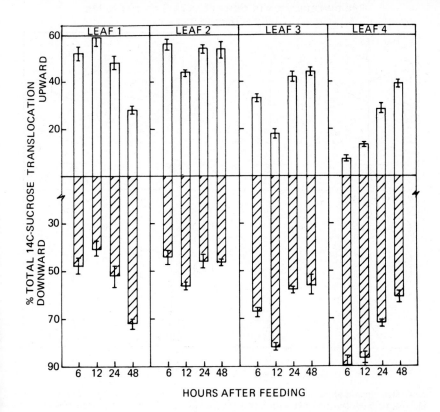

*Fig. 3.* Upward and downward transport of [14]C by successive leaves of okra. Each plant was fed with 5 μCi [14]C-sucrose. Each bar is the mean of 6 replicates ± the standard error.

Six hours after feeding, leaf 1 distributed the exported [14]C uniformly to the leaves, stem and root (Fig. 2A). With time, less radioactivity was recovered in the stem, while the root accumulated as much as 56 and 60% of the total exported [14]C at 24 and 48 h respectively. Initially at 6 h, leaf 4 distributed about 40% each to the leaves and the stem, and only 15% to the root. At 48 h, distribution to the root increased to 21% (Fig. 2D) while that to the leaves decreased to 21%.

While upward and downward translocation by leaf 1 was even at 6, 12 and 24 h, leaf 4 favoured upward transport with time (Fig. 3). The gross upward and downward translocation patterns were also reflected in the distribution to individual leaves and to individual stem internodes (Fig. 4).

*Effects of N and P on [14]C Retention and Distribution*
The 20 mg/l N, when compared with the 2 mg/l treatment, had no significant effect on plant weight (Fig. 5A), but the 200 mg/l treatment increased the weight of the leaves, root and stem. With P application, only the 20 mg/l level increased leaf weight.

The 20 and 200 mg/l N decreased retention of [14]C by the fed leaves (the

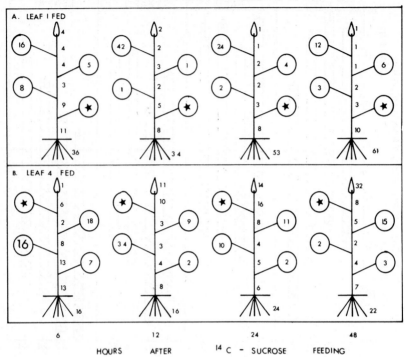

*Fig. 4.* Diagramatic representation of the distribution of $^{14}$C to leaves and stem internodes of okra plants. Each leaf was fed with 5 $\mu$Ci $^{14}$C-sucrose. Each value is the mean of 6 replicates $\pm$ the standard error.

youngest and the oldest) 48h after feeding $^{14}$C-sucrose (Fig. 5B). On the other hand, P at the 200 mg/l level increased retention by leaf 4 while it decreased retention by leaf 1 during the same period.

The high N level of 200 mg/l enhanced distribution by leaf 1 to the other leaves (Fig. 6A), accompanied by decreases in the distribution to the stem (Fig. 6B) and to the root (Fig. 6C). The high N level had no significant effect on distribution by leaf 4 to the stem and the root. On the other hand, the high P level decreased distribution by leaf 4 to the other leaves (Fig. 6A) while it increased that to the stem (Fig. 6B), with no significant effect on that to the root (Fig. 6C).

When leaf 1 was fed, N levels of 20 and 200 mg/l increased upward translocation (Table 1), an increase from 14% at 2 mg/l to 47% at the 200 mg/l N. The upward translocation enhancement by high levels of N was also reflected in the distribution pattern to individual leaves. When leaf 4 was fed, the 20 mg/l N decreased upward translocation. The medium and high levels of P decreased upward translocation by leaf 1 (Table 1). With leaf 4 feeding, upward translocation was increased by the 20 mg/l P level.

## Effects of Water Stress on $^{14}$C and $^{32}$P Distribution

When plants at the 7-leaf stage were subjected to wet and dry water regimes the

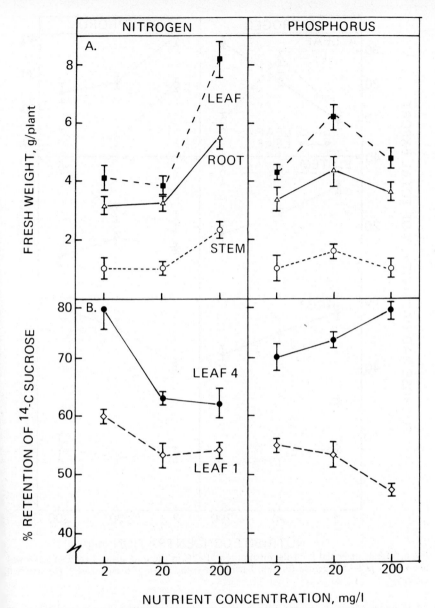

*Fig. 5.* Effects of N and P on (A) growth responses, and (B) $^{14}C$ retention. Leaf 1 or leaf 4 was fed with 5 $\mu$Ci $^{14}C$ for 48 h. Each point is the mean of 6 replicates $\pm$ the standard error.

relative leaf turgidities were 98 and 90% respectively. Water stress had no significant effect on plant weight or weights of individual tissues (Table 2). Water stress also had no significant effect on the distribution of $^{14}C$ by leaf 4, 6 and 24h after feeding (Fig. 7). When $^{32}P$ was fed, water stress increased the distribution of to other leaves (Fig. 7A), while it decreased that to the stem (Fig. 7B) at 6 h. At

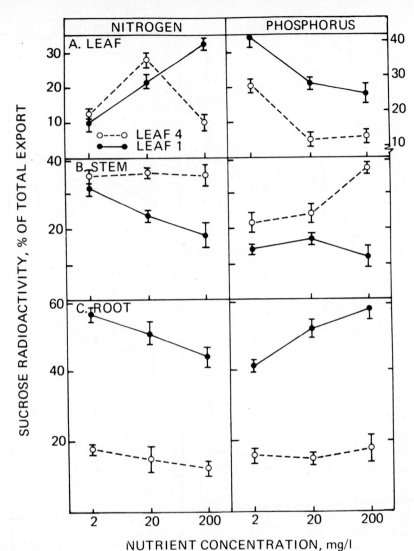

Fig. 6. Effects of N and P on the distribution of $^{14}C$ in the okra plant. Leaf 1 or leaf 4 was fed with 5 $\mu$Ci $^{14}C$-sucrose for 48h. Each point is the mean of 6 replicates $\pm$ the standard error.

24 h, water stress decreased distribution to the other leaves, but increased that to the root. Water stress generally favoured upward transport of P at 6 h, but downward transport at 24h (Table 3).

Leaf 1 exported $^{14}C$ most rapidly between 6 and 12 h during which there was a 20% export differential, compared with the period between 24 and 48h when the differential was only 10% (Fig. 1). In leaves 2 and 3 the period of most rapid export was between 12 and 24 h, while in leaf 4 there was a shift to the period between 24 and 48 h. Rapidity of transport is therefore directly related to lea

*Table 1.* Effects of N and P on the percentage upward and downward translocation of $^{14}C$ in okra plants.[a]

| Fed Leaf | Translocation direction | Nitrogen, mg/l | | | Phosphorus, mg/l | | |
|---|---|---|---|---|---|---|---|
| | | 2 | 20 | 200 | 2 | 20 | 200 |
| 1 | Upward | 14± 2 | 30± 3 | 47± 3 | 48± 2 | 38± 1 | 32± 6 |
| | Downward | 86± 4 | 70± 2 | 53± 3 | 52± 3 | 62± 1 | 68± 5 |
| 4 | Upward | 37± 3 | 23± 2 | 44± 5 | 38± 3 | 50± 2 | 37± 4 |
| | Downward | 63± 1 | 77± 6 | 56± 4 | 62± 3 | 50± 4 | 63± 6 |

a) Leaf 1 or leaf 4 was fed with 5 $\mu$Ci $^{14}C$-sucrose for 48h. Each value is the mean of 6 replicates $\pm$ the standard error.

*Table 2.* Dry weights of plant parts at two water regimes.[a]

| Treat-ment | g/plant | | | |
|---|---|---|---|---|
| | Leaves | Stem | Root | Total |
| Wet | 2.95 ±0.18 | 1.73 ±0.06 | 1.78 ±0.12 | 6.46 ±0.61 |
| Dry | 2.75 ±0.21 | 1.79 ±0.09 | 1.52 ±0.15 | 6.06 ±0.38 |

a) Each value is the mean of 6 replicates $\pm$ the standard error.

*Table 3.* Effects of water regimes on percentage upward and downward translocation of $^{14}C$ and $^{32}P$. [a]

| Time after feeding, hr | Translocation Direction | $^{14}C$ | | $^{32}P$ | |
|---|---|---|---|---|---|
| | | Wet | Dry | Wet | Dry |
| 6 | Upward | 37 ±1 | 32 ±2 | 26 ±1 | 45 ±2 |
| | Downward | 63 ±2 | 68 ±3 | 74 ±1 | 55 ±3 |
| 24 | Upward | 45 ±3 | 42 ±5 | 56 ±2 | 46 ±2 |
| | Downward | 55 ±2 | 58 ±3 | 44 ±1 | 54 ±3 |

a) 5 $\mu$Ci of $^{14}C$-sucrose or $NaH_2$ $^{32}PO_4$ was fed to leaf 4 when plants were at the 7-leaf stage. Each value is the mean of 8 replicates $\pm$ the standard error.

position and/or age. These patterns are in general agreement with those for most species (Thrower, 1962; Wardlaw, 1968).

In the okra plant, distribution of $^{14}C$ assimilates by leaf 1 was initially uniform to the leaves, stem and root (Fig. 2). With time, there was a shift in favour of the root. This is an indication of initial import by other leaves followed by a transfer to the root. Proximity to the root was therefore not a controlling factor. This is not in agreement with the general patterns reported for most species (Wardlaw, 1968). It is, however, similar to the unusual pattern of distribution in the tomato (Khan & Sagar, 1966). This pattern is further supported by data on upward and

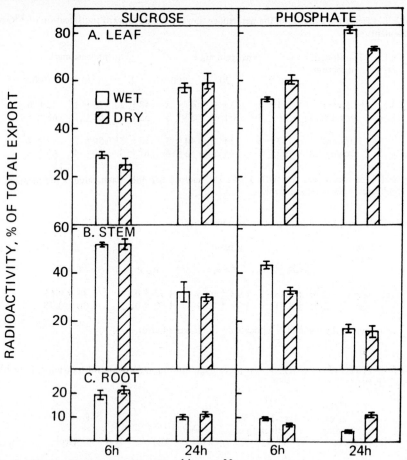

*Fig. 7.* Effects of soil water regimes on $^{14}C$ and $^{32}P$ translocation in the okra plant at the 7-leaf stage. Leaf 4 was fed with 5 $\mu$Ci $^{14}C$-sucrose or NaH$_2$$^{32}$PO$_4$. Each bar is the mean of 8 replicates $\pm$ the standard error.

downward translocation irrespective of the tissue (Fig. 3), which showed an initial even distribution of 50% upward and downward; but at 48 h, there was a 70% downward translocation. Wardlaw (1968) suggested that the upward movement from the lower leaves in the tomato (Khan & Sagar, 1966) could be related to redistribution of assimilates that have moved down the root system and subsequently back into the shoot through the xylem. This could not be determined from the work of Khan & Sagar (1966) since translocation was for a single period of 24 h. It is clear from the present study, however, that this is not the case since distribution to the root by leaf 1 was 36% at 6h, a value close to those of the leaves and the stem. Also at this early period, when individual leaves are considered leaf 4 imported the greatest percentage (16) of $^{14}C$ (Fig. 4). Leaf 2, immediately above the fed leaf 1, imported only about half as much $^{14}C$. The stem also imported a considerable percentage of the total $^{14}C$, an additional similar pattern to that in the tomato (Khan & Sagar, 1966). This suggests that the okra stem could be acting as a storage organ for assimilates for subsequent supplies to the leaves and the root, depending on temporal demands.

It is interesting to also note that the young leaf 4 initially exported a substantial percentage of its minimally exported $^{14}C$ downward (Fig. 3), but subsequently there was progressive upward translocation, increasing from 8% at 6 to 40% at 48 h. This is an indication of an initial downward translocation mainly to the leaves and the stem, which then pass the assimilates upward to the shoot apex. The initial downward movement was, again, reflected in transport to individual leaves (Fig. 4), which show about equal magnitudes of import by leaves 2 and 3 at 6 h, but at 12 h the lower leaf 2 imported as much as 34%, compared with 9% in the upper leaf 3. The stem internodes also exhibited the initial downward movement since internodes 1 and 2 each contained 13% at 6 h, but decreased with time in favour of the uppermost internode 5 just above the fed leaf.

The growth response of okra plants to high levels of applied N indicates a high growth requirement for N (Fig. 5A). Conversely, the lack of growth response to applied P is an indication of low P requirement. These patterns of growth response to applied N and P are generally similar to those of the fertilizer studies of Ahmed & Tulloch-Reid (1968) and Asif & Greig (1972). Both in the mature leaf 1 and in the young leaf 4 (Fig. 5B), decreases of $^{14}C$ retention due to high levels of N are in general agreement with the reports for potato (Anisimov, 1968) and taro (Izawa & Okamoto, 1961). Distribution was also influenced. For example, the 200 mg/l N level increased distribution by leaf 1 to other leaves by 210% (Fig. 6A), which was accompanied by decreases to the stem (Fig. 6B) and the root (Fig. 6C). The distribution of $^{14}C$ by leaf 4 was generally not influenced by high N level, probably because the young leaf is limited in its capacity to utilize root-applied N. The effects of P were generally opposite to those of N. The high P level of 200 mg/l increased $^{14}C$ retention by the fed young leaf 4 (Fig. 5B), thereby decreasing the amounts absorbed by the plant. But in the mature leaf 1, $^{14}C$ retention was decreased, implying increased absorption by the plant. This enhancement of absorption by leaf 1, due to high P level, could be a reflection of reduced demand for $^{14}C$ as a result of an induced efficient phosphorylation of sugar, and production of nucleotide phosphates (Adedipe & Fletcher, 1970 b; Bieleski, 1966). The depression of $^{14}C$ absorption by the young leaf 4, on the other hand, is an indication that the demand for external supply of P is low in an actively metabolizing leaf, so that the application of high levels of P would make for competitive inhibition of other aspects of metabolism. Anisimov (1966) also found that the application of P at high levels of N decreased translocation.

The influence of water stress on organic translocation is very much a subject of controversy (Hartt, 1967; Eaton & Ergle, 1948; Wardlaw, 1967). The controversy stems mainly from the fact that several reports relating water stress to the inhibition of translocation do not recognize the indirect effects of growth response on assimilate distribution. In the okra plant, relative turgidity of the leaves of water-stressed plants was 8% less than that of optimally-watered plants. When plants were subjected to a 2-cycle pulse of water stress, therefore, the leaves reflected differences in water contents without visible wilting. This 2-cycle stress led to no significant difference in plant weight or weight of different plant parts (Table 2). Results obtained from the present study are therefore a reflection of the direct effects of water stress, rather than indirect effects of water stress on plant growth.

That the export of both $^{14}C$ and $^{32}P$ from the fed leaf was not significantly influenced is an indication of marked drought-tolerance by the okra plant. When water stress is not protracted to bring about differences in growth, it also does not influence the translocation of $^{14}C$ both in term of distribution (to the leaves, stem and root), and direction (upward or downward). This is in agreement with the report by Wardlaw (1967) that when water stress acted directly on the leaf rather than indirectly through effects on growth, the velocity of movement through the conducting tissues was not sensitive to water stress. McWilliam (1968) has also reported that assimilates are readily moved from stem to root and buds during water stress-induced dormancy in perennial grasses. Although the inhibition of organic translocation has been ascribed to increased diffusion pressure deficit (Wiebe & Wihrheim, 1962) and decreased relative leaf turgidity (Plaut & Reinhold, 1965), differences in translocation patterns may have resulted from depressed growth due to diffusion pressure deficit differences of up to 12 atmosphere, and a relative leaf turgidity decrease of 28%. While water stress had no significant effect on the translocation of $^{14}C$, $^{32}P$ translocation was influenced. An initial decrease in $^{32}P$ distribution to the root, due to water stress, was followed at 24 h by increased distribution. This observation is in conflict with that of Wilson & McKell (1961) that water stress had no significant effect on translocation to the root of sunflower. The general effects of water stress on the distribution of $^{32}P$, compared with lack of influence on that of $^{14}C$, is probably related to the erratic translocation patterns of $^{32}P$ which result in or from rapid recycling (Adedipe & Fletcher, 1970a; Biddulph et al., 1958).

It is concluded that, in the vegetative okra plant, leaves do not export $^{14}C$-assimilates principally in terms of proximity to a given organ, but in response to demands by specific sinks which vary with time. High levels of applied N and P stimulate $^{14}C$ transport out of the fed leaf, but the effects of N on $^{14}C$ distribution are generally opposite to those of P. Soil water stress which does not result in a deleterious effect on growth also does not influence the translocation of $^{14}C$, but minimally influences that of $^{32}P$.

# REFERENCES

Adedipe, N.O. & R.A. Fletcher (1970a): Benzyladenine-directed transport of Carbon-14 and Phosphorus-32 in senescing bean plants. *J. Exptl. Bot.,* 21: 968-974.
Adedipe, N.O. & R.A. Fletcher (1970b): Retardation of bean leaf senescence by benzyladenine and its influence on phosphate metabolism. *Plant Physiol.,* 46: 614-617.
Adedipe, N.O. & D.P. Ormrod (1974): Effects of CCC and Phosfon on translocation patterns of $^{14}C$-sucrose in *Pisum sativum* L. *Z. Pflanzenphysiol.,* 71: 384-390.
Ahmed, N. & L.T. Tulloch-Reid (1968): Effects of fertilizer nitrogen, phosphorus, potassium and magnesium on yield and nutrient content of okra (*Hibiscus esculentus* L.). *Agron. J.,* 60: 353-355.
Anisimov, A.A. (1966): On the mechanism of the action of phosphorus on carbohydrate movement. *Soviet Plant Physiol.,* 13: 59-63.
Anisimov, A.A. (1968): Mechanisms by which nitrogen nutrition affects assimilate movement in potato. *Soviet Plant Physiol.,* 15: 8-12.
Asif, M.I. & J.K. Greig (1972): Effect of N, P and K fertilization on fruit yield, macro- and micronutrient levels, and nitrate accumulation in okra (*Abelmoschus esculentus* L. Moench). *J. Amer. Soc. Hort. Sci.,* 97: 440-442.

Barrs, H.D. & P.E. Weatherley (1962): A re-examination of the relative turgidity technique for estimating water deficits in leaves. *Aust. J. Biol. Sci.,* 15: 413-428.

Biddulph, O., S. Biddulph, R. Cory & H. Koontz (1958): Circulation patterns for phosphorus, sulfur and calcium in the bean plant. *Plant Physiol.,* 33: 293-300.

Bieleski, R.L. (1966): Accumulation of phosphorus, sulfate, and sucrose by excised phloem tissues. *Plant Physiol.,* 41: 447-454.

Eaton, F.M. & D.R. Ergle (1948): Carbohydrate accumulation in the cotton plant at low moisture levels. *Plant Physiol.,* 23: 169-187.

Hartt, C.E. (1967): Effect of moisture supply upon translocation and storage of $^{14}C$ in sugarcane. *Plant Physiol.,* 42: 338-346.

Hoagland, D.R. & D.I. Arnon (1950): Water culture method of growing plants without soil. *Calif. Agr. Expt. Stat. Circ. 347,* Univ. of Calif., Berkeley, U.S.A.

Izawa, G. & S. Okamoto (1961): Effect of mineral nutrition on content of organic constituents in taro plants during growth. *Soil and Plant Food,* 6: 127-135.

Khan, A.A. & G.R. Sagar (1966): Distribution of $^{14}C$-labelled products of photosynthesis during the commercial life of the tomato crop. *Ann. Bot. (N.S.),* 30: 727-743.

Khatamian, H., N.O. Adedipe & D.P. Ormrod (1973): Soil-plant-water aspects of ozone phytotoxicity in tomato plants. *Plant and Soil,* 38: 531-541.

Mason, T.G. & E.J. Maskell (1928): Studies on the transport of carbohydrates in the cotton plant. II. The factors determining the rate and the direction of movement of sugars. *Ann. Bot.,* 42: 571-636.

McGinty, R.A. & W.C. Barnes (1932): Observations on flower bud and pod development in okra. *Proc. Amer. Soc. Hort. Sci.,* 29: 509-513.

McWilliam, J.R. (1968): The nature of the perennial response in Mediterranean grasses. I. Senescence, summer dormancy and survival in *Phalaris. Aust. J. Agr. Res.,* 19: 397-409.

Neales, I.F. & Incoll, I.D. (1968): The control of leaf photosynthesis rate by the level of assimilate concentration in the leaf. A review of the hypothesis. *Bot. Rev.,* 34: 107-125.

Plaut, Z. & L. Reinhold (1965): The effect of water stress on $^{14}C$ sucrose transport in bean plants. *Aust. J. Biol. Sci.,* 18: 1143-1155.

Thrower, S.L. (1962): Translocation of labelled assimilates in soybean. II. The patterns of translocation in intact and defoliated plants. *Aust. J. Biol. Sci.,* 15: 629-649.

Wardlaw, I.F. (1967): The effect of water stress on translocation in relation to photosynthesis and growth. I. Effect during grain development in wheat. *Aust. J. Biol. Sci.,* 20: 25-39.

Wardlaw, I.F. (1968): The control and pattern of movement of carbohydrates in plants. *Bot. Rev.,* 34: 79-105.

Wiebe, H.H. & S.E. Wihrheim (1962): The influence of internal moisture deficit on translocation. *Plant Physiol.,* 37: suppl. L.

Wilson, A.M. & C.M. McKell (1961): Effects of soil moisture stress on absorption and translocation of phosphorus applied to leaves of sunflower. *Plant Physiol.,* 36: 762-765.

Zimmerman, M.H. (1960): Transport in the phloem. *Annu. Rev. Plant Physiol.,* 11: 167-190.

# ENVIRONMENTAL AND BIOLOGICAL CONTROL OF PHOTO-SYNTHESIS: GENERAL ASSESSMENT

I. ZELITCH

*Department of Biochemistry, The Connecticut Agricultural Experiment Station, New Haven, Conn., U.S.A. 06504*

## Introduction

I wish to thank the organizers, especially Dr. R. Marcelle, for all of their efforts in arranging this Symposium and for the hospitality during our stay. I think we are all honored to be able to participate indirectly in the life of this new and unusual University that is growing visibly around us even while we meet here.

The participants told us about their latest work and thus we learned about current research in many aspects of the environmental and biological control of photosynthesis. By subjecting their creative efforts to scrutiny and criticism-these scientists have helped to hasten progress in research on photosynthesis, the process primarily responsible for plant productivity.

Newspaper headlines during these meetings remind us constantly of the existence of a worldwide shortage of food, fiber, and wood, and there are strong indications that these shortages will increase in the future. Large increments in plant productivity are therefore needed to alleviate the widespread suffering that will continue to accompany this insufficient supply.

In many crops it is already clear that even the best technology and genetic engineering no longer produce greater yields. More knowledge is needed to permit further increases in net photosynthesis to be achieved, and presumably such characteristics will have to be introduced into cultivated plants by the use of new and innovative methods of plant breeding.

The participants come here from many countries and represent a variety of scientific disciplines and viewpoints. It has been an exciting week discussing and debating many important questions, and I appreciate the opportunity to offer a perspective about some of the subjects aired during this Symposium. It would not be possible to discuss all of the presentations, hence I have chosen to comment on only several of these diverse topics from the standpoint of possible research approaches to increasing net photosynthesis.

## The Relation between Net Photosynthetic $CO_2$ Fixation and Plant Productivity

Since 90 to 95% of the dry weight of plants is derived from photosynthetic $CO_2$ assimilation, it is not surprising that plant productivity is directly related to total net photosynthesis. Certain efficient photosynthetic species have come to be known as $C_4$-species and have slow rates of photorespiration. They fix $CO_2$ at rates two to three times faster than the less efficient, $C_3$, species at $25°$ to $35°C$

and high light intensities (Zelitch, 1971). Maximal crop growth rates are also usually about twice as great for the efficient species such as maize, sorghum, and sugarcane than for most other crop species. A few exceptions, such as sunflower and cattail (*Typha latifolia*), can be found to the rule that the $C_3$-species are less productive and have lower rates of net photosynthesis (Zelitch, 1971, p. 244; Gifford, 1974). A more detailed study of these few examples might allow a great deal to be learned about how some species overcome the handicap of a high rate of photorespiration.

When net photosynthesis and productivity are compared in the environments in which different crop species are commonly grown, the efficient photosynthetic species are much more productive. Table 1 shows that the average crop growth rate in the U.S.A. for maize silage, sorghum silage, and sugarcane is about twice as great as for spinach, tobacco, and hay. Thus these differences in plant productivity cannot be attributed to differences in the length of the growing season as suggested by Gifford (1974), but are related primarily to the great differences in rates of net photosynthesis normally encountered between these species.

Net $CO_2$ fixation is equal to the gross photosynthetic rate less the losses by respiration in the light and in darkness. Therefore net $CO_2$ uptake can be increased by improving the biochemical rate of carboxylation, the photosynthetic electron transport, or related processes, or by diminsihing the losses of carbon by respiration.

As indicated above, one can be confident that there is considerable room for increasing the productivity at least in the less efficient photosynthetic species so they become equal to that commonly found only in the $C_4$-species. Diminishing the stomatal and the internal physical and biochemical diffusive resistances that slow the flux of $CO_2$ from the atmosphere to the sites of fixation within the chloroplast would certainly increase net photosynthesis. As evidence of the importance of these internal diffusive resistances, which were often referred to at these meetings, it may be recalled that Hesketh (1963) showed that at higher than normal $CO_2$ concentrations and high irradiance, differences in net photosynthesis between maize and tobacco tended to disappear. This suggests that gross photosynthesis is not very different between these species, but that the internal diffusive resistances, including photorespiration, vary greatly in normal air.

## Some Speculations about Increasing Net Photosynthesis

It was pointed out at these meetings that the diffusive resistance of $CO_2$ into the chloroplast is probably larger than the resistance to enzymatic carboxylation. It seems likely that much of this transport resistance occurs at or near the envelope membrane that surrounds the chloroplast, and its magnitude is undoubtedly controlled by physical as well as biochemical factors. The effects of water deficit in decreasing photosynthesis, in addition to stomatal closure, may be due to an increase in the internal physical diffusive resistance. But the movement of substances into and out of chloroplasts also has an important indirect effect on the rate of $CO_2$ fixation.

Metabolites differ greatly in their ability to move through the spinach chloroplast envelope membranes (Table 2), and this undoubtedly will influence the pool sizes of different substances in the chloroplast. The concentration of various metabolites has been shown to regulate the activity of purified ribulose-1,5-di-

Table 1. Average crop growth rate of several species[a]

| Crop | Average market yield (lbs fresh weight/acre) | Estimated dry weight (lbs/acre) | Estimated growing season (weeks) | Average yield (lbs dry weight acre$^{-1}$ week$^{-1}$) | Average crop growth rate (gm dry weight m$^{-2}$ week$^{-1}$) |
|---|---|---|---|---|---|
| Maize silage | 23,600 | 7,080 | 17 | 417 | 47 |
| Sorghum | 21,600 | 6,480 | 17 | 381 | 43 |
| Sugarcane (cane) | 54,000 | 16,200 | 36 | 450 | 50 |
| Spinach | 5,800 | 580 | 5 | 116 | 13 |
| Tobacco (leaf) | 1,945 | 3,140[b] | 14 | 224 | 25 |
| Hay | 4,000 | 3,600 | 20 | 180 | 20 |

a) Data from Agricultural Statistics, 1969. U.S. Dept. of Agriculture. Taken from Table 9.1 in Zelitch (1971).
b) For tobacco, a yield of stalk was added equal to 90% of the yield of leaf, to give a total of 3696 lbs fresh weight per acre, or 3140 lbs dry weight.

phosphate carboxylase. For example, fructose-6-phosphate greatly stimulated the carboxylase while fructose-1,6-diphosphate was inhibitory (Buchanan, Schürmann & Kalberer, 1971). It was found that 6-phosphogluconate strongly inhibited the carboxylase at saturating bicarbonate concentrations (Chu & Bassham, 1972), but 6-phosphogluconate was an activator of the carboxylase at low levels of bicarbonate (Buchanan & Schürmann, 1973). This indicates how readily intermediates of the Calvin cycle might regulate $CO_2$ fixation, and suggests that changing the properties of the chloroplast envelope membrane could have important indirect effects on carboxylase activity.

*Table.2.* Rate of transport across the chloroplast envelope membrane by various substances[a]

| Fast Transport | Slow Transport |
| --- | --- |
| $HCO_3^-$ | Ribulose diphosphate |
| 3-Phosphoglycerate | Sedoheptulose diphosphate |
| Pentose monophosphates | Phosphoenolpyruvate |
| Triose phosphates | Fructose-6-phosphate |
| Orthophosphate | Glucose-6-phosphate |
| Malate | 6-Phosphogluconate |
| Succinate | Sucrose[b] |
| Glycolate | Glucose |
| Glycine | Fructose |
| Serine | ATP, ADP |
| Alanine | NADP, NAD |
| Aspartate | Pyrophosphate |
| Glutamate | $H^+$ |

a) Compiled from Chu & Bassham (1972); Heber (1974); Heldt & Rapley (1970); Heldt & Sauer (1971); Kelly & Gibbs (1973); Poincelot (1974); Schwenn, Lilley & Walker (1973); Walker (1973); Walker, Kosciukiewicz & Case (1973); Werdan & Heldt (1972).
b) Totally impermeable in intact chloroplasts.

This would seem to be a promising area of future research, and a beginning has already been made by Poincelot (1973) and Douce, Holtz & Benson (1973) who have described some of the properties of isolated chloroplast envelope membranes of spinach. Poincelot (1974) has also presented some of his initial studies on the penetration of various metabolites, including $HCO_3^-$, by isolated envelope membranes.

Evidence has been presented at these meetings that the accummulation of starch in soybean leaves and the failure of a rapid translocation of photosynthate is associated with decreases in net photosynthesis. Such results have in the past been attributed to the size of 'sinks', a concept which further describes these inhibitions of net photosynthesis but does nothing about explaining them. Perhaps in some instances, at least, the accumulation of metabolites within the chloroplast might decrease $CO_2$ fixation. This might occur by regulatory actions on the carboxylase enzyme as was suggested above, even before large amounts of carbohydrate accumulation can be detected.

There are some indications that the rate of photosynthetic electron transport in chloroplasts more likely limits $CO_2$ uptake at high irradiance than does ribulose-1,5-diphosphate carboxylase activity. How electron transport might be increased represents a challenging area of research about which little is now known.

The interesting point has been made at this Symposium that a study of wild plants that carry out photosynthesis effectively at extremes of temperature might teach us how these adaptations could be applied to crop species. Thus *Tidestroma* can grow in the desert even at $50°C$, but such species that perform well at high temperatures are inefficient in their net photosynthesis at $15°C$. Similar examples of increasing net photosynthesis up to about $35°C$, and poor rates at $15°C$ or lower, have been described for other cultivated $C_4$-species such as maize (Zelitch, 1971, p. 252). I believe that photosynthetically efficient species become relatively more efficient at higher temperatures because they do not have a greatly increasing rate of photorespiration (with its high $Q_{10}$), which normally masks the increasing gross photosynthesis that also occurs in $C_3$-species at higher temperatures (Zelitch, 1966). The reason why net photosynthesis is slow in $C_4$-plants at lower temperatures is not understood at present, and the indications offered at this meeting that the properties of membranes change greatly in response to alterations in temperature certainly deserve more study.

## Regulation of Dark Respiration

We have heard at these meetings that $CO_2$ enrichment of the atmosphere in a closed system, such as in a greenhouse, decreases the dark repiration, and also that there is relatively less dark respiration in 'shade' plants than in 'sun' plants. These examples demonstrate that dark respiration cannot be a fixed ratio of net photosynthesis, and suggest that dark respiration might be controlled and that some of it is not essential. I know of no compeling evidence that dark respiration does not occur rapidly in illuminated photosynthetic tissues, while there is ample evidence in the literature (Zelitch, 1971) that the tricarboxylic acid cycle functions rapidly during photosynthesis. Thus I assume that 'dark' respiration occurs equally well in the light in photosynthetic tissues.

The dark respiration takes place primarily, but not exclusively, in the mitochondria, and ATP is produced mainly by oxidative phosphorylation as a result of this respiration. It may be assumed that ATP production often limits growth in plants as it does in other living systems, but other products are also synthesized because of dark respiration and these substances are used in a variety of essential biosynthetic pathways.

The rate of dark respiration is usually about 10% of the rate of $CO_2$ fixation in bright light. Lower leaves are shaded and often have slow rates of photosynthesis, and respiration also occurs in stems, roots, and fruits that fix little or no $CO_2$. Thus the quantity of gross photosynthesis lost by dark respiration is considerable and varies from 29% to 71% in the examples shown in Table 3. It is therefore important to determine how much of the carbon lost by dark respiration is coupled to useful synthetic and growth processes and what part might be eliminated and bring about productivity gain.

A clear example of a wasteful role of dark respiration is shown in Table 4, in which net $CO_2$ exchange and increment dry weight increases were measured in two inbred maize varieties grown in a growth chamber. Net photosynthesis was similar in both varieties, but one variety showed about 50% faster growth which could be accounted for by its slower rate of dark respiration.

There are several possible biochemical explanations for a wasteful dark respiration. First, some dark respiration probably occurs outside the mitochrondria and is thus not coupled to rapid ATP production. Second, oxidative phosphorylation may not always be tightly coupled to respiration in the mitochrondria. Finally, it seems likely that some portion of mitochondrial respiration occurs by the inefficiently phosphorylating alternate pathway of electron transport that occurs in fungi (especially in 'poky', or slow growing, *Neurospora* mutants) and higher plants.

*Table 3.* Minimal quantity of gross photosynthesis lost by dark respiration (Taken from a larger list of references by Zelitch (1971), Table 9.3)[a]

| Species | Ratio of total dark respiration to gross photosynthesis % | Reference |
|---|---|---|
| Maize (upper leaves) | 29 | Lemon (1969) |
| Wheat | 48 | King & Evans (1967) |
| Alfalfa | 70 | King & Evans (1967) |
| Rice (different times during season) | 52, 71 | Tanaka et al. (1966) |

a) Gross photosynthetic assimilation was assumed to equal total net photosynthesis plus dark respiration in the light, but any losses resulting from photorespiration were neglected in all of these examples.

*Table 4.* Effect of dark respiration on growth of two maize varieties[a]

| Variety | Average net photosynthesis | Leaf dark respiration | Growth 20-30 days | Gross photosynthesis lost by dark respiration |
|---|---|---|---|---|
| | $mg\ CO_2/dm^2$ per hour | | g/plant per 10 days | % |
| Pa 83 | 14 | $1.2 \pm 0.1$ | $14.0 \pm 0.6$ | 26 |
| Wf 9 | 13 | $1.7 \pm 0.1$ | $9.4 \pm 0.8$ | 33 |

a) From Heichel (1971). The plants were raised in a growth chamber maintained at 12 hours light and 12 hours darkness at 25°C.

This alternate pathway (Figure 1) produces only one-third as much ATP for each pair of hydrogens oxidized as does the more conventional pathway (Schonbaum, Bonner & Storey, 1971; Lambowitz & Slayman, 1971; Lombowitz et al., 1972). The alternate pathway is insensitive to antimycin and cyanide, but it is specifically inhibited by salicylhydroxamic acid which has no effect on the conventional pathway. In isolated mitochondria from a number of plant species and tissues, under conditions of rapid respiration and phosphorylation, the alternate pathway contributed 1.0% to 100% of the respiration, with a value of 15% to 20% being encountered most often (Bahr & Bonner, 1973). These workers did not

256

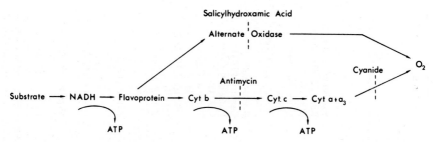

*Fig. 1.* The normal and alternate pathway of electron flow in mitochondria (Lambowitz & Slayman, 1971).

investigate mitochondria isolated from leaves, but it is well known that 50% or more of the dark respiration in leaves is cyanide-insensitive (Bonner & Wildman, 1946). Thus, eliminating the alternate pathway where it occurs, perhaps genetically, would be expected to result in greater productivity.

## The Magnitude of Photorespiration

It has been stated at this Symposium that photorespiratory $CO_2$ accounts for about 15% of net $CO_2$ fixation during photosynthesis in the $C_3$-species. This conclusion was based on the assumption that the properties of isolated ribulose-1,5-diphosphate carboxylase are the same as those of the enzyme *in vivo*, and that all of the glycolate is synthesized by the oxygenase activity that accompanies the carboxylase and gives rise to phosphoglycolate as one of the products. There are serious grounds for believing that both of these assumptions are questionable (Zelitch, 1975). Moreover, there is ample evidence in the literature (Table 5) to show that minimal rates of photorespiratory $CO_2$ evolution in many species accounts for at least 50% of net photosynthesis (not 15%). The results summarized in Table 5 were obtained by several different assays on soybean, sunflower, sugar beet, and tobacco, and at reasonable temperatures, 24° to 33.5° C.

Thus, a quantitative limitation can be set for biochemical reactions that are concerned with photorespiration in less efficient photosynthetic species. Reactions concerned with glycolate synthesis and oxidation must occur at rates at least 50% of net photosynthesis. For most species this is equivalent to approximately 10 mg $CO_2$ per $dm^2$·hr, or 114 $\mu$moles per g fresh wt·hr, or 76 $\mu$moles per mg chlorophyll·hr. Therefore photorespiration accounts for a major part of the diffusive resistance to $CO_2$ transport in $C_3$-species. Efficient photosynthetic species, including maize, sugar cane, and sorghum, show low or negligible rates of photorespiration by all assays (Zelitch, 1975).

Glycolate is probably the main photorespiratory substrate (Zelitch, 1971). Tobacco leaves synthesize sufficient glycolate to account for its role in rapid photorespiration, while maize leaves produce glycolate only about 10% as rapidly

*Table 5.* Minimal rates of photorespiration in species with rapid rates[a]

| Species | Method of assay | Temperature | Net photosynthesis in normal air | Photorespiration, % of net photosynthesis in normal air | Ratio photorespiration/dark respiration | References |
|---|---|---|---|---|---|---|
| | | °C | mgCO$_2$/dm$^2$·hr | | | |
| Soybean[b] | CO$_2$ release CO$_2$-free air | 26 | 35.2 | 46 | | Samish et al. (1972) |
| Soybean | Post-illumination CO$_2$ burst | 25 | 11.0 | 75 | | Bulley & Tregunna (1971) |
| Soybean | $^{14}$CO$_2$ release CO$_2$-free air | 24 | | | 2.3 | Laing & Forde (1971) |
| Soybean | CO$_2$ release CO$_2$-free air | 30 | 18 | 42 | 1.9 | Hofstra & Hesketh (1969) |
| Sunflower | Short time uptake $^{14}$CO$_2$ minus $^{12}$CO$_2$ | 25 | 25 | 60 | | Bravdo and Canvin (1973) |
| Sunflower | $^{14}$CO$_2$ release CO$_2$-free air | 25 | 28 | 27 | 3.5 | Ludwig & Canvin (1971) |
| Sugar beet | CO$_2$ release CO$_2$-free air | 25 | 25.2 | 47 | 8.5 | Terry & Ulrich (1973) |
| Sugar beet | $^{14}$CO$_2$ release CO$_2$-free air | 25 | 26 | 40 | 1.4 | Hofstra & Hesketh (1969) |
| Tobacco | $^{14}$CO$_2$ release CO$_2$-free air | 30 | 17-25 | | 1.5-6.0 | Zelitch & Day (1973) |
| Tobacco | Extrapolation net photosynthesis to 'zero' CO$_2$ | 25 | 11 | 55 | | Kisaki (1973) |
| Tobacco | Post-illumination CO$_2$ burst | 25 | 13.7 | 25 | 3.5 | Decker (1957) |
| Tobacco | CO$_2$ burst | 25.5 | 16.9 | 45 | | Decker (1959) |
| Tobacco | Post-illumination CO$_2$ burst | 33.5 | 14.8 | 66 | | Decker (1959) |

a) These values are minimal and are underestimates because photorespiration is assayed under conditions of high light intensity where the main flux of the gas (CO$_2$) is in the opposite direction. 'Dark' respiration contributes somewhat to the photorespiration ... discussed in Zelitch (1971).

258

(Zelitch, 1973, 1974). Photorespiration is undoubtedly slow in maize because glycolate is synthesized slowly. Perhaps maize synthesizes little glycolate because it is produced in the bundle sheath cells where synthesis is inhibited by the higher than usual $CO_2$ concentrations found there (Hatch, 1971).

It therefore appears that the $C_4$-species evolved an unnecessarily complicated biochemical mechanism for inhibiting glycolate biosynthesis. Moreover, it seems difficult to convert a $C_3$-species into a $C_4$-species by producing interspecific hybrids Björkman, 1973). This is not surprising since such a conversion of a $C_3$-plant to a $C_4$ would involve large changes in leaf morphology, chloroplast type, and enzyme activities. It would seem easier to accomplish a decreased rate of photorespiration in less efficient photosynthetic species by slowing glycolic acid synthesis more directly, and in this way one should expect to obtain large increases in net photosynthesis without invoking the $C_4$-system.

## The Biochemical Regulation of Photorespiration

Blocking the oxidation of glycolate and photorespiration in tobacco leaves with α-hydroxysulfonates increased net photosynthesis at $35°C$ several-fold (Zelitch, 1966). However, specific biochemical inhibitors of the biosynthesis of glycolate are potentially more interesting than inhibitors of glycolate oxidation. I recently found that glycidic acid (2,3-epoxypropionate), an epoxide similar in structure to glycolic acid, specifically inhibited glycolate synthesis about 50%, also inhibited photorespiration to a similar extent, and brought about an increase in net $CO_2$ fixation of about 50%.

These results demonstrate that the $C_4$-system is not essential in order to have low rates of photorespiration, and that $C_3$-tissues have an adequate photochemistry and carboxylation system to sustain more rapid rates of net photosynthesis when glycolate synthesis and photorespiration are blocked. Glycidate also inhibited glycolate synthesis in maize leaves, but net photosynthesis was not increased presumably because glycolate synthesis is already so slow in this tissue. Hence the results with biochemical inhibitors of photorespiration further confirm that large increases in net photosynthesis could be achieved in normal environments in many species by the biochemical or genetic regulation of glycolate synthesis, as already occurs naturally in species such as maize.

## Genetic Control of Photorespiration Within a Species

Zelitch & Day (1968) observed that a variety of tobacco (JWB Mutant) that grew slowly in a greenhouse environment had lower rates of net photosynthesis in normal air and a faster photorespiration than its fast growing sibling (JWB Wild). Since JWB Mutant involved a simple genetic change, this example demonstrated that genetic control, a pleiotropic effect in this instance, was capable of regulating photorespiration within a species. JWB Mutant tobacco also has a decreased chlorophyll content, an altered chloroplast structure, and a different rate of stomatal opening, hence changes in photorespiration in normal-appearing plants were sought. Some of our more recent results are discussed below.

It has been indicated at this Symposium that the ratio of glycolate production to photosynthetic $CO_2$ uptake has a fixed stoichiometry, thus implying that photorespiration is a constant fraction of net photosynthesis in a given environment. In addition to the contrary evidence with tobacco from our Station (Zelitch & Day, 1968, 1973), Wilson (1972) has described variations in photorespiration within populations of *Lolium,* and a difference in rate of photorespiration was claimed between two wheat varieties by Ghildyal & Sinha (1973).

More recently we have described the results of pedigree selections on siblings of several generations of normal-appearing Havana Seed tobacco plants with slower photorespiration and faster net photosynthesis than is common for this species. Superior plants, on selfing, produced about 25% of their progeny with slow photorespiration and fast net $CO_2$ uptake. The percentage was not increased in several successive generations. It was clearly established that some plants growing side by side showed about one-half the normal rate of photorespiration and similar rates of dark respiration, and this diminished photorespiration was accompanied by an increase in net photosynthesis of about 40%. Attempts are being made to increase the proportion of superior plants in a population by producing doubled haploid plants from selections first made for slow photorespiration and fast net photosynthesis in haploid plants obtained by anther culture (Kasperbauer & Collins, 1971). The recent advances in somatic cell genetics (Carlson, 1973), whereby large populations of haploid plant cells are treated in such a manner that only mutants with the desired phenotype (low rate of photorespiration or efficient dark respiration, for example) will survive, seems to offer the most promise of making more rapid progress in producing superior plants. From examples already obtained it seems clear that decreasing photorespiration in a population would be a desirable goal, and that its achievement should result in large increases in plant productivity.

## REFERENCES

Bahr, J.T. & W.D. Bonner, Jr. (1973): Cyanide-insensitive respiration. I. The steady states of shunk cabbage spadix and bean hypocotyl mitochondria. *J. Biol. Chem.,* 248: 3441-3445.
Björkman, O. (1973): Comparative studies on photosynthesis in higher plants. *Photophysiology,* 8: 1-63.
Bonner, J. & S.G. Wildman (1946): Enzymatic mechanisms in the respiration of spinach leaves. *Arch. Biochem.,* 10: 497-518.
Bravdo, B. & D.T. Canvin (1973): The use of [14]C as a tracer in photorespiration. *Plant Physiol.,* 51 Suppl.: 42 (Abstr.).
Buchanan, B.B. & P. Schürmann (1973): Regulation of ribulose 1,5-diphosphate carboxylase in the photosynthetic assimilation of carbon dioxide. *J. Biol. Chem.,* 248: 4956-4964.
Buchanan, B.B., P. Schürmann & P.P. Kalberer (1971): Ferredoxin-activated fructose diphosphatase of spinach chloroplasts. Resolution of the system, properties of the alkaline fructose diphosphatase component, and psysiological significance of the ferredoxin-linked activation. *J. biol. Chem.,* 246: 5952-5959.
Bulley, N.R. & E.B. Tregunna (1971): Photorespiration and the postillumination carbon dioxide burst. *Can. J. Bot.,* 49: 1277-1284.
Carlson, P.S. (1973): The use of protoplasts for genetic research. *Proc. Nat. Acad. Sci. U.S.A.,* 70: 598-602.
Chu, D.K. & J.A. Bassham (1972): Inhibition of ribulose 1,5-diphosphate carboxylase by 6-phosphogluconate. *Plant Physiol.,* 50: 224-227.
Decker, J.P. (1957): Further evidence of increased carbon dioxide production accompanying photosynthesis. *J. Sol. Energy Sci. Eng.,* 1: 30-33.

Decker, J.P. (1959): Comparative responses of carbon dioxide outburst and uptake in tobacco. *Plant Physiol.*, 34: 100-102.

Douce, R., R.B. Holtz & A.A. Benson (1973): Isolation and properties of the envelope of spinach chloroplasts. *J. Biol. Chem.*, 248: 7215-7222.

Ghildyal, M.C. & S.K. Sinha (1973): Variation in photorespiration in wheat genotypes using glycine-1-$^{14}$C decarboxylation technique. *Indian J. Exp. Biol.*, 11: 207-209.

Gifford, R.M. (1974): A comparison of potential photosynthesis, productivity and yield of plant species with differing photosynthetic metabolism. *Aust. J. Plant Physiol.*, 1: 107-117.

Hatch, M.D. (1971): The $C_4$-pathway of photosynthesis. Evidence for an intermediate pool of carbon dioxide and the identity of the donor $C_4$-dicarboxylic acid. *Biochem. J.*, 125: 425-432.

Heber, U. (1974): Metabolite exchange between chloroplasts and cytoplasm. *Annu. Rev. Plant Physiol.*, 25: 393-421.

Heichel, G. (1971): Confirming measurements of respiration and photosynthesis with dry matter accumulation. *Photosynthetica*, 5: 93-98.

Heldt, H.W. & L. Rapley (1970): Specific transport of inorganic phosphate, 3-phosphoglycerate and dihydroxyacetonephosphate, and of dicarboxylates across the inner membrane of spinach chloroplasts. *FEBS Lett.*, 10: 143-148.

Heldt, H.W. & F. Sauer (1971): The inner membrane of the chloroplast envelope as the site of specific metabolite transport. *Biochim. Biophys. Acta*, 234: 83-91.

Hesketh, J.D. (1963): Limitations to photosynthesis responsible for differences among species. *Crop Sci.*, 3: 107-110.

Hofstra, G. & J.D. Hesketh (1969): Effect of temperature on the gas exchange of leaves in the light and dark. *Planta (Berl.)*, 85: 228-237.

Kasperbauer, M.J. & G.B. Collins (1971): Reconstitution of diploids from anther-derived haploids in tobacco. *Crop Sci.*, 12: 98-101.

Kelly, G.J. & M. Gibbs (1973): A mechanism for the indirect transfer of photosynthetically reduced nicotinamide adenine dinucleotide phosphate from chloroplasts to the cytoplasm. *Plant Physiol.*, 52: 674-676.

King, R.W. & L.T. Evans (1967): Photosynthesis in artificial communities of wheat, lucerne, and subterranean clover plants. *Aust. J. Biol. Sci.*, 20: 623-635.

Kisaki, T. (1973): Effect of the age of tobacco leaves on photosynthesis and photorespiration. *Plant Cell Physiol.*, 14: 505-514.

Laing, W.A. & B.J. Forde (1971): Comparative photorespiration in Amaranthus, soybean and corn. *Planta (Berl.)*, 98: 221-231.

Lambowitz, A.M. & C.W. Slayman (1971): Cyanide-resistant respiration in *Neurospora crassa*. *J. Bacteriol.*, 108: 1087-1096.

Lambowitz, A.M., C.W. Slayman, C.L. Slayman & W.D. Bonner, Jr. (1972): Electron transport components of wild type and poky strains of *Neurospora crassa*. *J. Biol. Chem.*, 247: 1536-1545.

Lemon, E. (1969): Gaseous exchange in crop stands. In Eastin, J.D. et al. (Ed.): Physiological Aspects of Crop Yield. pp. 117-137. A.S.A. and C.S.S.A., Madison, Wis.

Ludwig, L.J. & D.T. Canvin (1971): The rate of photorespiration during photosynthesis and the relationship of the substrate of light respiration to the products of photosynthesis in sunflower leaves. *Plant Physiol.*, 48: 712-719.

Poincelot, R.P. (1973): Isolation and lipid composition of spinach chloroplast envelope membranes. *Arch. Biochem. Biophys.*, 159: 134-142.

Poincelot, R.P. (1974): Uptake of bicarbonate ion in darkness by isolated chloroplast envelope membranes and intact chloroplasts of spinach. *Plant Physiol.*, 54: 520-526.

Samish, Y.B., J.E. Pallas, Jr., G.M. Dornhoff & R.M. Shibles (1972): A re-evaluation of soybean leaf photorespiration. *Plant Physiol.*, 50: 28-30.

Schonbaum, G.R., W.D. Bonner, Jr. & B.T. Storey (1971): Specific inhibition of the cyanide-insensitive respiratory pathway in plant mitochondria by hydroxamic acids. *Plant Physiol.*, 47: 124-128.

Schwenn, J.D., R. McC. Lilley & D.A. Walker (1973): Inorganic pyrophosphatase and photosynthesis by isolated chloroplasts. I. Characterisation of chloroplast pyrophosphatase and its relation to the response to exogenous pyrophosphate. *Biochim. Biophys. Acta*, 325: 586-595.

Tanaka, A., K. Kawano & J. Yamaguchi (1966): Photosynthesis, respiration, and plant type of the tropical rice plant. *Int. Rice Res. Inst. Techn. Bull.* 7.

Terry, N. & A. Ulrich (1973): Effects of potassium deficiency on the photosynthesis and respiration of leaves of sugar beet. *Plant Physiol.,* 51: 783-786.

Walker, D.A. (1973): Photosynthetic induction phenomena and the light activation of ribulose diphosphate carboxylase. *New Phytol.,* 72: 209-235.

Walker, D.A., K. Kosciukiewicz & C. Case (1973): Photosynthesis by isolated chloroplasts: some factors affecting induction in $CO_2$-dependent oxygen evolution. *New Phytol.,* 72: 237-247.

Werdan, K. & H.W. Heldt (1972): Accumulation of bicarbonate in intact chloroplasts following a pH gradient. *Biochim. Biophys. Acta,* 283: 430-441.

Wilson, D. (1972): Variation in photorespiration in *Lolium. J. Exp. Bot.,* 23: 517-524.

Zelitch, I. (1966): Increased rate of net photosynthetic carbon dioxide uptake caused by the inhibition of glycolate oxidase. *Plant Physiol.,* 41: 1623-1631.

Zelitch, I. (1971): Photosynthesis, Photorespiration, and Plant Productivity. 347 p. Academic Press, New York.

Zelitch, I. (1973): Alternate pathways of glycolate synthesis in tobacco and maize leaves in relation to photorespiration. *Plant Physiol.,* 51: 299-305.

Zelitch, I. (1974): The effect of glycidate, an inhibitor of glycolate synthesis, on photorespiration and net photosynthesis. *Arch. Biochem. Biophys.,* 163: 367-377.

Zelitch, I. (1975): Pathways of carbon fixation in green plants. *Annu. Rev. Biochem.,* 44: In press.

Zelitch, I. & P.R. Day (1968): Variations in photorespiration. The effect of genetic differences in photorespiration on net photosynthesis in tobacco. *Plant Physiol.,* 43: 1838-1844.

Zelitch, I. & P.R. Day (1973): The effect on net photosynthesis of pedigree selection for low and high rates of photorespiration in tobacco. *Plant Physiol.,* 52: 33-37.

Special Session

on

Crassulacean Acid

Metabolism

# THE LABELLING OF THE CARBOXYL CARBON ATOMS OF MALATE IN *KALANCHOË CRENATA* LEAVES

J.W. BRADBEER*, W. COCKBURN** & S.L. RANSON***

*Department of Plant Sciences, University of London King's College.
**Department of Botany, University of Leicester.
***Department of Plant Biology, University of Newcastle.

*Abstract*

Evidence is provided which establishes the correctness of the finding that, when malate was synthesized by darkened acidifying CAM leaves in the presence of $^{14}CO_2$, the distribution of label between carbon atoms 4 and 1 of the malate approximated to a ratio of 2 : 1. This result was not an artefact of the degradation procedure as suggested by Sutton & Osmond (1972). As the explanation, that this distribution of label resulted from a double carboxylation process, has not found support from recent work in other laboratories, it would seem advisable to seek an alternative hypothesis.

## Introduction

Bradbeer (1954) reported that when *Kalanchoë crenata* leaves fixed $^{14}CO_2$ in the dark the radioactivity found in malate (COO$^-$-CHOH-CH$_2$-COO$^-$) was distributed between carbon atoms 4 and 1 in a 2 : 1 ratio. Furthermore the 2 : 1 ratio was maintained when the leaves were allowed to starve for many hours after the incorporation of $^{14}CO_2$. The thesis also contained the proposal that the 2 : 1 ratio arose as a result of two sequential carboxylation reactions being responsible for the labelling of the malate. The discovery of ribulosebisphosphate carboxylase (Quayle et al., 1954; Weissbach et al., 1954) permitted the development of the hypothesis that the first carboxylation involved this enzyme while the second carboxylation involved phosphoenolpyruvate carboxylase (Walker, 1957). Thus, as shown in Fig. 1, the first carboxylation would give one unlabelled and one labelled molecule of 3-phosphoglycerate, the label being in carbon atom 1. The conversion of the 3-phosphoglycerate to phosphoenolpyruvate followed by the second carboxylation and the formation of malate would give one malate labelled in carbon atoms 1 and 4 and one malate labelled only in carbon atom 4, thus accounting for the 2 : 1 ratio of label. As Stiller (1956) had obtained similar results and had reached a similar conclusion, a joint report, which contained the scheme shown in Fig. 1, was published (Bradbeer et al., 1958).

Jolchine (1959) confirmed these findings for *Bryophyllum daigremontianum* and further investigations at Newcastle established that the 2 : 1 ratio was normally found for malate synthesized by acidifying CAM material but that, in the absence of acidification, divergence from the 2 : 1 ratio occurred (Avadhani, 1957; Bradbeer, 1963; Cockburn, 1965). More recently Sutton & Osmond (1972)

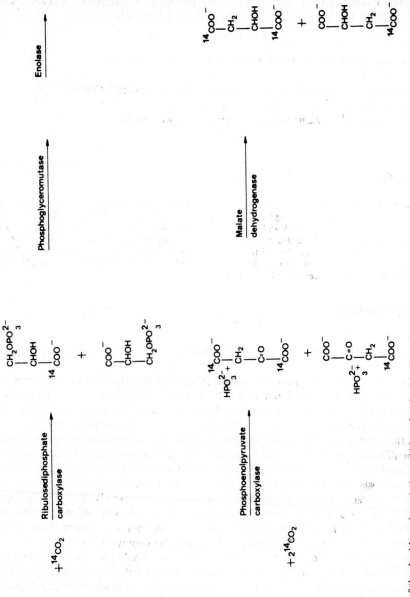

*Fig 1* Diagram of the double carboxylation pathway for the labelling of malate.

found that acidifying Crassulacean leaves gave ratios for the labelling of the 4 and 1 carboxyls of malate substantially in excess of the 2 : 1 ratio. To account for the discrepancy between their results and the previously reported ones they suggested that the 2 : 1 ratio represented an artefact arising from the misuse of the degradation of malate by *Lactobacillus arabinosus*. In this paper evidence is provided to establish the reliability of the *Lactobacillus* degradation technique and of the data obtained by its use.

## Materials and Methods

Mature leaves of *Kalanchoë crenata* Haw. were cut in the late afternoon and usually given a further period of illumination to ensure vigorous acidification in the subsequent dark period. $^{14}CO_2$ was generated from Ba $^{14}CO_3$ and supplied to the leaves after they had been in the dark for at least 30 min. After killing, the leaves were exhaustively extracted with 80% ethanol and water. The labelled malate was isolated chromatographically on Whatman n⁰, 3MM paper with *tert*-amyl alcohol: formic acid: water (3:1:3) and n-propanol: ammonia: water (3:1:1) as the developing solvents.

The malate was degraded stepwise in a series of reactions, each yielding $CO_2$ from one of the carbon atoms. Degradations were carried out either in Warburg flasks, in which case inactive L-malate was added to the labelled samples to give about 20 $\mu$moles per degradation, or in sealed 50 ml wide-necked Erlenmeyer flasks with centre wells, with 400 $\mu$moles of L-malate per degradation. $CO_2$ was trapped by 0.4 ml of 10% sodium hydroxide solution in the centre well of the reaction flask, and after transfer to another container, the carbonate was precipitated by the addition of 1 ml of 10% barium chloride solution. The barium carbonate was collected by a procedure which involved centrifugation and washing successively with hot $CO_2$-free water and ethanol. The barium carbonate was counted as an infinitely thick sample on an aluminium planchette under a conventional end-window Geiger-Muller tube. Corrections were applied for background and the result of blanks which were carried out at all stages to test the contamination with $CO_2$ from the reagents and the atmosphere. The sodium hydroxide solutions used were in all cases carbonate free.

Carbon atom 4 was removed by incubation of the malate with *Lactobacillus arabinous* for 4 hr at 35°C in a modification of Nossal's method (1952). 400 $\mu$moles of malate required organisms harvested from 40 ml of culture medium in a total volume of 6 ml. The residual lactate was isolated by a simple chromatographic procedure with silica gel. The lactate was decarboxylated, yielding carbon atom 1 as $CO_2$, by mixing with 4 ml of 5N sulphuric acid and 2 ml of a 2% potassium permanganate solution in a reaction flask and allowing it to stand at room temperature for 90 min. Although the degradation of the remaining acetate is not discussed in the present paper details about the degradation of the acetate are given elsewhere (Bradbeer, 1963).

The degradations with 20 $\mu$moles of L-malate were scaled down versions of those with 400 $\mu$moles. No differences were found between the results of degradations carried out with the large and small amounts of L-malate.

## Results and Discussion

Fig. 2 shows the release of $CO_2$ from 19.8 µmoles of L-malate which were decarboxylated in a manometric flask under similar conditions to those used for the degradation. This amount of L-malate released 425 µl of $CO_2$ which is 96% of the expected recovery. Half of the $CO_2$ was released within 20 min of the commencement of the reaction which was essentially complete after 80 min. Fig. 3 shows that there was a linear relationship between $CO_2$ release and the amount of L-malate provided. Out of 10 determinations of the $CO_2$ release from L-malate standards (obtained from Koch-Light Laboratories Ltd., Colnbrook, Bucks, U.K.) the $CO_2$ production was found to average at 97% of the expected values. The data in Figs. 2 and 3, which were obtained prior to the use of this technique for the degradation of radioactive malate, established that this technique was a suitable degradation procedure.

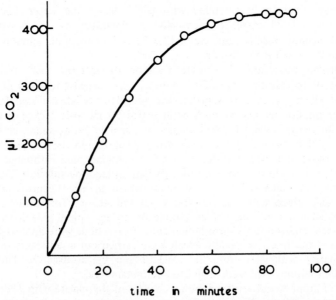

*Fig. 2.* The release of $CO_2$ from 19.8 µmoles of L-malate by *Lactobacillus arabinosus* in a Warburg flask at 35°C.

At the time that the first degradations were carried out specifically labelled malate was not available for use in checking the accuracy of the degradation results. However colleagues in our laboratory used the same technique to degrade malate which had been labelled by $^{14}CO_2$ in the dark in tissues which do not show CAM. The data have been reported previously (Bradbeer et al., 1958) but are presented again with some CAM results in Table 1. *Kalanchoë crenata* roots contained malate in which the radioactivity was equally distributed between carbons 1 and 4 while both fumitory leaves and sunflower seedlings contained malate which was predominantly labelled in carbon atom 4. Thus in the roots there was completed equilibration between the carboxyls while in the other tissues

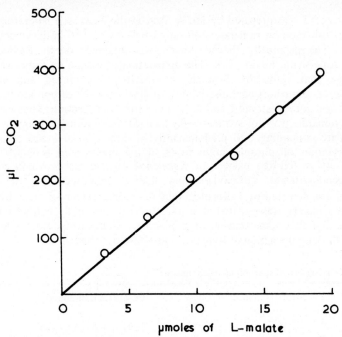

*Fig. 3.* The total $CO_2$ released from different amounts of L-malate by *Lactobacillus arabinosus* in Warburg flasks at 35°C.

equilibration was slight. These results showed us that the 2:1 labelling of the malate carboxyls was not an artefact of the degradation procedure.

*Table 1.* Distribution of label in malate from a number of tissues after dark $^{14}CO_2$ fixation.

| Plant material | Fixation conditions | $^{14}C$ distribution | | | | Reference |
|---|---|---|---|---|---|---|
| | | 1 | 2 | 3 | 4 | |
| | | \multicolumn{4}{c}{COO⁻-CHOH-CH₂-COO⁻} | |

| Plant material | Fixation conditions | 1 | 2 | 3 | 4 | Reference |
|---|---|---|---|---|---|---|
| *Kalanchoë crenata* leaves | 90 min at 19°C | 30 | | 0 | 70 | Bradbeer (1954) |
| | 9 hr at 19°C | 37 | | | 63 | Bradbeer (1954) |
| *Kalanchoë crenata* roots | 1 hr at 20°C | 51 | | 0 | 49 | Avadhani (1957) |
| | 6 hr at 20°C | 48 | | 0 | 52 | Avadhani (1957) |
| *Fumaria officinalis* leaves | 5 min at 20°C | 4 | | 0 | 96 | Bradbeer (1957) |
| *Helianthus annuus* seedlings | 45 sec at 24°C | 12 | | 0 | 88 | Bradbeer (1957) |
| | 3 hr at 24°C | 5 | | 0 | 95 | Bradbeer (1957) |
| | 12 hr at 24°C | 5 | | 0 | 95 | Bradbeer (1957) |

Cockburn (1965) eventually obtained specifically labelled L-malate for a test of the degradation technique. L-malate labelled exclusively and uniformly in carbon atoms 1 and 4 and L-malate labelled exclusively and uniformly in carbon atoms 2 and 3 was prepared by Mr. A. Peat in the Newcastle laboratory by the action of fumarase on fumarate uniformly labelled with $^{14}$C in the corresponding positions. The specifically labelled fumarate was obtained from the Radiochemical Centre, Amersham, Bucks., U.K. The preparation of malate by fumarase has the advantage that only the L-isomer of malate is produced and also that equilibration, ensuring equal distribution of label between carbon atoms 1 and 4 and between carbon atoms 2 and 3, occurs during the reaction. Thus even if the original fumarate was not symmetrically labelled the resulting L-malate would be.

L-malate containing $^{14}$C predominantly in carbon atom 4 was prepared by β-carboxylation of phosphoenolpyruvate in the presence of NaH $^{14}$CO$_3$ and NADH$_2$ by a cell-free extract of *Kalanchoë crenata* leaves which contained phosphoenolpyruvate carboxylase and malate dehydrogenase. As labelled fumarate was detected by radioautography of chromatograms of the extract from which the malate was isolated it is evident that the crude extract contained fumarase and that some transfer of radioactivity from carbon atom 4 to carbon atom 1, through the action of fumarase, must have occurred.

Table 2. Degradation of specially-labelled malate.

| Known distribution of label | $^{14}$C distribution found[1] | | | |
|---|---|---|---|---|
| | 1 | 2 | 3 | 4 |
| | COO⁻-CHOH-CH$_2$-COO⁻ | | | |
| [1,4-$^{14}$C]-malate | 51 | 0 | 0 | 49 |
| [2,3-$^{14}$C]-malate | 0.7 | 45 | 55 | 0 |
| predominantly [4-$^{14}$C] malate | 5 | 0 | 0 | 95 |

1) Each result is the mean of three degradations.

Table 2 shows the results of the degradation of the specifically labelled malate. In this case the stepwise degradation of the labelled malate was taken to completion as described by Bradbeer (1963). The L-malate labelled uniformly in carbon atoms 1 and 4 gave a result which was within experimental error of the expected one. L-malate labelled uniformly in carbon atoms 2 and 3 showed only a trace of activity in the carboxyl carbon atoms. The label was apparently not quite uniformly distributed between carbon atoms 2 and 3, but this discrepancy is of no concern in the topic discussed in this paper. For malate predominantly labelled in carbon atom 4, the 5% of label in carbon atom 1 probably represents some randomization through fumarase as discussed above. If any of the carbon 4 label is not released as CO$_2$ it is probable that it would amount to 2% or less of the amount of label in carbon atom 4. It may be concluded, on the basis of the data in Tables 1 and 2 and Figs. 2 and 3, that the 2:1 distribution of label between the carboxyl carbons of malate is a true representation of the labelling of malate and cannot be accounted for as an artefact of the degradation technique.

Sutton & Osmond (1972) have shown that when the *Lactobacillus* degradation was misused by the provision of either too little malate, (0.2 μmoles/assay) or too much malate (40-100 μmoles/assay), under their conditions erroneous results were obtained. In the case of very low amounts of malate a substantial amount of the radioactivity is likely to be incorporated within the organism and results will be distorted. In the case of excessively high amounts of malate the degradations will not be completed. They also found, with all quantities of malate, that no more than 90% of the carbon atom 4 was released as $^{14}CO_2$. This latter discrepancy is more disturbing, but as in our hands *Lactobacillus arabinosus* released at least 95% of carbon atom 4 as $CO_2$, and more probably about 98%, any error from this source is quite small. We have found the results of the *Lactobacillus arabinosus* degradation technique to be as accurate as Sutton & Osmond found for the NADP-malic enzyme degradation. In all cases we have used the *Lactobacillus arabinosus* technique with the amounts of L-malate and organism within the proper range for the effective operation of the method.

The reason for the difference between the labelling of malate found by us and that found by Sutton & Osmond clearly resides in the CAM plant and not in discrepancies in the degradation. Sutton & Osmond (1972) found that when four CAM species fixed $^{14}CO_2$ in the dark for one min or less over 90% of the malate label was in carbon atom 4, the remainder being in carbon atom 1. Extension of the labelling period from 5 min to 2 h resulted in a change towards an approximately 2:1 ratio of labelling. These results may be contrasted with those obtained at Newcastle where a 2:1 labelling pattern for malate was always found with acidifying leaves and was even obtained when *Kalanchoë crenata* leaves were exposed to $^{14}CO_2$ for 4 sec in the dark.

In the next paper Cockburn gives further data about the labelling of malate in CAM plants. The labelling found by Sutton and Osmond does not disprove the double carboxylation hypothesis however, since the slow labelling of carbon atom 1 of malate could result from the presence of appreciable unlabelled pools of intermediates on a pathway from 3-phosphoglycerate to phosphoenolpyruvate. However, the following papers by Cockburn, Kluge, Osmond & Sutton all contain some evidence which does not appear to be consistent with the occurrence of a double carboxylation pathway in acidifying CAM plants. As not one piece of evidence seems to be conclusive the double carboxylation pathway cannot finally be disproved. However, it would seem advisable to seek an alternative hypothesis to account for the labelling of malate.

# REFERENCES

Avadhani, P.N. (1957): Studies on the fixation of carbon dioxide in leaves and roots of *Kalanchoë* and roots of *Ricinus*. Ph. D. thesis. University of Durham.

Bradbeer, C. (1957): Some interrelations of fat, carbohydrate and organic acid metabolism in etiolated *Helianthus annuus* seedlings. Ph.D. thesis. University of Durham.

Bradbeer, J.W. (1954): Some carbohydrate and acid conversions in *Kalanchoë crenata*. Ph.D. thesis. University of Durham.

Bradbeer, J.W. (1963): Physiological studies on acid metabolism in green plants. IX. The distribution of $^{14}C$ in malate of darkened *Kalanchoë* leaf fragments after infiltration with labelled pyruvate. *Proc. Roy. Soc. B.*, 157: 279-289.

Bradbeer, J.W., S.L. Ranson & Mary Stiller (1958): Malate synthesis in Crassulacean leaves. I. The distribution of $C^{14}$ in malate of leaves exposed to $C^{14}O_2$ in the dark. *Plant Physiol.*, 33: 66-70.

Cockburn, W. (1965): Problems related to acid synthesis in higher green plants. Ph.D. thesis. University of Newcastle.

Jolchine, G. (1959): Sur la distribution du $^{14}C$ dans les molécules d'acide malique synthétisées par fixation de $^{14}CO_2$ dans les feuilles de *Bryophyllum daigremontianum* Berger. *Bull. Soc. Chim. biol.*, 41: 227-234.

Nossal, P.M. (1952): Estimation of L-malate and fumarate by malic decarboxylase of *Lactobacillus arabinosus*. *Biochem. J.*, 50: 349-355.

Quayle, J.R., R.C. Fuller, A.A. Benson & M. Calvin (1954): Enzymatic carboxylation of ribulose diphosphate. *J. Am. Chem. Soc.*, 76: 3610-3611.

Stiller, M. (1956): Experiments on the relationship between succulence and acid accumulation in *Bryophyllum* with an addendum on the mechanism of acid accumulation. M.S. thesis. Purdue University.

Sutton, B.G. & C.B. Osmond (1972): Dark fixation of $CO_2$ by Crassulacean plants. Evidence for a single carboxylation step. *Plant Physiol.*, 50: 360-365.

Walker, D.A. (1957): Physiological studies on acid metabolism. 4. Phosphoenolpyruvic carboxylase activity in extracts of Crassulacean plants. *Biochem. J.*, 67: 73-79.

Weissbach, A., P.Z. Smyrniotis & B.L. Horrecker (1954): Pentose phosphate and $CO_2$ fixation with spinach extracts. *J. Am. Chem. Soc.*, 76: 3611-3612.

# THE PATHWAY OF MALATE SYNTHESIS IN CRASSULACEAN ACID METABOLISM

W. COCKBURN & A. McAULAY

*School of Biological Sciences, University of Leicester, Leicester, LE1 7RH, England*

*Abstract*

Malate synthesized in darkness from $^{14}CO_2$ by tissues which exhibit crassulacean acid metabolism often contains a 2:1 distribution of label between carbon atoms 4 and 1. The occurrence of this distribution of label and its significance in relation to the pathway of malate synthesis are studied by degradation of malate accumulated under a variety of experimental conditions by a range of plant species – some which exhibit crassulacean acid metabolism and some which do not. The data so obtained indicates transfer of label from carbon 4 to carbon 1 of malate (equilibration) but does not allow a distinction to be made between the two possibilities of malate synthesis via a single or a double carboxylation pathway followed in both cases by equilibration of carboxyl label. Mass spectrometric detection and quantification of the isotopic species of malate synthesized from $^{13}CO_2$ by actively acidifying crassulacean tissue indicated unequivocally that the malate was synthesized via a single carboxylation pathway.

The 2:1 distribution of malate carboxyl label is accounted for in terms of equilibration of the carboxyl carbon label of malate initially labelled exclusively in carbon 4 during movement from the site of malate synthesis to storage at a metabolically inert site.

## Introduction

It has been frequently found that malate synthesised from $^{14}CO_2$ in the dark by tissues which exhibit crassulacean acid metabolism, contains approximately a 2:1 distribution of $^{14}C$ between carbons 4 and 1 (Bradbeer, Ranson & Stiller 1958; Varner & Burell 1950; Stiller 1959; and Cockburn 1965). This labelling pattern persisted throughout long experimental periods and Bradbeer et al. (1958) postulated that this militated against the production of the labelling pattern by equilibration of malate synthesised by a single carboxylation reaction and therefore initially labelled exclusively in carbon atom 4. They suggested instead that malate was synthesized by a unique pathway which includes two sequential carboxylations (Fig. 1). The first involves the ribulose diphosphate carboxylase reaction which, if $^{14}CO_2$ is utilized, results in the synthesis of equal quantities of labelled and unlabelled 3-phosphoglycerate. In the second carboxylation, phosphoenolpyruvate derived from the 3-phosphoglycerate is carboxylated to yield oxaloacetate which is reduced to malate. Being derived from the two species of 3-phosphoglycerate the oxaloacetate and malate produced would comprise equal quantities of molecules containing $^{14}C$ either exclusively in carbon 4 or in both carbon 4 and carbon 1. Degradation of such a mixture of molecules would give the characteristic 2:1 distribution of label. More recently (Cockburn, 1965;

*Acknowledgements:* We are grateful to Professor S.L. Ranson for his advice and encouragement and to the Science Research Council for financial support.

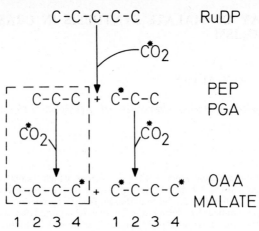

**Fig. 1.** Possible pathways of malate synthesis. The entire diagram represents the carbon skeletons of intermediates of the double carboxylation pathway and the section within the dotted line represents those of the single carboxylation pathway. RuDP = ribulose diphosphate, PEP = phosphoenolpyruvate, PGA = 3-phosphoglycerate, OAA = oxaloacetate.

and Sutton & Osmond, 1972) have obtained labelling patterns other than the 2:1 distribution in crassulacean malate. Sutton & Osmond (1972) suggested that the 2:1 distribution was an artefact of the *lactobacillus* degradation technique used by the earlier workers and concluded on this basis that crassulacean tissues synthesised malate by a pathway involving a single carboxylation. The work described here was undertaken to clarify the situation regarding the occurrence and significance of the 2:1 distribution of label 1) by further investigations of the labelling pattern in malate produced from $^{14}CO_2$ under a variety of conditions by a range of species, and 2) by the investigation of the possible contribution of a double carboxylation pathway to malate synthesis by a method which does not depend on the degradation of malate.

## Materials and Methods

*Plant Materials* were grown in the Botanic Gardens of the Universities of Leicester and Newcastle or were bought in local markets.

*Measurement of Acid Content* was made by a titration of a boiling water extract of the tissue against NaOH to the phenolphthalein end point and results were expressed as milliequivalents of acid per 100 g F.W. of plant material.

*Carbon Dioxide Feeding.* Isotopically labelled $CO_2$ was generated from sodium or barium $^{13}C$ or $^{14}C$ carbonate by the addition of lactic acid and the gas so produced was flushed by $CO_2$-free air into an evacuated darkened vessel containing the plant material. Sufficient $CO_2$ was supplied to allow an acid content increase of approximately 80 milliequivalents per 100 g F.W., which, in addition to respiratory $CO_2$ released by the tissues, would ensure that acid synthesis was not $CO_2$-limited.

*Malate Isolation and Purification* from boiling water extracts of the tissues was achieved by Dowex-1(formate) column chromatography utilising formic acid gradient elution or by chromatography on Whatman 3 mm paper using tertiary amyl alcohol, formic acid, water (3:1:1 V/V) as solvent. Detection of malic acid was by its radioactivity or by the use of bromocresolpurple spray reagent (0.04% W/V, pH 8.0).

*Malate Degradation* was performed as described by Cockburn (1965). Carbon 4 was released as $CO_2$ from radioactive malate (adjusted to 100 $\mu$ moles with unlabelled malate) by malate-adapted *Lactobacillus arabinosus* 17/4 and trapped in NaOH solution. Carbon 1 of the original malate was released as $CO_2$ by acid permanganate oxidation of the lactic acid remaining after lactobacillus treatment and was absorbed in NaOH solution.

The amount of radioactivity in malate atoms 2 and 3 (assessed by comparison of the sum of the amounts measured in atoms 2 and 3 with the total radioactive content as determined by persulphate oxidation) was found to be negligible in all cases.

*Estimation of Radioactivity.* $^{14}CO_2$ trapped in NaOH solution was released by the addition of acid and was reabsorbed in a 1.0 M solution of Hyamine-10X in methanol which was then assayed for $^{14}C$ by liquid scintillation techniques.

*Degradation of Specifically Labelled Malate.* Malate labelled specifically and equally in carbon atoms 1 and 4 gave results within 1% of the expected 1:1 distribution and the degradation of malate labelled equally and exclusively in atoms 2 and 3 resulted in 0.7% of the total activity being erroneously attributed to carbon 1 and none to carbon 4. Degradation of malic acid prepared from phosphoenolpyruvate and $^{14}CO_2$ by phosphoenolpyruvate carboxylase contaminated with fumarase and therefore predominantly but not exclusively labelled in carbon 4 gave an indicated 95% of the total label in carbon 4.

*Gas Chromatography/Mass Spectrometry.* Dimethylmalate was prepared by diazomethane treatment of malate and tri-trimethylsilyl malate was prepared by warming malate to 60° C in bis trimethylsilyl acetamide. Mass spectrometry linked to gas chromatography was carried out by the Physico Chemical Measurements Unit, Harwell, U.K.

*Detection and Quantitatification of Isotopic Species of Malate.* Calculations were made by the method of Biemann (1962) from the mass spectra of samples of malate from *Kalanchoë daigremontiana* leaf material supplied in darkness with $^{12}CO_2$ or $^{13}CO_2$ (87.6% enriched). In the case of dimethylmalate calculations were made from the m/e 113, 114 and 115 peaks in the spectrum which correspond to fragments formed by the loss of the elements of water and a methoxy group and in the case of tri-trimethylsilyl malate from the m/e 350, 351 and 352 peaks which result from the loss of a methyl group from one of the trimethylsilyl groups.

# Results and Discussion

The distribution of label in malic acid isolated from leaves of *Kalanchoë crenata, Ananas comosus, Spinacea oleracea, Pelargonium zonale,* from leaves and stems of *Kleinia articulata* and *Sedum praealtum* and from stems of *Opuntia monacantha* supplied in darkness with $^{14}CO_2$ for periods of 5 minutes, 30 minutes and 6 h are

*Table 1.* The percentage of carboxyl carbon label in carbon atom 4 of malate synthesized in the dark by a variety of plant species.

| Species | % of carboxyl carbon label in carbon atom 4 of malate synthesized in the dark by a variety of plant species | | | |
|---|---|---|---|---|
| | incubation period | | | |
| | 10 sec | 5 min | 30 min | 6 h |
| *Sedum praealtum* | - | 68 | 62 | 63 |
| *Opuntia monacantha* | - | 66 | 59 | 51 |
| *Kleinia articulata* | - | 63 | 57 | 51 |
| *Ananas comosus* | - | 67 | 60 | 64 |
| *Kalanchoë crenata* | 83 | 69 | 60 | 60 |
| *Spinacea oleracea* | - | 58 | 52 | 50 |
| *Pelargonium zonale* | - | - | 51 | - |

shown in Table 1. In all species the highest percentage of label was found in carbon atom 4 after 5 minutes incubation. Malate isolated from *Kalanchoë, Ananas, Kleinia, Sedum* and *Opuntia* contained approximately two thirds of the total label in carbon atom 4 and the remainder in carbon atom 1. No significant amount of label was detected in atoms 2 or 3. The percentage of total label in carbon atom 4 consistently decreased with increase in incubation period, equal distribution of label between carbon atoms 1 and 4 being approached in malate synthesized by spinach during 30 minutes' incubation. Changes in the distribution of label in malate from the species known to exhibit CAM took place more slowly. It was most rapid in *Kleinia* and *Opuntia* followed by *Sedum, Kalanchoë* and *Ananas.*

The present results differ from those of Bradbeer et al. (1958) in that the characteristic labelling pattern was not retained for the long periods (up to 154 h) observed by these workers. Instead, as has also been reported by Sutton & Osmond, the percentage of carboxyl label in carbon atom 1 increased with time of incubation. Such an occurrence could be the result of an increasing synthesis of malate labelled only in carbon atom 1, although this is considered unlikely since never more than 50% of the total carboxyl label was observed in carbon atom 1. The preferred interpretation is that there has been interchange of malate carboxyl groups, as would be the case if malate were converted to the symmetrical fumarate molecule then reconverted to malate via the enzyme fumarase. Such an equilibration process, acting upon malate synthesized by the double carboxylation pathway, would result, eventually, in the conversion of the malate species labelled exclusively in carbon atom 4 to a mixture of equal numbers of molecules labelled either exclusively in carbon atom 4 or carbon atom 1. The distribution of label in molecules labelled in both carbons 1 and 4 (the other type of molecule produced by the double carboxylation pathway) would remain unchanged and degradation of the mixture, following complete equilibration, would indicate that 50% of the total carboxyl label was in carbon 4 and 50% in carbon 1.

The occurrence of equilibration of malate carboxyl label is not necessarily antagonistic to the double carboxylation hypothesis. Although the degradation results do indicate the operation in 'CAM-type' plants of a mechanism potentially capable of the transient production of the 2:1 distribution of label from malate initially labelled exclusively in carbon atom 4, they are also perfectly consistent with the equilibration of malate upon which the 2:1 labelling pattern was imposed at synthesis.

The time of incubation of 'CAM-type' plants during which complete equilibration was approached was shortest in *Kleinia* and *Opuntia* and longest in *Kalanchoë* and *Ananas*. The rate at which such equilibration proceeds may reflect species-related differences in the rapidity of interchange between metabolically active and inactive pools of malate (Bradbeer et al., 1958; Cockburn, 1965) and also variations in rate of the reactions responsible for equilibration. However, regardless of the identity of the reactions involved, the rate of equilibration of label will also depend upon the amount of malate participating in the reaction. Thus, leaves in different phases of acid accumulation might be expected to have different rates of carboxyl label equilibration, simply as a result of differences in malate content. Since the capacity to accumulate acid also varies with the physiological state of the leaf, e.g. its age (Ranson & Thomas, 1960; Bradbeer, 1962), it is clear that considerable variation in rates of equilibration of malate label are to be expected even in different parts of an individual plant. Thus, although variations in equilibration rate between species may well occur, the present differences observed between *Kleinia* and *Opuntia* on the one hand and *Ananas, Sedum* and *Kalanchoë* on the other, may not be related to genuine inter-specific differences but, instead, may simply reflect the physiological state of the plant material used.

A dramatic effect of the physiological state of the leaves on the labelling pattern of malate is shown in the distribution of label in malate synthesized by leaves of *Kalanchoë crenata* in which malate accumulation was complete. An exposure to $^{14}CO_2$ of only 10 seconds duration resulted in the synthesis of malate containing equal amounts of label in the carboxyl carbons. This may be interpreted as indicating that access of newly synthesized malate to the storage pool is restricted in fully-acidified tissue. It may also indicate that the rate of equilibration outside the storage pool is increased under these conditions. No evidence of a large increase in the rate in previously accumulated malate — presumably located within the storage pool — has been observed when acid accumulation is complete.

The observation of malate containing more than 66% of the total label in carbon 4 (Table 1, Sutton & Osmond, 1972) can not be taken as unequivocal evidence against double carboxylation since even exclusive operation of the double carboxylation pathway would, following the supply of $^{14}CO_2$, initially produce singly labelled malate from the pool of three carbon intermediates present in the tissue and associated with the pathway. Only later (providing $^{14}CO_2$ is still being utilized) would the 2:1 labelling pattern become established. Furthermore, simultaneously operation of double and single carboxylation pathways could operate concurrently resulting in more than 66% of the total being detected in carbon 4. The situation is further complicated by the transfer of label from carbon 4 to carbon 1 and it is clear that malate degradation experiments are not capable of distinguishing between the two (not mutually exclusive) possibilities of 1) malate synthesis by single carboxylation followed by equilibration of label or 2) malate synthesis by double carboxylation followed by equilibration.

Since only the double carboxylation pathway synthesizes malate molecules containing two isotopically labelled atoms the detection and estimation of this species of malate in malate synthesized from isotopically labelled $CO_2$ in the dark allows the extent of participation of the double carboxylation pathway in crassulacean malate synthesis to be uniquivocally determined. Transfer of label from carbon 4 to carbon 1 in this case has no effect on the interpretability of results since a singly labelled molecule remains singly labelled and a doubly labelled remains doubly labelled.

Table 2. Mole % composition of dimethyl and trimethylsilyl derivatives of malate accumulated as a result of 12 h exposure of leaf tissue of *Kalanchoë daigremontiana* to $^{13}CO_2$ in darkness.

| | | Malate species | | |
|---|---|---|---|---|
| | | Double $^{12}C$ | Single $^{13}C$ | Double $^{13}C$ |
| Mole % composition | Dimethyl malate | 74.5 | 25.5 | ⟨0.5 |
| | Trimethyl-silyl malate | 72.7 | 27.3 | ⟨0.1 |

Table 2 shows the relative amounts of singly, doubly and unlabelled malate in malate synthesized by actively acidifying *Kalanchoë daigremontiana* leaf tissue from $^{13}CO_2$ during a 12 h incubation at 15°C in darkness. The tissue accumulated 6.0 milliequivalents of acid per 100 g FW during this period. The data was calculated from combined gas chromatography/mass spectrometry of TMS and dimethyl derivatives of malate from the leaf tissue supplied with $^{13}CO_2$ and from identical controls supplied with $^{12}CO_2$. The substantial proportion of singly labelled species leave no doubt of the incorporation of the isotope and the virtually complete absence of the doubly labelled species clearly indicates that the synthesis of malate has been by a reaction sequence involving a single carboxylation.

We are left without an explanation for the 2:1 labelling pattern which so often occurs in crassulacean malate. The most probable origin of this distribution of label appears to involve the transfer of label from carbon atom 4 to carbon atom 1 of malate initially exclusively labelled in carbon atom 4. The persistence of the labelling pattern, once formed, has been attributed to the storage of malate at some site in which the transfer of label from carbon 4 to carbon 1 is slow. If this is indeed the case then the transfer of label from carbon 4 to 1 which results in the 2:1 labelling pattern can only occur during the time elapsing between the synthesis of malate and its entry into the 'storage site'. It is conceivable that the time involved in transport from synthetic to storage site could be similar even under a variety of experimental conditions and could be of sufficient duration to allow the transfer of about 33% of the carboxyl label to carbon 1 — thus producing the 2:1 distribution of label. The exactness of timing required to yield this effect would be reduced if, as seems likely, there is for a given rate of the reactions causing equilibration an exponential relationship between the proportion of the total label in carbon 4 and the rate of transfer of label to carbon atom 1. Allowing this assumption it can be calculated that if the conversion of malate containing 100% of the total carboxyl label in carbon 4 to malate containing 75% requires one half-life of equilibration then the change from 70% to 60% would also occupy one half-life and the change from 60% to 55% two half lives. Thus the extremes of total equilibration or total lack of equilibration are less likely to be observed than some intermediate stage. If we accept the suggestion made above that equilibration occurs most rapidly during movement of malate from its sites of synthesis to the site of storage and that the time taken to achieve this transport may not vary widely then the possibility of random equilibration of carboxyl label leading to relatively precise and persistent distribution of label becomes acceptable.

## REFERENCES

Biemann, K. (1962): Mass spectrometry; organic chemical applications. Pp. 223-227. McGraw-Hill, New York.
Bradbeer, J.W., S.L. Ranson & M.L. Stiller (1958): Malate synthesis in Crassulacean leaves. I. The distribution of $^{14}C$ in malate of leaves exposed to $^{14}CO_2$ in the dark. Plant Physiol., 33: 67-70.
Bradbeer, P. (1962): The distribution and variation in the intensity of carboxylating mechanisms in plants. Ph.D. thesis, University of Durham.
Cockburn, W. (1965): Problems relating to acid synthesis in higher green plants. Ph.D. thesis, University of Newcastle-upon-Tyne, England.
Jolchine, G. (1959): Sur la distribution du $^{14}C$ dans les molecules d'acide malique synthetisées par fixation de $^{14}CO_2$ dans les feuilles de Bryophyllum daigremontianum Berger. Bull. Soc. Chim. Biol., 41: 227-234.
Ranson, S.L. & M. Thomas (1960): Crassulacean acid metabolism. Ann. Rev. Plant Physiol., 11: 81-110.
Stiller, M.L. (1959): The mechanism of malate synthesis in Crassulacean leaves. Ph.D. thesis, Purdue University, Lafayette.
Sutton, B.G. & C.B. Osmond (1972): Dark fixation of $CO_2$ by Crassulacean plants. Evidence for a single carboxylation step. Plant Physiol., 50: 360-365.
Varner, J.E. & Burrell, R.C. (1950): Use of $^{14}C$ in the study of the acid metabolism of Bryophyllum calcinum. Arch. Biochem., 25: 250-257.

# MALATE SYNTHESIS IN CRASSULACEAN ACID METABOLISM (CAM) VIA A DOUBLE $CO_2$ DARK FIXATION?

M. KLUGE, L. BLEY & R. SCHMID

*Botanical Institutes of the Technical Universities Darmstadt and Munich (BRD)*

## Abstract

Malate labelled after $^{14}CO_2$ dark fixation exhibits often a label ratio of 1:2 between the carbon atoms one and four of malate $^{14}C$. This typical label distribution was explained by Bradbeer, Ranson & Stiller (1958) in terms of a two step $CO_2$ fixation.

It has been assumed by these authors that the primary step involves $CO_2$ fixation via RuDP carboxylase. 3-PGA produced in this reaction is converted into PEP which serves as substrate of the secondary carboxylating step ($\beta$-carboxylation of PEP via PEP carboxylase).

The results presented in this paper provide some evidence that a double $CO_2$ fixation might not operate *in vivo* in the malate synthesis during CAM. $^{14}CO_2$ dark fixation and synthesis of $^{14}C$-malate in leaf tissue slices obtained from the CAM plants *Kalanchoe daigremontiana* and *Sedum praealtum* were clearly stimulated by exogenously applied unlabelled PEP or 3-PGA. However, no alteration in the label ratio between the carboxyls of malate $-^{14}C$ could be observed under these conditions. This suggests that malate molecules deriving from the exogenous $CO_2$ acceptors show the same label distribution as malate deriving from endogenous PEP (i.e. about 1:2 between $C_1$ and $C_4$), although they were definitely not the product of a double $CO_2$ fixation. So the asymmetrical label distribution in malate $^{14}CO_2$ produced in CAM cannot be regarded as evidence for a two step mechanism in $CO_2$ dark fixation. However, PEP prelabeled by RuDP carboxylase catalysed $^{14}CO_2$ fixation is used in malate synthesis in the light. This is indicated by the increase of label in $C_1 - C_3$ of malate with increasing light intensity during $CO_2$ fixation. In presence of DCMU, however, the label distribution in malate remained the same as in the dark.

## Introduction

It is now generally accepted that the substantial malate synthesis during the night which characterizes the CAM of succulents (for details see reviews: Ranson & Thomas, 1960; Beevers et al., 1966; Ting 1971) proceeds via a $CO_2$ dark fixation. Above that there is overwhelming evidence that PEP carboxylase is involved in this event as a key enzyme. $\beta$-carboxylation of phosphoenolpyruvate via PEP-carboxylase according to equ. 1 should label exclusively the $C_4$-carboxyl of the resulting malate, if $^{14}CO_2$ is used as substrate of this reaction.

$$PEP + CO_2 \xrightarrow{\text{PEP–C}} \text{Oxaloacetate} \xrightarrow{\text{MDH}} \text{Malate} \qquad \text{(equ. 1)}$$

Supported by Deutsche Forschungsgemeinschaft. Abbreviations are used as follows: CAM = Crassulacean Acid Metabolism; fw = fresh weight; MDH = malate dehydrogenase; OAA = Oxaloacetate; PEP = phosphoenolpyruvate; PEP-C = phosphoenolpyruvate carboxylase; 3-PGA = 3-phosphoglyceric acid; R-5-P = Ribulose-5-Phosphate; RudP = Ribulose-1,5-Diphosphate.

A label distribution like this can easily be found in malate $-^{14}C$ synthesized by PEP carboxylase and malate dehydrogenase *in vitro*. However, when Bradbeer et al. (1958) isolated malate $-^{14}C$ from CAM plants labelled by $^{14}CO_2$ dark fixation, they found the label distribution to be asymmetric with one part being localized in carbon 1 and two parts in carbon 4 of malate. A label distribution like this, however, is difficult to explain if a simple β-carboxylation of PEP according to equ. 1 is the mechanism of the $CO_2$ dark fixation in CAM. To overcome this difficulty, Bradbeer et al. (1958) postulated a metabolic pathway outlined in equ. 2 which includes two carboxylating steps switched in series:

$$
\text{RuDP} + {}^{*}CO_2 \longrightarrow
\begin{array}{cc} {}^{*}C_1 & C_1 \\ | & | \\ C_2 & + C_2 \\ | & | \\ C_3 & C_3 \end{array}
\rightarrow \rightarrow
\begin{array}{cc} {}^{*}C_1 & C_1 \\ | & | \\ C_2 & + C_2 \\ | & | \\ C_3 & C_3 \end{array}
\xrightarrow{\;2\;{}^{*}CO_2\;}
\begin{array}{cc} {}^{*}C_1 & C_1 \\ | & | \\ C_2 & + C_2 \\ | & | \\ C_3 & C_3 \\ | & | \\ {}^{*}C_4 & {}^{*}C_4 \end{array}
\qquad \text{(equ. 2)}
$$

| 3-PGA | PEP | Oxaloacetate or Malate |

The first step is assumed by Bradbeer to be identical with the carboxylation of RuDP via RuDP-carboxylase, producing 3-PGA which is converted into PEP. These PEP molecules are used in the secondary carboxylation step where PEP carboxylase is involved.

Bradbeer's hypothesis explained the asymmetry in the label distribution of malate so elegantly and convincingly, that no serious doubts about its validity arose during many years since it has been published. Only recently, Sutton & Osmond (1972) showed a label distribution in malate$-^{14}C$ with more than 90% of label in $C_4$ after short term $^{14}CO_2$ fixation. The result indicates a mechanism of $CO_2$ fixation according to equ. 1 rather than a double fixation. These authors suspected the method used by Bradbeer et al. (1958) in degrading malate to pretend a 1:2 ratio between $C_1$ and $C_4$ label. In view of these discrepancies, we started some investigations which should provide clear evidence either in favour or against Bradbeer's postulation of a double $CO_2$ dark fixation in CAM. Some of our results have been published earlier (Kluge et al., 1974).

## Materials and Methods

Leaf slices of CAM plants (*Bryophyllum daigremontianum, Sedum praealtum*) or intact leaf material (*Tillandsia usneoides*) were used in our experiments. All methods used has been described in detail by Kluge et al. (1974); see also Kluge et al. (1973), Kluge & Heininger (1973).

## Results and Discussion

Our experiments based on the following argumentation: if the asymmetrical label distribution as found by Bradbeer et al. in malate-$^{14}$C after $^{14}CO_2$ dark fixation is caused by a double $CO_2$ fixation (equ.2), it can be predicted that the provision of exogenous unlabelled PEP will increase the proportion of label localized in $C_4$ with respect to $C_1$. The reason for this is that malate molecules labelled only in $C_4$ will derive from exogenously provided unlabelled PEP. These $C_4$ labelled malate molecules will reduce the proportion of malate molecules prelabelled in $C_1$ during the $^{14}CO_2$ fixation by RuDP carboxylase (see fig. 1).

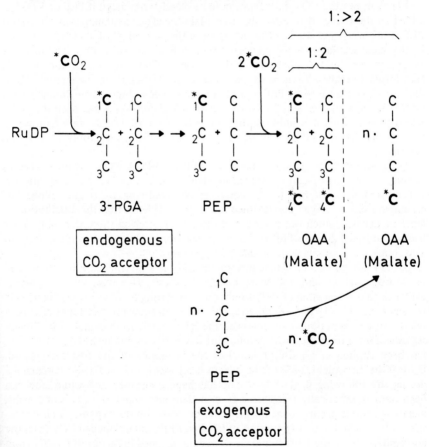

*Fig. 1.* Shifting in the label distribution between $C_1$ and $C_4$ of malate after $^{14}CO_2$ dark fixation which is to be expected if the double fixation hypothesis is valid and exogenously applied PEP is used in $^{14}CO_2$ dark fixation (for details see text).

However, if exogenous PEP is used as $CO_2$ acceptor in $^{14}CO_2$ dark fixation without the label ratio of 1:2 between $C_1$ and $C_4$ of malate-$^{14}$C being shifted in favour of $C_4$, it can be concluded that malate molecules deriving from unlabelled

PEP must have also an asymmetric label ratio near 1:2. So it would evident that the RuDP carboxylase reaction is not necessary to label the $C_1$ of malate. This allows to assume the PEP carboxylase reaction being the only carboxylating step during $CO_2$ dark fixation in CAM.

We tested our argumentation using slices of leaf tissue obtained from CAM plants. The tissue slices were suspended in buffer and fed with $NaH^{14}CO_3$ in the dark (30-60 min.) with or without PEP or 3-PGA added. Substantial amounts of $^{14}CO_2$ were fixed by the tissue in the dark and incorporated with 75% in malate, 15% in citrate/isocitrate and 10% in aspartate and glutamate, which is a label distribution nearly identical to what can be found in intact leaves *in vivo*.

Fig. 2 shows the $CO_2$ dark fixation to be clearly stimulated if PEP or 3-PGA is added to the tissue. It is reasonable to explain the effect assuming that exogenous PEP is taken up and used as $CO_2$ acceptor in the $\beta$-carboxylation. 3-PGA may act in the same way after being converted into PEP by the cells. This assumption is evidenced by the fact that radiocarbon from 3-PGA-$^{14}C$ fed to the tissue is transferred to malate by the cells (see Kluge et al., 1974). Enhancement of $^{14}CO_2$ dark fixation after feeding PEP or 3-PGA has also been observed by Rouhani et al. (1973) in isolated mesophyll cells of *Sedum telephium*. As it has been shown by Kluge et al. (1974) it is unlikely that the stimulation of $^{14}CO_2$ dark fixation by exogenous PEP or 3-PGA has other reason than increased supply of $CO_2$ acceptors.

Thus, although exogenous, unlabelled PEP or 3-PGA were obviously used as carbon skeletons in $\beta$-carboxylation, no significant changes in the label ratio between $C_1$ and $C_4$ of malate-$^{14}C$ could be observed (Fig. 2). If one follows our argumentation displayed above, these results give evidence that the distribution of label in malate (which was according to Bradbeer often in the order of 1:2) has been established by the PEP carboxylase reaction and events occuring after it. This, however, conflicts with Bradbeer's hypothesis.

It shall not be excluded that in some special cases prelabelled PEP can be used in $\beta$-carboxylation and following malate synthesis. However, we can provide evidence that prelabeling of PEP is only possible during or after $^{14}CO_2$ fixation in the light, i.e. $^{14}CO_2$ photosynthesis. This has been shown in two types of experiments. In one experiment we followed the label distribution in malate in *Tillandsia usneoides* during the night, a following day and a second night (fig. 3). $^{14}CO_2$ has been applied in a pulse (1 hour) at the beginning of the first dark period. During the first night, 63% of the malate label was found in $C_4$. So it remained during the following day while malate has been consumed and while label has been photosynthetically transferred from malate into starch. In the second night, malate synthesis occurred again, this time, however, on the expense of photosynthetically labelled starch (provision of labelled PEP) and unlabelled $CO_2$ from the atmosphere. Now the label found $C_4$ of malate dropped in favour of $C_1-C_3$ from 60 to 40%.

In other experiments, we allowed tissue slices of *Sedum praealtum* to fix $^{14}CO_2$ in the light under increasing light intensities. Under this conditions, in addition to the products of photosynthesis also some malate gained radioactivity (fig. 4). The proportion of label found in $C_4$ of malate decreased with increasing light intensity. If the photosynthesis was inhibited by DCMU, the label distribution in malate-$^{14}C$, however, could not be influenced by the light intensity. Here

284

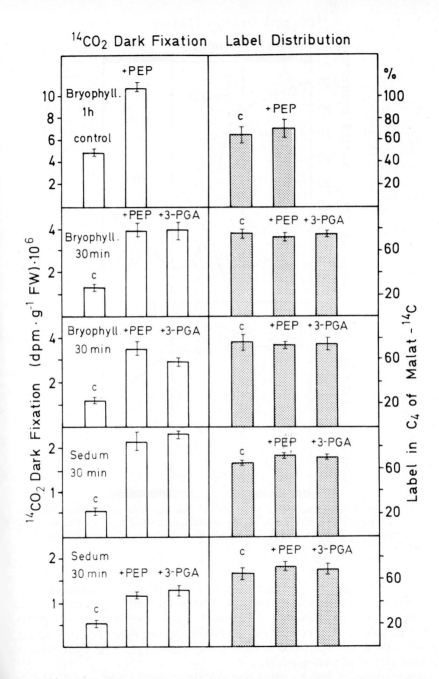

*Fig. 2.* $^{14}CO_2$ dark fixation and label distribution of malate in tissue slices with or without addition of exogenous $CO_2$ acceptors. Each column represents the arith. mean with standard deviation of five parallel treatments.

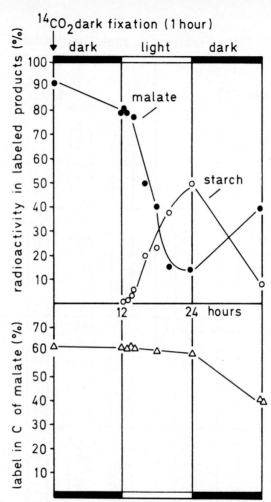

*Fig. 3.* Conversion of malate-$^{14}$C into starch-$^{14}$C and vice versa in *Tillandsia usneoides* labelled by a pulse $^{14}CO_2$ dark fixation at the beginning of the first dark period (upper part). Label found in $C_4$ of malate during the experiment. Temperature 20°C; 35.000 lux during the light.

it remained the same as in the dark control, i.e. 70% of label being localized in $C_4$. The influence of light intensity and of DCMU on the label distribution in malate indicates clearly the derivation of PEP used in malate synthesis from a photosynthetic product. This product may be 3-PGA, deriving either directly from the photosynthetical carboxylation of RuDP or indirectly via glycolysis from starch as aan end product of photosynthesis.

Summarizing, it shall be stated that our results. seems to disprove Bradbeer's hypothesis. However, we emphasize that we are not able to offer a conclusive alternative hypothesis which allows to explain the asymmetry of the label

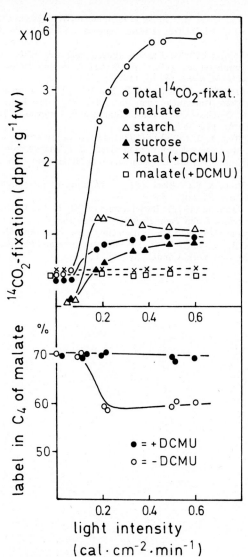

*Fig. 4.* $^{14}CO_2$ fixation in the light and labelled products in tissue slices of *Sedum praealtum* suspended in buffer (upper part). Label found in $C_4$ of malate (lower part).

distribution in CAM generated malate. A possible explanation has been offered by Ting (1971) who argued in favour of randomization of label, originally localized in $C_4$, controlled by compartmentation.

## REFERENCES

Beevers, H., M.L. Stiller & V.S. Butt (1966): Metabolism of organic acids. In Steward, F.C. (Ed.): Plant Physiology, Vol. IV B, Academic Press, New York & London.

Bradbeer, J.W., S.L. Ranson & M.L. Stiller (1958): Malate Synthesis in Crassulacean leaves. I. The distribution of $C^{14}$ in malate of leaves exposed to $C^{14}O_2$ in the dark. *Plant Physiol.,* 33: 66-70.

Kluge, M. & B. Heininger (1973): Untersuchungen über den Efflux von Malat aus den Vacuolen der assimillierenden Zellen von *Bryophyllum* und mögliche Einflüsse dieses Vorganges auf den CAM. *Planta (Berl.),* 113: 333-343.

Kluge, M., O.L. Lange, M. v. Eichmann & R. Schmid (1973): Diurnaler Säuerhytmus bei *Tillandsia usneoides:* Untersuchungen über den Weg des Kohlenstoffs sowie die Abhängigkeit des $CO_2$-Gaswechsels von Lichtintensität, Temperatur und Wassergehalt der Pflanze, *Planta (Berl.),* 112- 357-372.

Kluge, M., Ch. Kriebitsch & D. v. Willert (1974): Dark fixation of $CO_2$ in Crassulacean Acid Metabolism: Are two carboxylation steps involved? *Z. Pflanzenphysiol.,* 72: 460-463.

Ranson, S.C. & M. Thomas (1960): Crassulacean Acid Metabolism. *Ann. Rev. Plant Physiol.,* 11: 81-110.

Rouhani, J., H.M. Vines & C.C. Black (1973): Isolation of mesophyll cells from *Sedum telephium* leaves *Plant Physiol.,* 51: 97-103.

Sutton, B.G. & C.B. Osmond (1972): Dark fixation of $CO_2$ by Crassulacean Plants. Evidence for a single carboxylation step. *Plant Physiol,* 50: 360-365.

Ting, L. (1971): Non autotrophic $CO_2$ fixation and Crassulacean Acid Metabolism. In Hatch, M.D., Osmond, C.B., Slatyer, R.O. (Eds.): Photosynthesis and Photorespiration. 169-185. Wiley Interscience, New York, London, Sydney, Toronto.

# PHOTOSYNTHETIC EFFICIENCY OF CAM PLANTS IN RELATION TO $C_3$ AND $C_4$ PLANTS

S.R. SZAREK[1] & I.P. TING

*Department of Biology and The Philip L. Boyd Deep Canyon Desert Research Center, University of California, Riverside, California 92502 U.S.A.*

## Abstract

An investigation of CAM with the use of the gas transfer equations, $T = D\Delta\!\smallsmile\!/R_a + R_s$ and $P = D\,'\!\Delta\!\smallsmile\!/R_a + R_s + R_m$, suggests that CAM is adaptive to xeric conditions and similar to the $C_4$ pathway. Transpiration ratios (TR) are low and on the order of 150 whereas for $C_4$-plants and $C_3$-plants they average 300 and 600 respectively. Carbon assimilation efficiency (CAE) estimates, the ratio of $P$ to $P_{max}$ where $P_{max} = D\,'\!\Delta\!\smallsmile\!/(R_a + R_s)$ or the estimated $CO_2$ assimilation rate at $R_m = 0$, in the dark are about 0.7-0.9 similarly to $C_4$ plants but unlike $C_3$ estimates of 0.2 to 0.3. We have concluded that the CAM pathway for plants in an arid, desert environment has a two-fold interpretation: primarily, immediately after periods of rainfall and consequent high plant water potentials, nocturnal acidification rates are highest. Minimum TR of 25-70 and maximum CAE values of 0.75-0.82 occur. The level of acid metabolism decreases with decreasing water potential. Secondly, during periods of drought and water potentials of $-15$ to $-20$ bars a reduced, but steady level of acidification occurs by the reassimilation of endogenous $CO_2$. Perhaps, further, these plants can shift from the CAM option to the $C_3$ option under extended, favorable conditions.

## Introduction

Crassulacean acid metabolism (CAM) is the massive (100-200 $\mu$eq g$^{-1}$ fresh weight) diurnal fluctuation of organic acids (viz., malic acid), resulting from a dark or nighttime carboxylation via P-enolpyruvate carboxylase followed by the subsequent decarboxylation of malic acid during the day period, and finally the concomitant photosynthetic $CO_2$ fixation of endogenously produced $CO_2$ by ribulose diphosphate carboxylase. Other criteria to define CAM metabolism are stomata, largely open at night and closed during the day, and some succulence. But it is the diurnal fluctuation of malic acid which is diagnostic for CAM.

Recently, Black (1973) has clearly stated the belief that CAM is one of three known photosynthetic options, i.e., the $C_3$-option, the $C_4$-option, and the CAM-option. This concept, coupled with the recent observation (Osmond et al, 1973) that CAM plants may shift to the $C_3$-option under certain environmental condi-

[1] Present address: Department of Botany and Microbiology, Arizona State University, Tempe, Arizona 85281 U.S.A.

*Acknowledgement:* This investigation received financial support from the Dessert Biome of the United States/International Biological Program through a National Science Foundation Grant GB 15886.

*Abbreviations:* $C_3$: reductive pentose phosphate cycle; $C_4$: dicarboxylic acid cycle; CAE: carbon assimilation efficiency; CAM: Crassulacean acid metabolism; TR: transpiration ratio; $\Psi$: water potential.

tions, allows for ecological and physiological interpretation of succulent plant metabolism.

We have used the parameters of the gas transfer equations to compare the three photosynthetic groups of plants. Transpiration (T) is conveniently described as: $T = D\Delta\sim/R_a + R_s$, where $T = g\ H_2O$ loss dm$^{-2}$ hr$^{-1}$; $D$ = diffusion coefficient for $H_2O$; $\Delta\sim$ = vapor pressure difference between leaf source and atmosphere sink; $R_a$ = boundary layer resistance to diffusion in sec cm$^{-1}$; $R_s$ = stomatal resistance to diffusion in sec cm$^{-1}$. Similarly, $CO_2$ assimilation can be described as: $P = D'\Delta\sim'/R_a + R_s + R_m$, where $P = CO_2$ uptake in g $CO_2$ dm$^{-2}$ hr$^{-1}$; $D'$ = diffusion coefficient for $CO_2$; $\Delta\sim'$ = $CO_2$ difference between atmosphere source and leaf sink (here we assume 59.3 mg $CO_2$ cm$^{-3}$ x 10$^5$); $R_a$ = boundary layer resistance to diffusion in sec cm$^{-1}$; $R_s$ = stomatal resistance to diffusion in sec cm$^{-1}$; $R_m$ = a residual resistance which could include diffusion and transport of $CO_2$ after $R_s$, and all biochemical events associated with $CO_2$ assimilation.

Given the above definitions of the terms of the gas transfer equations, $R_s$ and $R_m$ can be evaluated among the three photosynthetic option plant groups. In addition, two ratios based on the above equations have been useful. Firstly, the transpiration ratio (TR), basically the ratio of equation 1 to equation 2, i.e., T/P, gives an indication of the amount of water used per carbon dioxide assimilated. Usual figures for $C_3$, $C_4$, and CAM are 600, 300, and about 150 respectively. A major complication of TR is that the term $\Delta\sim$ of T, and hence TR, is tightly coupled to the environment and therefore TR is expected to vary considerably. To overcome this problem in part we have previously defined a term E (for photosynthetic efficiency) or CAE (for carbon assimilation efficiency in CAM plants) which is less a function of the environment and more an intrinsic plant parameter. E or CAE is defined as the ratio of actual carbon dioxide assimilation (P) to the maximum carbon dioxide assimilation, $(P_{max})$ where $P_{max}$ is defined as carbon dioxide assimilation at $R_m = 0$.

In this paper we will attempt to place the CAM photosynthetic option in perspective with the $C_3$ and $C_4$ photosynthetic options using data obtained from the gas transfer equations.

## Materials and Methods

Plants of *Opuntia basilaris* Engelm. & Bigel. and *Opuntia bigelovii* Engelm. used in this study were in natural stands at the Boyd Deep Canyon Desert Research Center, California.

The primary climatic characteristics of the Center are: (a) annual average rainfall of 9.8 cm; (b) high summer daytime temperatures, generally less than 47° C; (c) moderate winter daytime temperatures, generally above 15° C with minimum temperatures generally greater than 4° C; and (d) high solar radiation throughout most of the year, with occasional cloud cover (Szarek, 1974). Since the study was conducted in a desert environment, the results are particularly useful toward expanding the recorded observations of stress environment physiology of CAM plants.

Methodologies and equipment were adopted which could be efficiently utilized for analyses of in situ acid metabolism and gas exchange. Stem tissue acidity was determined by titration. Plant and soil water potential ($\psi$) were measured by thermocouple psychrometry. Transpiration was measured with a diffusion resistance hygrometer, and carbon dioxide – $^{14}C$ assimilation was measured with a portable $^{14}CO_2$ porometer. The transpiration ratio (TR) was calculated by dividing the instaneous rate of transpiration by the instaneous rate of $^{14}CO_2$ assimilation, and the maximum $CO_2$ assimilation was calculated with the assumption that the $CO_2$ concentration difference between the atmosphere and site of carboxylation was constant and 59.3 mg $CO_2/cm^3$ x $10^5$. All procedures were described elsewhere (Szarek & Ting, 1974).

The nocturnal acidification rate is the theoretical $CO_2$ assimilation rate required to maintain the observed level of nocturnal acid synthesis. This rate was calculated assuming: (a) malic acid synthesis is linear throughout the nighttime; and (b) a stoichiometric relationship exists between acidification and $CO_2$ uptake.

## Results and Discussion

### Response to precipitation
Precipitation was the major environmental parameter influencing the physiological activity of *Opuntia* spp., by effects on plant $\psi$. Water was absorbed within 24 to 36 hr after rainfall. Plant water status followed soil water status in the top 30 cm of the soil profile, until soil $\psi$ decreased to below −40 bars. Soil $\psi$ did not directly regulate acid metabolism. Plant $\psi$ was high following rainfalls as small as 6 mm, and values of −1 to −3 bars were commonly measured. The nocturnal $^{14}CO_2$ assimilation of *O. basilaris* reached maximum rates, i.e., 10 to 15 mg $CO_2/dm^2 \cdot hr$, when plant $\psi$ was high, and stomatal resistance ranged from 2 to 3 sec/cm. During such periods maximum nighttime acidity levels commonly reached 75-95 $\mu$eg/g fresh wt. The day/night difference of stem tissue titratable acidity approached 100 to 150% increases following rainfall. During periods of enhanced nocturnal acidification the efficiency of gas exchange was highest.

### TR and CAE estimates
The patterns of annual variation of TR and CAE are presented in Fig. 1. During a period of rainfall from January through March, mean TR values ranged from 50 to 200, with the single lowest value being 25. After the termination of rainfall TR increased during April and remained in excess of 1000 until a midsummer rainfall occurring in August. Mean TR values during August ranged from 135 to 215, with the single lowest value being 70. TR subsequently increased again and remained very high throughout the rainless late fall and early winter. The annual TR of *O. basilaris* showed significant variation, although mean TR values were within the range of the values reported for the three photosynthetic groups of plants (Black, 1973). Typical values from the literature are 450 to 600 for $C_3$ plants, 250 to 350 for $C_4$ plants and 25 to 150 for CAM plants (Table 1). The large variation of nocturnal TR reported above indicates that during periods of rainfall and high plant $\psi$ the efficiency of water use by CAM plants is highest, and TR is below the values of $C_3$ and $C_4$ plants. However, during periods of decreasing plant $\psi$ without rainfall, CAM plant TR is above the level of $C_3$ and $C_4$ plants. These

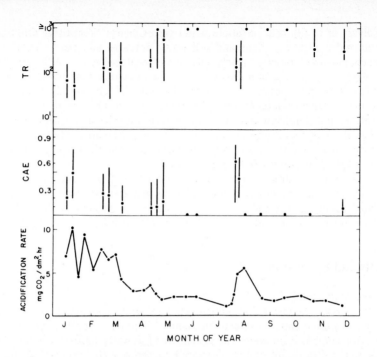

*Fig. 1.* (Upper) Annual variation in nocturnal transpiration ratio (TR). The mean value (closed circle) and range of the values (bar) are presented for each sampling period during 1973. Data from *O. basilaris*. (Middle) Annual variation in nocturnal carbon assimilation efficiency (CAE). The mean value (closed circle) and range of the values (bar) are presented for each sampling period during 1973. Data from *O. basilaris*. (Lower) Annual variation in the nocturnal acidification rate. The mean value (closed circle) of twelfe acidity samples are presented for each sampling period during 1973. Data from *O. basilaris*.

data seem to be consistent with Neales (1973) who found an increase in TR from 32 to 256 when *Agave americana* was transferred from 15° C nights to 36° C nights with day temperatures constant at 25° C.

The annual variation of CAE presented in Fig. 1 followed the same pattern, in response to rainfall and plant $\psi$. Early in the year, mean CAE values ranged between 0.2 to 0.5, with the single highest value being 0.75. CAE decreased in April and remained near zero values until the August rainfall, when mean values ranged between 0.4 to 0.6, with the single highest value being 0.82. CAE subsequently decreased throughout late fall and early winter. The annual CAE of *O. basilaris* has previously been shown to correlate with the annual variation of the mesophyll resistance to $CO_2$ transfer and assimilation (Szarek & Ting, 1974).

*Acidification rate and $CO_2$ recycling*

The nocturnal acidification persisted throughout the year (Fig. 1), despite the significant annual variations in TR and CAE. During the months from January until March the acidification rate declined from 10 to 6 mg $CO_2$/$dm^2$·hr. During the rainless late fall and early winter the acidification rate again decreased to

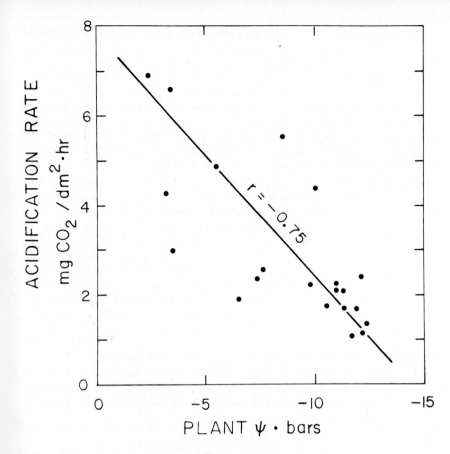

*Fig. 2.* Effect of plant water potential ($\psi$) on the nocturnal acidification rate. Each acidification rate is the mean value of twelve samples, and each $\psi$ value is the mean value of three samples taken during sampling periods of 1973. Data from *O. basilaris.*

values near 2 mg $CO_2/dm^2 \cdot$hr. Rainfall influenced the level of acid metabolism by affecting plant $\psi$. In *O. basilaris* the relationship between acidification rate and plant $\psi$ appears to be nearly linear (Fig. 2). High rates of nocturnal acidification occurred when plant $\psi$ was high, and lower rates occurred when plant $\psi$ was lowest.

In *Opuntia* spp. atmospheric gas exchange during periods of drought is greatly reduced by continuous stomatal closure and the highly impervious cutinized stem. The cuticular diffusion resistances to water are very high, averaging over 600 sec cm$^{-1}$ and 1000 sec cm$^{-1}$ for *O. basilaris* and *O. bigelovii,* respectively (Ting, et al., 1972). Such high gas diffusion resistance must also restrict the loss of endogenous $CO_2$. The persistence of metabolic activity (Fig.1) appears to be due to the continuous day/night recycling of endogenous $CO_2$ (Szarek, et al., 1973).

In the dark, $CO_2$ is produced by respiration, and consumed by acidification. Net dark respiration occurs over a wide range of temperatures in *O. basilaris,* with the temperature optimum between 35 to 40°C (Szarek & Ting, 1974). Moreover, the rate of respiration of stem tissue equilibrated in solutions of -25 bars osmotic potentials was maintained at 60% of the control rate. Further, tissue collected from plants prior to and after precipitation showed substantial respiration despite $\psi$ of -10 to -12 bars. These results suggest respiration continues at a moderate level despite in situ conditions of water stress, since plant $\psi$ are usually greater than -16 to -20 bars in both *Opuntia* spp. Nocturnal acidification in *O. basilaris* persisted at a low level, i.e., 2 mg $CO_2$/ $dm^2 \cdot hr$. despite the cessation of atmospheric gas exchange (Szarek & Ting, 1974). The persistent consumption of endogenous $CO_2$ by acidification suggests continuous enzymatic activity, i.e., phosphoenolpyruvate carboxylase, throughout such periods of stomatal closure.

In the light, $CO_2$ produced by both deacidification and photorespiration and consumed by photosyntheisis, is nevertheless, conserved by stomatal closure. Previously, $^{14}CO_2$ assimilated in the dark by CAM plants was shown to be conserved and subsequently appear in photosynthetic products (Kunitake & Saltman, 1958), and in *Opuntia ramosissima* 80% of the dark fixed $^{14}CO_2$ appeared as $^{14}C$-products (Ting & Dugger, 1968). Since the rate of deacidification in *O. basilaris* is 40% greater than the rate of acidification (Szarek, 1974), $CO_2$ production resulting solely from deacidification would be near 2.3 mg $CO_2$/$dm^2 \cdot hr$. during periods of drought. Photorespiration affects the net carbon balance of *O. basilaris* by net $CO_2$ evolution with stem temperatures in excess of 25°C (Szarek, 1974). Additionally, stem temperatures and stem tissue acidity levels influence photosynthetic activity since net $O_2$ evolution occurs in the light with temperatures below 20°C and acidity levels less than 95 $\mu$eq/g fresh wt.

The metabolic activity of *O. bigelovii*, as estimated by nocturnal acidification persisted throughout three years of severe experimental drought conditions (Fig.-3). The acidification rate during June, 1974 was similar to the rate measured in

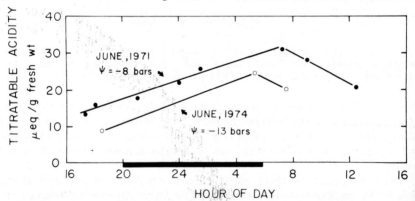

*Fig. 3.* The circadian variation in titratable acidity preceding and following a three year experimental drought. Data for June, 1971 (closed circle) and data for June, 1974 (open circle). Each acidity sample presented for the two sampling periods. Each plant water potential ($\psi$) determination is the mean value of two samples taken at sunrise from the apical region of the plant. Data from *O bigelovii*.

June, 1971, although the maximum acidity level had decreased from 30 to 24 $\mu$eq/g fresh wt. During this time the plants of *O. bigelovii* had been cut off at ground level and supported above the desert surface by ring stands and clamps

After three years without any water the plant $\psi$ at the apex had decreased only to -13 bars, a remarkable conservation of water since initial plant $\psi$ was -8 bars during June, 1971. This *Opuntia* sp. responds to severe drought by „drying out" from the bottom of the plant to the top. Individual stems also „dried out" from their point of attachment to their growing apex. The total number of living, green stems had been reduced from 40 to 5 at the end of the three year experiment, although metabolic activity persisted in the remaining stems at a moderate level.

*Table 1.* Summary of gas exchange parameters for CAM, $C_4$, and $C_3$ plants.

| | $P^{a)}$ | $R_s(min)^{b)}$ | $R_m(min)^{b)}$ | TR(ave) | E or CAE (max) |
|---|---|---|---|---|---|
| CAM$^{c)}$ | | | | | |
| dark | 10-15 | 2-10 | 1-2 | 25-150 | 0.7-0.9 |
| light | 3-20 | $\rangle 6.0$ | 20-40 | 150-600 | 0.2-0.3 |
| $C_4$ | 30-60 | $\rangle 1.0$ | 1.5 | 250-350 | 0.7-0.9 |
| $C_3$ | 20-40 | $\langle 1.0$ | 3.0 | 450-600 | 0.2-0.3 |

a) Expected range of $CO_2$ assimilation rates in mg $dm^{-2}$ $hr^{-1}$.
b) Stomatal or leaf resistance in sec $cm^{-1}$
c) CAM estimates taken from our own data plus literature, and $C_3$ and $C_4$ data summarized from available literature.

*Comparison with $C_3$ and $C_4$ plants.*

Table 1 summarizes a comparison between $C_3$-, $C_4$-, and CAM-photosynthetic option plants. The CAM data are taken from our own studies and from literature citations whereas the $C_3$ and $C_4$ data are largely derived from the literature. The generally higher $R_s$ estimates for CAM in comparison with $C_3$ and $C_4$ probably account for the generally lower carbon dioxide assimilation rates of CAM plants in both dark and light since our estimates of $R_m$ in the dark are in the range of $C_4$-plant $R_m$ estimates. However, the combination of resistances and environment results in a lower transpiration ratio for CAM than for $C_3$ and $C_4$. Yet in the light CAM can be as high as $C_3$-plants. Hence the overall low TR of CAM plants is in large part a function of the lower evaporative demand of night when most transpiration and photosynthesis occur.

A variety of data are now available in the literature allowing calculation of E for $C_3$ and $C_4$ plants. For example, data of Ludlow (1970) and Ludlow and Wilson (1972) summarized in Gifford (1974) for the $C_3$ plants *Phaseolus atropurpurens* and *Vigna luteola* give 0.32 and 0.3 respectively. In a comparative study, Slatyer (1971) reported resistance data for $C_3$ *Atriplex hastata* and $C_4$ *A. spongiosa* which give E estimates of 0.3 and 0.79 respectively. Data of Bull (1969) for a $C_4$ *Saccharum* species show 0.83, for $C_4$ *Amaranthus viridis* 0.86, and for *Zea mays* 0.72. Data of Gifford and Musgrave cited in Gifford (1974) show 0.65 for $C_4$ *Zea mays*. Björkman et al (1972) reported data resulting in a near maximum E of 1 for $C_4$ *Tidestromia oblongifolia*.

A study by Neales et al (1968) with pineapple (a CAM plant) showed an E value of 0.93 in the light and a CAE of 0.72 in the dark. In our own *Opuntia* study, our estimates of CAE ranged from 0.2 to 0.82.

The CAE estimates for CAM plants in the dark are in the range of the $C_4$ plant estimates for E. Since this ratio is largely independent of the environment, we suspect that the similarity results in part because of the similar assimilating pathway, i.e., initially through P-enolpyruvate carboxylase (Ting, 1971). In the light, the CAE ratio for *Opuntia* aligns to a level comparable to $C_3$-option plants. We interpret this to mean that $CO_2$ assimilation in the light is through the $C_3$ pathway (Kluge, 1971).

The latter should not be confused with the hypothesis that CAM plants can shift to the $C_3$-option (Osmond et al., 1973). Here $CO_2$ assimilation would be predominantly in the light and little or no diurnal malic acid fluctuation would occur. Our *Opuntia basilaris* plants of Deep Canyon presumably are always in the CAM-option since the isotopic discrimination estimate was -12 o/oo (data obtained by Dr. J.C. Lerman).

*Adaptive significance of CAM*

Under the arid conditions of the Boyd Desert Research Center, CAM plants appear to respond only to moisture. In other situations, CAM metabolism may respond to salt treatment (Winter, 1973), thermoperiod (Ting, 1971), or photoperiod (Queiroz, 1974). Except in the latter, the response which may be either an enhancement of CAM activity or a shift from CAM to the $C_3$-option, water status may be the regulating feature. We visualize under extreme drought, CAM plants will be hermatically sealed and recycle endogenous $CO_2$ through the CAM pathway. Little $CO_2$ loss allows for survival. Secondly, in arid environments during favorable periods, exogenous $CO_2$ is assimilated via the CAM pathway. And finally, during extended mesic periods in arid or arid-tropical environments CAM plants may shift to the $C_3$ photosynthetic option (Osmond et al, 1973).

Hence, the adaptive significance of the CAM photosynthetic option has a two-fold interpretation. Firstly, and primarily, the nocturnal assimilation of $CO_2$ into organic acids accounts for the high efficiency of water use and carbon assimilation immediately after periods of optimum environmental and plant conditions. Secondly, CAM contributes to the continuous day/night recycling of endogenous $CO_2$, which may be efficiently conserved by strict stomatal regulation and high cuticular diffusion resistances. The recycling of endogenous $CO_2$ maintains a moderate level of metabolic activity during long periods of severe drought and enables immediate physiological responses to small amounts of rainfall. We conclude that the CAM photosynthetic option is an adaptation to water stressed environments in that there is a minimal amount of water loss per carbon dioxide assimilated. This increased carbon assimilation efficiency relative to water loss is reflected in the low TR estimates and high CAE values.

# REFERENCES

Björkman, O., R.W. Pearcy, A.T. Harrison & H.A. Mooney (1972): Photosynthetic adaptation to high temperatures: a field study in Death Valley, California. *Science,* 175: 786-789.

Black, C.C. (1973): Photosynthetic carbon fixation in relation to net $CO_2$ uptake. *Annu. Rev. Plant Physiol.,* 24: 253-286.

Bull, T.A. (1969): Photosynthetic efficiencies and photorespiration in Calvin cycle and $C_4$-di carboxylic plants. *Crop Sci.,* 9: 726-729.

Gifford, R.M. (1974): A comparison of potential photosynthesis, productivity and yield c plant species with differing photosynthetic metabolism. *Aust. J. Plant. Physiol.,* 1 107-117.

Kluge, M. (1971): Studies on $CO_2$ fixation by succulent plants in the light. In Hatch, M.D., Osmond, C.B., & R.O. Slatyer (Eds.): Photosynthesis and Photorespiration. Pp. 283-287. Wiley – Interscience. N.Y.

Kunitake, G., & P.D. Saltman (1958): Dark fixation of $CO_2$ by succulent leaves: conservation of the dark fixed $CO_2$ under diurnal conditions *Plant Physiol.*, 33: 400-403.

Neales, T.F. (1973): The effect of night temperature on $CO_2$ assimilation, transpiration, and water use efficiency in *Agave Americana* L. *Aust. J. Biol. Sci.*, 26: 705-714.

Neales, T.F., A.A. Patterson & V.J. Hartney (1968): Physiological adaptation to drought in the carbon assimilation and water loss of xerophytes. *Nature*, 219: 469-472.

Osmond, C.B., W.G. Allaway, B.G. Sutton, J.H. Troughton, O. Queiroz, U. Luttge, K. Winter (1973): Carbon isotope discrimination in photosynthesis in CAM plants. *Nature*, 246: 41-42.

Queiroz, O. (1974): Circadian rhythms and metabolic patterns. *Annu. Rev. Plant Physiol.*, 25: 115-134.

Slatyer, R.O. (1971): Relationship between plant growth and leaf photosynthesis in $C_3$ and $C_4$ species of *Atriplex*. In Hatch, M.D., Osmond, C.B., & Slatyer, R.O. (Eds.): Photosynthesis and Photorespiration, pp. 76-81, Wiley – Interscience, N.Y.

Szarek, S.R. (1974): Physiological mechanisms of drought adaptation in *Opuntia basilaris* Engelm. & Bigel. Ph.D. Thesis. University of California, Riverside.

Szarek, S.R., H.B. Johnson & I.P. Ting (1973): Drought adaptation in *Opuntia basilaris*. Significance of recycling carbon through Crassulacean acid metabolism. *Plant Physiol.*, 52: 539-541.

Szarek, S.R. & I.P. Ting (1974): Respiration and gas exchange in stem tissue of *Opuntia basilaris*. *Plant Physiol.*, 54: 829-834.

Szarek, S.R. & I.P. Ting (1974): Seasonal patterns of acid metabolism and gas exchange in *Opuntia basilaris*. *Plant Physiol.*, 54: 76-81.

Ting, I.P. (1971): Nonautotrophic $CO_2$ fixation and Crassulacean Acid Metabolism. In Hatch, M.D., Osmond, C.B. & Slatyer, R.O. (Eds.): Photosynthesis and Photorespiration. Pp. 169-185. Wiley – Interscience N.Y.

Ting, I.P. & W.M. Dugger (1968): Non-autotrophic carbon dioxide metabolism in cacti. *Bot. Gaz.*, 129: 9-15.

Ting, I.P., H.B. Johnson & S.R. Szarek (1972): Net $CO_2$ fixation in Crassulacean Acid Metabolism plants. In Black, C.C. (ed.): Net carbon dioxide assimilation in higher plants. *Proc. Symp. S. Sect. Am. Soc. Plant Physiol.*, Pp. 26-53. North Carolina.

Winter, K. (1973): $CO_2$-Fixierungsreaktionen bei der Salzpflange *Mesembryanthemum crystallinum* unter variierten Aussenbedingungen. *Planta* (Berl.), 114: 75-85.

# THE GAS EXCHANGE PATTERNS OF CAM PLANTS

T.F. NEALES

*Botany Department, University of Melbourne, Parkville 3052, Victoria, Australia.*

*Abstract*

The gas exchange patterns of CAM plants are discussed with particular reference to the diurnal changes in the $CO_2$ flux. A model is proposed in which the variations of this pattern between species is seen as a gradation from the 'non-CAM' ($C_3$ or $C_4$ plants) to the 'super-CAM'; in the latter of which all the gas exchange takes place at night. The category of 'weak-CAM' is proposed for such plants as *Portulacaria afra* and *Kleinia articulata*. The 'full-CAM' pattern is that found in *Agave americana, Opuntia stricta* and *K.daigremontiana* and many other species. The dependence of the predominance of nocturnal $CO_2$ assimilation on night temperature and drought is discussed. Whilst salinity is shown to affect titratable acidity and the $\delta^{13}C$ ratio in the Australian halophyte, *Disphyma australe,* no net uptake of $CO_2$ at night in this plant was observed, with or without salt. The CAM pattern in thick-leaved tropical orchids is confirmed by $\delta^{13}C$ values.

## Introduction

*History of attitudes towards CAM*

It has always seemed strange that the intensity with which the biochemistry of CAM plants was studied in the late 1940's and 1950's did not then result in realisation of CAM for what it is — an alternative variant in the mode of carbon acquisition, or photosynthesis, by some plants.

Before 1955, most of the studies of CAM were directed towards the under-standing of the biochemistry of acid metabolism. This was true of the important schools of Wolf (Leipzig), Thomas et al. (Newcatsle) and of Vickery & Pucher (New Haven). In 1946, Thomas (1949) and Wolf (1949) both showed that exogenous $CO_2$ was a metabolite in the formation of malate in CAM plants; they rediscovered and explained what they named the 'de Saussure (1804) effect'. Microbiologists had earlier (Wood & Werkman, 1936) shown $CO_2$ to be a meta-bolite in organic acid formation and also, at this time, Krebs & Johnson (1937), in Cambridge, were demonstrating the central importance of organic acids in the cellular oxidation of carbohydrate in animal tissues. It is of little wonder, there-fore, that organic acid formation was the focus of such attention in CAM plants.

However, the extent to which the 'dark' assimilation of $CO_2$ contributed to the carbon balance of CAM plants did not become clear until quantitative studies were done on the diurnal rhythm of $CO_2$ fluxes in CAM plants. Such studies awaited the advent of the infra red gas analyzer (IRGA or URAS). The first of such studies was that of Gregory et al. (1954), using *Kalanchoe blossfieldiana,* but

*Acknowledgements:* Robin Duke provided competent technical assistance. The results of Fig. 6 were obtained in conjunction with V.J. Hartney, and those of Table 3 by cooperation with Dr. C.S. Hew. The $\delta^{13}C$ analysis were made by Mrs. K. Card at the instigation of Dr. J. Troughton. Grateful acknowledgement is also made to the Australian Research Grants Committee for support.

their emphasis was on the interrelationship of $CO_2$ fluxes and the photoperiodic induction of flowering. Neurnberk (1961) showed the extent of the distribution among various families of the ability to assimilate $CO_2$ in the dark, and Joshi et al. (1965) introduced the innovation of measuring the rhythms of both $CO_2$ and water vapour exchange in the pineapple. The CAM pattern of gas exchange was then associated, as had long been inferred, with high water-use-efficiencies. The magnitude and diurnal rhythms of $CO_2$ and water vapour exchange in a range of CAM plants became clear after the papers of Kluge & Fischer (1967) and Neales et al. (1968).

Table 1. Maximum fluxes of $CO_2$ into CAM leaves (mg $CO_2$ $dm^{-2}h^{-1}$).

|  | By day | By night |
|---|---|---|
| Agave americana | 6.70 | 11.80 |
| Ananas comosus | 0.94 | 0.48 |
| Aeonium howarthii | 4.00 | 3.2 |

Measurements of water vapour flux also supported previous findings (Nishida, 1963) of the 'inverted' stomatal rhythm of these plants. The quantitative measurements of $CO_2$ flux established clearly the considerable degree to which 'dark' $CO_2$ assimilation contributed to the total daily carbon balance of the plant, and it was also evident that the nocturnal $CO_2$ influx was less sensitive to drought than the daytime $CO_2$ influx. For instance, in a droughted plant of *Agave americana* 96% of the total daily $CO_2$ influx took place in the 8h night.

It is also of note that the comparatively recent rise in the interest in the biology of CAM followed the discovery and elucidation of the $C_4$ variant of photosynthesis (Kortschak et al., 1965; Hatch & Slack, 1966). This event gave a great stimulus to the study of the comparative physiology and biochemistry of photosynthesis.

## Gas Exchange Patterns in CAM Plants

*Defining the scope of this paper*
It is my assignment to discuss the diurnal patterns of gas exchange of CAM plants. I wish to qualify and restrict this area as follows. I will confine myself to a consideration of the fluxes of $CO_2$ and water vapour, neglecting the oxygen flux, because this is the subject of a later address by Dr. Marcelle. Furthermore, I shall neglect the interesting areas of the endogenous rhythms of $CO_2$ fluxes in CAM plants and will also only deal in passing with the evidence that this flux is also greatly affected by the photoperiodic regime. Both of these are topics that will doubtless be expertly discussed by Dr. Queiroz in this symposium.

The topics I will deal with are — a, the gas exchange characteristics of a hypothetical 'super-CAM' plant; b. the actual variants of this that we and other have observed in various plants; c. environmental factors (apart from photoperiod) that modify CAM behaviour.

The discussion will stop short of attempting to attribute variations in th observed gas exchange to biochemical events as, again, these topics are due to be dealt with by other contributors. ·

## Models: Non-CAM and Super-CAM

One framework, admittedley teleological, within which to place this discussion, is a hypothetical model of what might be the most extreme form of CAM possible — here called 'Super-CAM'.

An obvious essential feature of the $C_3$ or $C_4$ photosynthesis characteristics of land plants is that these modes of carbon assimilation require, in nature, that a flux of light quanta and $CO_2$ molecules be present together at the chloroplast in order that the photoreduction of $CO_2$ can take place. The storage of light-produced assimilatory power is very limited, and the $CO_2$ assimilation of leaves of $C_3$ and $C_4$ plants ceases abruptly when the light source is removed. Consequent upon this, and the normal stomatal rhythm of these plants, is that if such plants are placed in a 'Square-wave' (lights on/lights off) environment, then a 'Square-wave' pattern (Non-CAM) of gas exchange results (Fig. 1A).

We may regard CAM as essentially the possession by some plants of the ability to separate, between day and night, the requirement for light and the requirement for $CO_2$. Coupled with this ability, one must add the day-closed/night-open 'abnormal' rhythm of stomatal aperture. A 'Super-CAM' plant is here defined as one in which this separation is complete: the day time gas exchange is nil or very low, and the leaf being sealed from the atmosphere; whereas at night all the gas exchange of the plant takes place. Fig. 1D illustrates the diurnal gas exchange pattern of such a plant. Whilst this model of 'Super-CAM' would have the maximum water use efficiency, it does also have certain constraints. The first is that by day the dissipation of the net radiation flux as latent heat is not possible, hence high leaf temperatures are to be expected, and indeed are found, in *Opuntia* for instance. Also it assumes that sufficient photo-formation of the source of precursor (starch) of the dark $CO_2$ acceptor (phosphoenolpyruvate) will take place by day. High water use efficiency would also depend upon low night-time transpiration whilst the stomata are open. This is entirely possible because at night wind speeds are lower than by day and also the water vapour pressure gradient from leaf to atmosphere is much smaller than by day, due to lower leaf and atmosphere temperatures.

## Variants of Gas Exchange Patterns in Succulent Plants.

In this section are considered the variants that we have observed in the patterns of gas exchange, that have been found in well watered plants, which are succulent and might, therefore, be expected to have CAM characteristics.

a. 'Non-CAM'

Fig. 2 shows the 'Non-CAM' pattern of gas exchange that was found in the succulent halophyte *Disphyma australe* ('pigface') when grown in Hoagland's solution plus 0.025 NaCl. The effects of salinity on the $CO_2$ assimilation of this plant are considered later. This 'Non-CAM' pattern was also found in *Mesembryanthemum crystallinum* and *Salicornia quinquifolia*.

b. 'Weak-CAM'

Two succulent plants that we have examined (*Portulacaria afra* and *Kleinia articulata*) have diurnal patterns of gas exchange, which are similar to 'non-CAM'

# CAM GAS EXCHANGE PATTERNS

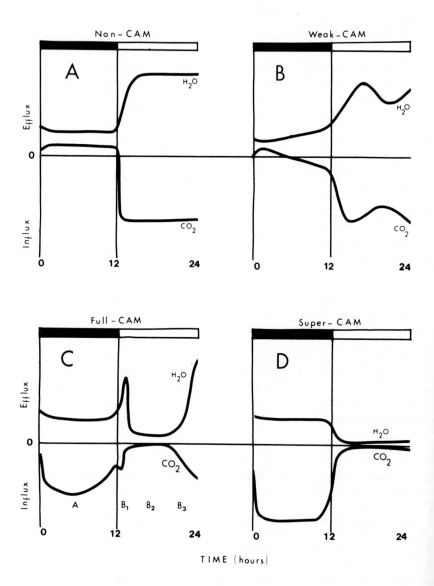

*Fig. 1.* CAM Gas Exchange Patterns (see text for discussion).

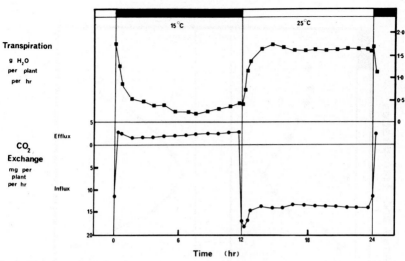

Fig. 2. Water Vapour and $CO_2$ fluxes of *Disphyma australe*. (Day/Night temperature 25/15°C. Irradiance c.100wm⁻² ).

patterns but consistently show a small noctural uptake $CO_2$ This 'Weak-CAM' pattern of *Portulacaria afra* is shown in Fig.3. These plants may possibly be, in an evolutionary as well as a physiological sense, intermediate between $C_3$ and CAM-type plants. Leaves of this plant had $\delta^{13}C$ values (3 replicates) of $-24.6$, $-25.2$ and $-26.1\%o$.

c: 'Full-CAM'

The type of gas exchange pattern that is most typical of CAM plants is that in which the dark assimilation of $CO_2$ contributes significantly to the carbon balance of the plant. This is here called the 'Full-CAM' pattern of behaviour. This is shown quite typically in *Agave americana* (Fig. 4), plants of which had $\delta^{13}C$ values (3 replicates) of $-14.2$, $-14.3$ and $-14.5\%o$.

The following plants have shown similar patterns

| | |
|---|---|
| Cactaceae | *Opuntia stricta* |
| Bromeliaceae | *Ananas comosus* (Pineapple) |
| | *Bilbergia pyrimaldalis* |
| Crassulaceae | *Aeonium arborium* |
| | *Aeonium howarthii* |
| Liliaceae | *Aloe distans* |
| Agavaceae | *Agave americana* |

The published patterns of $CO_2$ exchange of *Bryophyllum daigremontianum* (Osmond & Allaway, 1974; Kluge & Fischer, 1967), *Tillandsia usneoides*(Kluge et al., 1973) and the thick-leaved orchid *Oncidium lanceanum* (Wong, 1973), are similar. It should be emphasized that these patterns are obtained with well-watered plants in conditions of night temperature below 20° C.

The 'Full-CAM' pattern of $CO_2$ assimilation has several phases (Fig. 1C): A

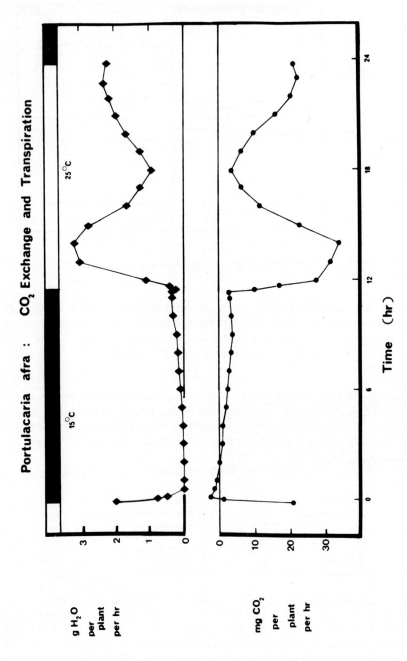

*Fig. 3.* Water Vapour and CO$_2$ fluxes of *Portulacaia afra*. (Conditions as for Fig. 2).

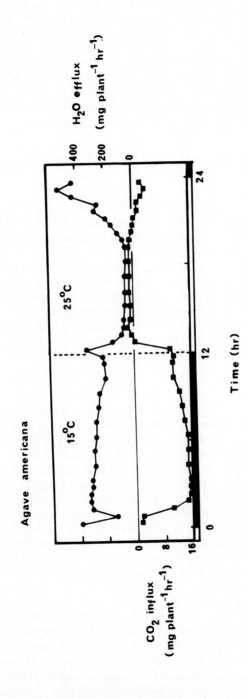

*Fig. 4.* Water Vapour and CO₂ fluxes of *Agave americana*. (Day/Night temperature 25/15°C. Irradiance c.120wm⁻².)

(Dark): a 'Dark' rate of $CO_2$ assimilation comparable to, or in excess of, the light rate. This influx reaches an early maximum and then decreases. If very long nights are given, this $CO_2$ influx in the dark ceases after a time, dependent upon the previous light regime, when a respiratory efflux results. B(Light): B1.An initial influx which rapidly diminishes as stomatal closure takes place. B2. A period of very low $CO_2$ influx and $H_2O$ efflux. This may last for periods of many hours, during which stomatal resistance is high. B3. A phase of increasing $CO_2$ influx and water vapour efflux which is accompanied by stomatal opening. This phase is absent, or much reduced, in droughted plants.

## Environmental Factors Affecting the Gas Exchange Pattern

*Night Temperature*
The accumulation of titratable acidity in the leaves of CAM plants has long been known to be decreased by high, and favoured by low, night temperatures. As anticipated (Fig. 5), such treatments applied to *Agave americana* also decrease, or eliminate, the nocturnal assimilation of $CO_2$ (Neales, 1973). This perhaps substantiates the claim that CAM is an adaptation to environments of cool nights. It is also evident that, when the nocturnal assimilation of $CO_2$ is inhibited, this causes the increase in daytime gas exchange and the plant approximates to the 'Non-CAM' pattern of behaviour (Fig. 5). Kluge et al. (1973) have elegantly shown the same effect of temperature on the $CO_2$ assimilation of *Tillandsia usneoides,* in which the nocturnal component of $CO_2$ uptake was greatly diminished by an increase in night temperature from $20°$ C to $25°$ C and eliminated at $30°$ C.

*The Effects of Drought*
The daytime gas exchange of both *Bryophyllum daigremontianum* (Kluge & Fischer, 1967) and *Agave americana* (Neales, Hartney & Patterson, 1968) are greatly reduced in plants deprived of water (Fig. 6). The nocturnal phase of gas exchange is much less affected, thus the positive carbon balance of the plant is maintained and water loss greatly reduced. Under drought, 98% of the total net $CO_2$ assimilation of an *Agave americana* plant (in a 16hr/8hr light/dark regime) takes place in the 8hr dark period, (Fig. 6), when the plant approaches the 'Super-CAM' condition (Fig. 1D). This suppression of daytime gas exchange in droughted plants has been observed by Bartholomew (1973) in the Crassulacean plant, *Dudleya farinosa,* growing in the field, and Szarek et al. (1973) have shown that in *Opuntia basilaris* under drought, gas exchange by day and night is minimal with very high diffusive resistances (c.600 s $cm^{-1}$) throughout the day/night cycle.

*The Effect of Salinity on CAM*
The induction of a CAM-type response by NaCl given to the rooting medium was discovered by Winter & Willert (1972) in *Mesembryanthemum crystallinum* and also in *Carpobrotus edulis,* (Winter,1973). In 1973 we tried to repeat this effect with *Disphyma australa,* a plant closely related to *C.edulis.* However, we failed to observe any induction of nocturnal $CO_2$ assimilation (Fig. 2) and the $\delta^{13}C$ values of plants grown for one month in Hoagland's Solution, in which the concentration

Agave americana

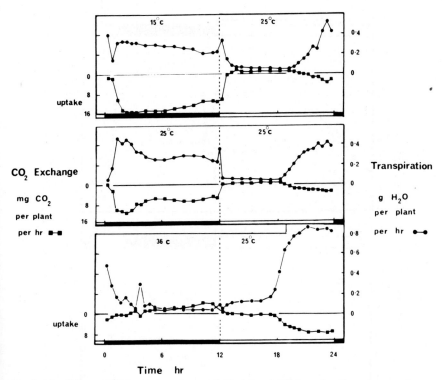

*Fig. 5.* The Effect of Night Temperatures of 15, 25 and 35°C on the diurnal gas exchange patterns of *Agave americana.*

of NaCl ranged from 0 to $10^{-1}$ M, varied between $-33.8‰$ to $-30.7‰$ (Table 2). The type of CAM activity that has been induced by Winter & Willert (1972) appears to be similar to that here classified as 'Weak-CAM' (Fig. 1).

The $\delta^{13}C$ values of *Disphyma australa* indicate that CAM is only unduced in this plant by NaCl concentrations that greatly restrict growth (Table 2). Gas exchange measurements failed to show nocturnal $CO_2$ assimilation in any of the NaCl treatments used.

## CAM in Orchids

Some species of orchids have long been known to exhibit CAM activity in terms of the accumulation of titratable acidity at night, of the nocturnal net assimilation of $CO_2$, and the possession of the inverted stomatal rhythm (Neurnbergk, 1961; McWilliams, 1970). Furthermore, the thick-leaved orchids have been shown to possess greater CAM activity than the thin-leaved (McWilliams, 1970; Wong & Hew, 1973). In co-operation with Dr. Hew, from Singapore, and Dr. John Troughton, in New Zealand, we obtained leaf specimens of 5 thick-leaved and 5 thin-leaved orchids from Singapore, and measured the $\delta^{13}C$ values of this leaf tissue. The value of the thin-leaved orchids ranged between $-22.6‰$ and $-28.1‰$, whilst

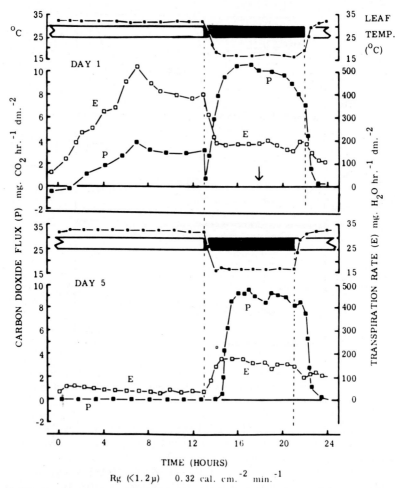

Fig. 6. The Gas Exchange Patterns of *Agave americana* when supplied with water, and 4 days after the roots were deprived of water.

the range for the thick-leaved varieties was from −14.9‰ to −16.2‰ (Table 3).

This clearly substantiates the previous observations that the thick-leaved orchids have CAM activity. Moreover, this occurred at Singapore, where the minimum night temperatures are c.22° C. As here confirmed for the Orchidaceae, the Bromeliacea also appear to have CAM and 'Non-CAM' forms (Medina & Troughton, 1974).

## Discussion

In CAM plants the patterns of $CO_2$ exchange over a diurnal cycle vary both with

*Table 2.* The Effects of Salinity on *Disphyma australe.*

| Expt. No. | Growth medium | NaCl added (M) | Titratable acidity ($\mu$ mole. g FW$^{-1}$) Morning | Titratable acidity ($\mu$ mole. g FW$^{-1}$) Evening | $\delta^{13}C$[b] (‰) |
|---|---|---|---|---|---|
| 1 | Soil | 0 | 7.1 | 3.9 | -31.2 |
|  | Soil | 0.1 | 8.9 | 4.0 | -26.4 |
|  | Soil | 0.5[a] | 10.1 | 4.0 | -25.6 |
| 2 | Nutrient soln. | 0 | - | - | -31.7 |
|  | Nutrient soln. | .025 | - | - | -32.1 |
|  | Nutrient soln. | .050 | - | - | -34.6 |
|  | Nutrient soln. | .100 | - | - | -31.3 |
|  | Nutrient soln. | .500[a] | - | - | -24.4 |
| 3 | Nutrient soln. | 0 | - | - | -33.8 |
|  | Nutrient soln. | $10^{-5}$ | - | - | -31.1 |
|  | Nutrient soln. | $10^{-4}$ | - | - | -30.5 |
|  | Nutrient soln. | $10^{-3}$ | - | - | -31.0 |
|  | Nutrient soln. | $10^{-2}$ | - | - | -30.1 |
|  | Nutrient soln. | $10^{-1}$ | - | - | -30.7 |

a) 0.5 M NaCl severely inhibited growth.
b) Measured by Dr. J.H. Troughton & K. Card.


*Table 3.* Carbon Isotope Ratios ($\delta^{13}C$ ‰) of some Orchids.

|  | $\delta^{13}C$ ‰ CAM | $\delta^{13}C$ ‰ 'Non-CAM' |
|---|---|---|
| Orchids[a] | $-15.3 \pm 0.2$ (n = 5) | $-26.6 \pm 1.1$ (n = 5) |
| Leaf thickness | 2.5 to 1.5 mm | 0.5 to 0.3 mm |
| Bromeliads[b] | $-13.9 \pm 1.2$ (n = 13) | $-25.3 \pm 1.8$ (n = 14) |

a) Grown on Singapore Island (Hew, Neales & Troughton, unpublished).
b) Grown at Caracas, Venezuela (from: Medina & Troughton, 1974).


genotype and environment. The Orchids and Bromeliads (Medina & Troughton, 1974) are two families that clearly have CAM and 'Non-CAM' forms. The carbon isotopic ration of plant material (Osmond et al., 1973) and of plant extracts (Lerman & Queiroz, 1974) is a valuable indication of the current mode of carbon assimilation. The CAM type of metabolism is remarkable for its sensitivity to the environmental conditions of night temperature, photoperiod, drought and salinity. The impression is gained that the underlying metabolic processes are more capable of induction and repression by environmental factors than are the metabolic processes and structures associated with $C_4$ photosynthesis. The mechanism of this induction or repression of CAM continues to be an interesting field of investigation.

# REFERENCES

Bartholomew, B. (1973): Drought response in the gas exchange of *Dudleya farinosa. Photosynthetica,* 7: 114-120.

Gregory, F.G., I. Spear & K.V. Thimann (1954): The interrelation between $CO_2$ metabolism and photoperiodism in *Kalanchoe. Plant Physiol.,* 29: 220-229.

Hatch, M.D. & C.R. Slack (1966): Photosynthesis by sugar cane leaves – a new carboxylation reaction and the pathway of sugar formation. *Biochem. J.,* 101: 103-111.

Joshi, M.C., J.S. Boyer & P.J. Kramer (1965): Growth, $CO_2$ exchange, transpiration and transpiration ratio of pineapple. *Bot. Gaz.,* 126: 174-179.

Kluge, M. & K. Fischer (1967): Uber Zusammenhange dem $CO_2$ – Austauch und der Abgabe von Wasserdampf durch *Bryophyllum daigremontianum* Berg. *Planta (Berl.),* 77: 212-223.

Kluge, M., D.L. Lange, M. Eichmann & R. Schmid (1973): Diurnaler Saurhythmus bei *Tillandsia usneoides. Planta (Berl.),* 112: 357-372.

Kortschak, H.P., C.E. Hartt & G.O. Burr (1965): Carbon dioxide fixation in Sugar cane leaves. *Plant Physiol.,* 40: 209-213.

Krebs, H.A. & W.A. Johnson (1937): The role of citric acid in intermediate metabolism in animal tissues. *Enzymologia,* 4: 148-156.

Lerman, J.C. & O. Queiroz (1974). Carbon fixation and isotope discrimination by a Crassulacean Plant: dependence upon photoperiod. *Science,* 183: 1207-1209.

McWilliams, E.L. (1970): Comparative rates of dark $CO_2$ uptake and acidification in the Bromeliaceae, Orchidaceae and Euphorbiaceae. *Bot. Gaz.,* 131: 285-290.

Medina, E. & J.H. Troughton (1974): Dark $CO_2$ fixation and the carbon isotope ratio in Bromeliacea. *Pl. Science Letters,* 2: 357-362.

Neales, T.F. (1973): The effect of night temperature on $CO_2$ assimilation, transpiration and water use efficiency in *Agave americana. Aust. J. Biol. Sci.,* 26: 705-714.

Neales, T.F., V.J. Hartney & A.A. Patterson (1968): Physiological adaptation to drought in the carbon assimilation and water loss of xerophytes. *Nature,* 219: 469-472.

Neurnbergk, E.L. (1961): Endogener rhytmus and $CO_2$ – stoffwechsel bei pflanzen mit diurnalem saurerhythmus. *Planta (Berl.),* 56: 20-70.

Nishida, K. (1963): Studies on stomatal movement of Crassulacean plants in relation to acid metabolism. *Physiol. Plantarum,* 16: 281-298.

Osmond, C.B. & W.G. Allaway (1974): Patterns of $14CO_2$ fixation during photosynthesis in *Kalanchoe daigremontiana. Aust. J. Plant Physiol.,* 2: (in press).

Osmond, C.B. et al. (1973): Carbon isotope discrimination in Photosynthesis in CAM plants. *Nature,* 246: 41-42.

Saussure, T. de (1804): Recherches chimique sur la vegetation. Paris 1804.

Szarek, S.R., H.B. Johnson & I.P. Ting (1973): Drought adaptation in *Opuntia basilaris. Plant Physiol.,* 52: 539-541.

Thomas, M. (1949): Physiological studies in acid metabolism in green plants. I. *New Phytol.,* 48: 390-420.

Vickery, H.B. & J.K. Pucher (1940). Organic acids in plants. *Annu. Rev. Biochem.,* 9: 529-544.

Winter, K. & D.J. von Willert (1972): NaCl-induzierter Crassulaceensaurestoff wechsel bei *Mesembryanthemem crystallinum. Z. Pflanzenphysiol.,* 67: 166-170.

Winter, K. (1973): NaCl- induzierter Crassulaceensaurestoff wechsel bei einer Weiteren Aizoacee: *Carpobrotus edulis. Planta (Berl.),* 115: 187-188.

Wong, S.C. (1973): M.Sc. Thesis, U. of Nanyang, Singapore.

Wong, S.C. & C.S. Hew (1973): Photosynthesis and photorespiration in some thin-leaved orchids. *J. Singapore Nat. Acad. Sci.,* 3: 150-157.

Wolf, J. (1949): Beitrage Zur Kenntuis des Saurestoffwechsels sukkulenter Crassulacea VI *Planta (Berl.),* 37: 510-534.

Wood, H.G. & C.H. Werkman (1936): The utilisation of $CO_2$ in the dissimilation of glycerol by the propionic acid bacteria. *Biochem. J.,* 30: 48-53.

# ENVIRONMENTAL CONTROL OF PHOTOSYNTHETIC OPTIONS IN CRASSULACEAN PLANTS *

C.B. OSMOND

*Department of Environmental Biology, Australian National University, Canberra, Australia*

## Abstract

Photosynthetic metabolism in plants capable of Crassulacean acid metabolism (CAM) may involve the net fixation of $CO_2$ in the preceeding dark period or the net fixation of $CO_2$ in the light. The properties of these two carboxylation reactions have been examined and evidence suggesting that dark fixation is essentially a $C_4$-like process and light fixation is essentially a $C_3$-like photosynthetic process is presented.

A number of environmental conditions directly influence the contribution of $C_3$-like and $C_4$-like photosynthetic options to the net carbon gain of these plants. These responses, the indicator significance of the $\delta^{13}C$ value in CAM plants, and the relationship to growth and persistence of these species are discussed.

Plants capable of Crassulacean acid metabolism may fix $CO_2$ from the surrounding air at substantial rates in the light or in the dark. This paper reviews evidence which suggests that CAM plants are functionally similar to $C_3$-plants during steady state fixation of exogenous $CO_2$ in the light and that the same plants are functionally similar to $C_4$-plants during $CO_2$ fixation in the dark. Environmental conditions readily modify the contribution of light and dark $CO_2$ fixation to net carbon gain in these plants so it is clear that the carboxylation reactions may be considered as optional photosynthetic processes.

These options seem to be unique to plants capable of CAM for there is no convincing evidence to date that conventional $C_3$-plants have the capacity for net carbon uptake via the $C_4$-acids. Likewise, there is no convincing evidence that $C_4$-plants have the capacity for direct carbon uptake via the carbon reduction cycle without the prior involvement of the $C_4$-acids. The increasing number of claims for intermediate forms between $C_3$ and $C_4$-plants have yet to be critically assessed.

## Optional Carboxylation Reactions in CAM Plants

A number of studies have established that the activity of PEP carboxylase and RuDP carboxylase in extracts of CAM plants is sufficient to account for the observed rates of $CO_2$ fixation in the light or in the dark (about 0.5 - 1.0 $\mu$moles. min$^{-1}$.mg$^{-1}$ chlorophyll) (Kluge & Osmond, 1972; Osmond & Allaway, 1974). At

---

*Abbreviations:* CAM = Crassulacean acid metabolism; PEP = phosphoenol pyruvate; RuDP = ribulose-1,5-diphosphate; 3-PGA = 3-phosphoglycerate.

*The author is grateful to Mrs. K.A. Card and Dr. J.H. Troughton DSIR, Lower Hutt, New Zealand, for $\delta^{13}C$ analyses cited in the text. Mr. Graham White collected some of the *Opuntia* samples.

the same time, these plants have sufficient activity of malic enzyme or PEP carboxykinase to catalyse the decarboxylation of malate formed during dark $CO_2$ fixation (Kluge & Osmond, 1972; Dittrich et al., 1973). There is no evidence that these enzymes are compartmented in different cells, as is the case in $C_4$-plants (Rouhani et al., 1973) so that the regulation of the activity of the two carboxylases and the decarboxylase is a problem of some importance. The *in vivo* data cited below suggest that these activities are strictly regulated.

*Products and labelling patterns during $^{14}CO_2$ fixation*

a. *$CO_2$ fixation in the dark*
Malate is the principal product of $^{14}CO_2$ fixation in the dark in all CAM species investigated. Early reports indicated that the distribution of $^{14}C$ within malate was constant with time and was approximately 66% in the 4-C carboxyl and 34% in the 1-C carboxyl (Bradbeer, Ranson & Stiller, 1958). It was proposed that this labeling pattern could be explained in terms of a primary carboxylation of RuDP and a secondary carboxylation of PEP derived from 3-PGA, the product of the first carboxylation reaction. More recent experiments, using a different degradation method, have not confirmed these observations but show rather, that the distribution of label within malate does not remain constant with time and is initially almost exclusively confined to the 4-C carboxyl carbon (Sutton & Osmond, 1972). Table 1 shows the proportion of radioactivity in the 4-C carboxyl

Table 1. Distribution of $^{14}C$ within malate-$^{14}C$ isolated from CAM plants after 1 min dark $^{14}CO_2$ fixation (Ambient temperature 20-25°C). CAM activity shown as percentage increase in malate content during 12 hr dark.

| Species | $^{14}C$ in 4-C carboxyl (percent) | CAM activity (percent) |
|---------|-----------------------------------|------------------------|
| *Bryophyllum tubiflorum*[*] | 89, 93 | 320 |
| *Kalanchoe fedtschenkoi* | 88, 87 | 150 |
| *Kalanchoe beharensis* | 87, 88 | 460 |
| *Sedum praealtum* | 89, 90 | 760 |
| *Sedum morganianum* | 89, 89 | 240 |
| *Opuntia inermis* | 90, 89 | 260 |
| *Trichocerus* sp. | 51, 79 | 260 |
| *Agave americana* | 89, 89 | 390 |
| *Aloe* sp. | 63, 75 | 220 |

[*] Data of Sutton & Osmond (1972) for comparison.

after 1 min dark $^{14}CO_2$ fixation by a range of CAM plants exposed to $CO_2$ concentrations between approximately 400 and 600 ppm. With two exceptions these data and those cited above indicate that the fixation of $CO_2$ by CAM plants in the dark involves the primary carboxylation of PEP to yield 4-C labelled oxaloacetate and subsequently malate. The two exceptions in this table were tissues in which the specific activity of malate was exceptionally low and difficulties were experienced with the *in vitro* degradation by NADP malic enzyme.

## b. $CO_2$ fixation in the light

During steady state $CO_2$ fixation in the late light period a variable proportion of [14]C may be fixed into malic acid. The primary products of [14]$CO_2$ fixation are 3-PGA and phosphorylated compounds (Kluge, 1969) and pulse chase experiments show that malate behaves as an end-product rather than as an intermediate of $CO_2$ fixation in the light (Osmond et al., 1973). Degradation of the malate-[14]C formed during these experiments shows that this malate may have arisen by a secondary carboxylation of PEP derived from 3-PGA as suggested by Bradbeer, Ranson & Stiller (1958). Table 2 shows that the distribution of [14]C in malate formed during steady state fixation in the light is 60% in the 4-C carboxyl, a value

*Table 2.* Distribution of [14]C within malate isolated from *Kalanchoe daigremontiana* following 15 sec [14]$CO_2$ fixation in the light or dark (Osmond & Allaway, 1974).

| Phase of $CO_2$ fixation | [14]C in 4-C carboxyl (percent) |
|---|---|
| Steady state, light | 57, 63 |
| Initial burst, light | 78, 76, 79, 79 |
| Steady state, dark | 90, 92 |

similar to that obtained by Kluge (1975) with CAM tissues exposed to high light intensities. This distribution declines to about 30% in a 5 min chase in [12]$CO_2$ suggesting the malate becomes uniformly labeled in a relatively short time. CAM plants show a characteristic initial burst of $CO_2$ fixation on illumination and malate is the principal labelled product of 15 sec [14]$CO_2$ fixation at this time. This malate contains a higher proportion of [14]C in the 4-C carboxyl suggesting that perhaps some dark $CO_2$ fixation may persist into the light phase (Osmond & Allaway, 1974).

Avadhani, Osmond & Tan (1971) obtained different results in some experiments with *Sedum praealtum.* Dark fixation malate contained about 60% of the [14]C in the 4-C carboxyl and malate labelled in the light in leaves after 24 hr illumination was initially nearly 100% labelled in the 4-C carboxyl. Subsequent experience has thrown doubt on the *Lactobacillus* degradation technique used in these experiments and these results have not been confirmed by *in vitro* techniques (Table 1).

### Kinetics of net $CO_2$ exchange and response to $O_2$ tension

Björkman & Osmond (1974) have recently examined the interaction between $CO_2$ and $O_2$ concentration, leaf temperature and light intensity during the steady state fixation of $CO_2$ by *Kalanchoe daigremontiana* in the light and in the dark. This work was done because there is good evidence, *in vivo* and *in vitro,* that $CO_2$ fixation via RuDP carboxylase in $C_3$-plants is substantially inhibited by $O_2$ tensions in the region of 21% whereas $CO_2$ fixation via PEP carboxylase in $C_4$-plants is not sensitive to $O_2$ tension between 2 and 21%. It was reasoned that such experiments might provide additional information as to the primary carboxylation reactions of CAM plants in the light and dark.

*Table 3.* Effects of $O_2$ and $CO_2$ concentration on net $CO_2$ fixation by leaves of *Kalanchoe daigremontiana* in the light and dark. (Data of Björkman & Osmond, 1974).

| Light intensity | Leaf temperature | Intercellular $CO_2$ concentration | Ambient $O_2$ concentration | Rate of $CO_2$ fixation |
|---|---|---|---|---|
| (nano Einsteins) | (°C) | ($\mu$Bars) | (percent) | ($\mu$moles cm$^{-2}$ sec$^{-1}$) |
| zero | 16° | 287 | 20 | 0.27 |
| | | | 4 | 0.27 |
| | | | 36 | 0.30 |
| 40 | 16° | 250 | 20 | 0.45 |
| | | 181 | 4 | 0.80 |
| | | 287 | 36 | 0.29 |
| 40 | 29° | 0 | 20 | -0.19 |
| | | | 4 | -0.02 |
| | | 70 | 20 | 0.11 |
| | | | 4 | 0.28 |
| | | 370 | 20 | 0.80 |
| | | | 4 | 1.06 |
| | | 650 | 20 | 1.30 |
| | | | 4 | 1.51 |

Table 3 summarizes some of the data from these experiments. $CO_2$ fixation in the dark at 16°C was insensitive to $O_2$ tension between 2 and 36%. The dark fixation process had a high affinity for $CO_2$; half maximum velocity was calculated to correspond to 3 $\mu$M $CO_2$ in the intercellular spaces. On the other hand, $CO_2$ fixation in the light was markedly sensitive to $O_2$ tension over a wide range of leaf temperatures. The inhibition due to oxygen was partially but not fully reversed by increasing intercellular $CO_2$ concentration to 650 $\mu$Bars or higher levels in other experiments. $CO_2$ fixation in the light had a much lower affinity for $CO_2$; calculated half maximal velocity at 9 $\mu$M (16°C) and 37 $\mu$M (29°C). The $CO_2$ compensation point under light saturating conditions was approximately 50 $\mu$Bars at 20% $O_2$ and declined to zero in 4% $O_2$. A post-illumination burst was frequently observed, but this was insensitive to $O_2$ tension. These data are consistent with involvement of RuDP carboxylase as the primary step in $CO_2$ fixation in the light. They are consistent with the $O_2$ inhibition of this carboxylase, associated with the generation of substrates for photorespiration but suggest some interesting differences from $C_3$-plants ($O_2$ insensitive post-illumination burst; incomplete reversal of $O_2$ inhibition by high $CO_2$) which require further study.

The data also indicate that PEP carboxylase is the primary means of $CO_2$ fixation in the dark. There is no evidence for the participation of an $O_2$ sensitive carboxylation process (such as RuDP carboxylase) in dark $CO_2$ fixation.

*Other evidence*

$C_3$-plants and $C_4$-plants may be readily distinguished on the basis of the $\delta^{13}C$

value. The $\delta^{13}C$ values of $C_3$-plants (-20 to -30‰) are due, in a large measure, to the discrimination characteristics of the primary RuDP carboxylase in these plants (Whelan et al., 1973). In the same way, the $\delta^{13}C$ values of $C_4$-plants (-10 to -15‰) reflect the slight discrimination of PEP carboxylase *in vitro*. The $\delta^{13}$ values of CAM plants may range between those characteristic of $C_3$-plants and of $C_4$-plants (Allaway et al., 1974; Bender et al., 1973). As discussed below, changes in the $\delta^{13}C$ value of a CAM plant are most readily interpreted on the basis of $C_3$-like $CO_2$ fixation in the light (via RuDP carboxylase) and $C_4$-like $CO_2$ fixation in the dark (via PEP carboxylase) (Osmond et al., 1973).

Brownell & Crossland (1974) have provided very interesting, indirect evidence for the similarity between $C_4$ photosynthesis and dark $CO_2$ fixation in CAM plants. These authors had previously shown that the requirement for sodium as a micronutrient ion was restricted to plants with the $C_4$ pathway of photosynthesis. They found that *Bryophyllum* exhibited a growth response to sodium as a micronutrient only when it was forced to depend largely on dark $CO_2$ fixation for carbon assimilation. When plants were grown on long days in full sunlight no sodium requirement was evident. This intriguing experiment was interpreted to indicate substantial similarity between dark $CO_2$ fixation in CAM and $C_4$ photosynthesis.

## The $\delta^{13}C$ Values as an Index of Carboxylation Options

That the contribution of dark $CO_2$ fixation and light fixation to total carbon gain in CAM plants is under direct environmental control may be deduced from a wide range of studies in the literature. The role of day/night temperature regime (Neales 1973a, b), daylength (Queiroz & Morel, 1974) and water status (Kluge & Fischer, 1967; Neales et al., 1968; Bartholomew, 1973) have recently been reappraised. In addition, Winter & Von Willert (1972) have shown that in certain Azoids the addition of NaCl or other salts to the growth medium induces water-stress and CAM in plants which do not normally show a diurnal fluctuation in malic acid content. These studies provide a wealth of possibilities for laboratory experiment and a basis for assessing the physiological strategies of these plants under natural conditions.

Recognizing the significance of these environmental interactions and the clear distinction between the pathways of net fixation of exogenous $CO_2$ in the light and dark, an obvious question was posed. The $\delta^{13}C$ values for CAM plants recorded in the literature range between those characteristic of $C_3$-plants on the one hand and $C_4$-plants on the other (Allaway et al., 1974; Bender et al., 1973). Does this variability occur within a single species of CAM plant? Can the environmental conditions be manipulated to alter the contribution of $C_3$-like and $C_4$-like carboxylation and so bring about predictable changes in the $\delta^{13}C$ value of the plant? Preliminary experiments involving a wide range of environmental conditions indicated an affirmative answer to both of these questions (Osmond et al., 1973) and more detailed studies have followed.

Neales (1973a,b) showed that elevated temperatures at night totally abolished fixation of exogenous $CO_2$ in the dark in *Agave* and *Ananas*. Attempts to grow *Kalanchoe daigremontiana* with night temperatures above $30°C$ have been unsuccessful but one series of experiments in the CSIRO Canberra Phytotron with $20°C$ day $25°C$ night yielded satisfactory results. The elevated night temperature substantially reduced $CO_2$ fixation in the dark (Allaway et al., 1974) and malate synthesis, principally derived from the refixation of endogenous $CO_2$ was about half that of plants growing on a $30°C$ day/$15°C$ night (Table 4). In plants grown

*Table 4.* Effect of daylength and temperature regime on malate accumulation and $\delta^{13}C$ value (whole plant) in *Kalanchoe daigremontiana*. Malate accumulation shows as difference between titratable acidity at end of day and end of night. $\delta^{13}C$ value of clonal starting material -20.1‰ (Data taken from Allaway et al., 1974; $\delta^{13}C$ values are for total shoot carbon in each case.)

| Growth conditions | $\Delta$ Malate ($\mu$ equiv/g fresh wt) | $\delta^{13}C$ value (‰) |
|---|---|---|
| 25 weeks; 8 hr day $30°$, 16 hr night $15°$ | 74 | -18.5 |
| 25 weeks; 8 hr day $20°$, 16 hr night $25°$ | 45 | -23.5 |
| 14 weeks; 16 hr day $30°$, 8 hr night | 120 | -17.6 |
| 14 weeks; 16 hr day $20°$, 8 hr night $25°$ | 80 | -19.6 |

on short and long days, the $\delta^{13}C$ value (total plant carbon) was less negative in the $30°C/15°C$ treatments, than in the $25°C/20°C$ treatments, consistent with the reduced contribution of dark fixed carbon in the latter (Table 4).

Lerman & Queiroz (1974) have found comparable changes in the $\delta^{13}C$ value (total plant carbon) in *K. blossfeldiana,* a species in which CAM can be induced by short days, without any change in day/night temperature regime (reviewed by Queiroz & Morel, 1974). These authors have made a most important additional observation. When either of the above species of *Kalanchoe* show CAM metabolism, the $\delta^{13}C$ value of the soluble extract from the plant tissue is in the vicinity of -10‰, whereas that of the isoluble material is considerably more negative. (Lerman & Queiroz, 1974; Lerman et al., 1974). These data are strong evidence for the direct participation of PEP carboxylase in dark $CO_2$ fixation, but also emphasize the need for careful fractionation studies in CAM plants. The $\delta^{13}C$ value (total plant carbon) may underestimate the shift from $C_3 + C_4$ like metabolism to almost exclusively $C_3$ or $C_3$-like metabolism. At the same time $\delta^{13}C$ values for soluble products of about -10‰ suggest that the precursors of the dark carboxylation reactions (starch and PEP) have similarly low $\delta^{13}C$ values and tha the whole of the acidification/deacidification cycle must be isolated from other metabolic events to some extent.

## Water availability

It has been appreciated for some time that plants with CAM are predominantly associated with arid regions and that the ability to gain $CO_2$ at night when the evaporative demand is low, represents a considerable improvement in water use efficiency. Kluge & Fischer (1967) and Bartholomew (1973) showed that an early response to removal of water in CAM plants was the cessation of $CO_2$ fixation in the light and a reduction in the rate of dark $CO_2$ fixation. Droughting thus increase the dependence of the CAM plant on dark $CO_2$ fixation for net carbon gain and studies with *Opuntia* spp. in the field indicate that these CAM plants depend solely on dark $CO_2$ fixation (Szarek et al., 1974; Szarek & Ting, 1974). In line with these observations, when water was withheld from *K. daigremontiana* growing in the CSIRO Canberra Phytotron, it was found that the $\delta^{13}C$ value (total plant carbon) declined from control values of -18.5‰ to -13.5‰ and from -23.0‰ to -17‰ in two experiments (Allaway et al., 1974).

These substantial shifts in the $\delta^{13}C$ value in response to water stress were obtained in a second experiment using rather different conditions. *K. daigremontiana* was grown under continuous light (40 Wm$^{-2}$, 25°) and under short-day

*Table 5.* Effect of continuous light and water stress on malate accumulation and $\delta^{13}C$ value in *Kalanchoe daigremontiana*. $\delta^{13}C$ value of clonal starting material -22.5‰ $\delta^{13}C$ value of controls in Treatment A, C3- 15.7‰ - C4 -18.0‰ and in Treatment B, C3, -29.0‰ C4, -15.7‰ ($\delta^{13}C$ values are for total shoot carbon in each case.)

| Growth treatment | Growth period (days) | Δ Malate ($\mu$ equiv/gfwt) | $\delta^{13}C$ value (‰) |
|---|---|---|---|
| A. Continuous light 25° | 107 | 17 | -26.5 |
| Transfer A → B* | 16 | 141 | -21.6 |
| water withheld | 40 | - | -17.6 |
| B.  6 hr light, 30° 18 hr dark 15° | 107 | 138 | -19.9 |
| Transfer B → A* | 16 | 23 | -22.9 |
|  | 40 | - | -24.3 |

* Values for youngest leaves only, similar values obtained with older parts of the plant.

conditions (6 hr light 143 Wm$^{-2}$, 25° C/18 hr dark 15° C) in artificially lit growth cabinets. $C_3$ and $C_4$ species of *Atriplex* were grown in the same cabinets during part of the experiment to provide controls should the $\delta^{13}C$ value of conventional plants shift due to these growth conditions. Water was witheld from the plants in continuous light (treatment A, Table 5) and in an 82 day period the $\delta^{13}C$ value (total shoot carbon) declined from -26.9‰ to -23.3‰. Over the same period the

$\delta^{13}C$ value (total shoot carbon) of the short day plants (treatment B, Table 5) declined from -24.5‰ to -17.4‰

Reciprocal changes were made between these two extreme treatments and the results are shown in Table 5. It is clear that the $\delta^{13}C$ value of plants grown in continuous light changed quite rapidly in the direction of an increased contribution from dark $CO_2$ fixation following transfer from treatment A to B. A similar but opposite trend was observed when plants were transferred to treatment A from B. These plants commenced to grow rapidly and the new tissues formed under continuous light showed $\delta^{13}C$ values consistent with a large contribution from $CO_2$ fixation mediated by RuDP carboxylase. The change observed on transfer from treatment A to B is not so readily explained, for the $\delta^{13}C$ value became rather less negative over a period in which no significant growth occurred. These data suggest that a substantial part of the plant carbon may have been recycled via PEP carboxylase or that a substantial amount of new (soluble) material of low $\delta^{13}C$ value had been added. Although these changes are obviously in the directions predicted by the known environmental effects on the $C_3$-like and $C_4$-like photosynthetic options in CAM plants, it is clear that more detailed analyses of the $\delta^{13}C$ value and percentage composition of different plant fractions are required.

It should be noted that two conditions are required for $\delta^{13}C$ values in the vicinity of -10‰. First, carbon should be fixed via PEP carboxylase, an enzyme which shows little discrimination against the heavier isotopes of carbon (Whelan et al., 1973) and second, the carbon so fixed should not exchange with exogenous $CO_2$ during subsequent metabolism. In $C_4$-plants both requirements are fulfilled as a matter of course. In CAM plants the second requirement is not so frequently met and the exchange of $CO_2$ between internal and external pools during deacidification is frequently observed. This exchange is minimised during water stress when stomatal resistance is maximal, and this may account in part for the large changes in $\delta^{13}C$ value observed under these conditions. These two requirements are implicit in the model of Lerman et al. (1974).

*Photosynthetic options under natural conditions*

As discussed above, CAM plants growing in arid habitats may be entirely dependent on the dark $CO_2$ fixation, $C_4$-like photosynthetic option. It is not surprising then that most of the $\delta^{13}C$ values for established CAM plants growing under these conditions are in the vicinity of -10 to -15‰ (Troughton & Berry, private communication). *Opuntia inermis,* an introduced species in eastern Australia, has been examined over a wide range of environmental conditions. The $\delta^{13}C$ value showed remarkably little variation from -12‰ in samples collected in all but the southernmost part of its present distribution (figure 1). Only in the more mesic southern part do the $\delta^{13}C$ values indicate that *O. inermis* may fix $CO_2$ in the light via the $C_3$-like photosynthetic option. Many of the coastal habitats in the north of the range are equally mesic but there the plant is restricted to rocky areas and may experience frequent periods of water stress. Laboratory studies with *K. daigremontiana* suggest that the $\delta^{13}C$ value of $CO_2$ incorporated in the light may decline in subsequent periods of water stress, presumably because carbon is recycled through PEP carboxylase (Szarek et al., 1973). The periodic

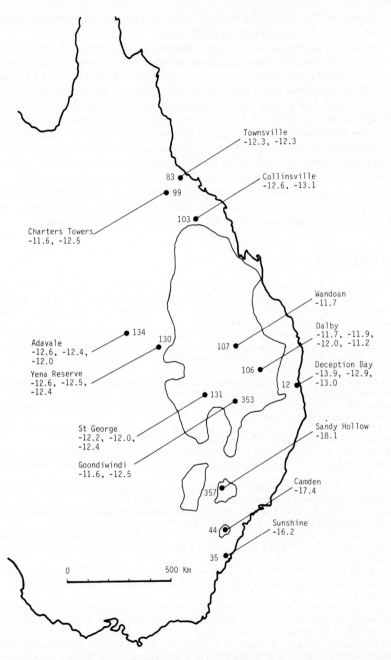

*Fig. 1.* $\delta^{13}C$ values (‰) obtained with samples of *Opuntia inermis* throughout its range in Eastern Australia. The number following the locality refers to the climate diagram (Walter & Leitz, 1961) most nearly corresponding to the sample site. The outline shows the approximate region in which this species constituted a serious pest prior to eradication in the period 1927-1933.

contribution of $CO_2$ fixation in the light under favourable conditions may thus be difficult to identify by means of the $\delta 13C$ value.

It would be interesting to compare these data with those for tropical CAM species such as *Ananas*. These plants are most productive under irrigated conditions with warm night temperatures, as is *Kalanchoe* (Allaway et al., 1974). It seems likely that some CAM plants grow best when conditions favour $CO_2$ fixation in the light but growth rates even then do not approach those of $C_3$-plants, although water use efficiency may be comparable. There is little doubt that CAM plants can persist when $CO_2$ fixation is restricted to the dark. Water use efficiency is undoubtedly better than that of $C_3$-plants under these conditions (Neales, 1973b), but growth rates are likely to be extremely low. Thus parallel evolution of the $C_4$-acid carboxylation-decarboxylation sequence has, on the one hand allowed $C_4$-plants to exploit opportune conditions in the arid with little loss of growth potential, and on the other, has permitted CAM plants to persist in more unfavourable environments only at the expense of much reduced growth.

## REFERENCES

Allaway, W.G., C.B. Osmond & J.H. Troughton (1974): Environmental regulation of growth, photosynthetic pathway and carbon isotope discrimination ratio in plants capable of crassulacean acid metbolism. Proc. Intern. Conf. Mechanisms of Regulation of Plant Growth. *Bull. Roy Soc. N.Z.* 12: 195-202.

Avadhani, P.N., C.B. Osmond & K.K. Tan (1971): Crassulacean acid metabolism and the $C_4$ pathway of photosynthesis in succulent plants. In Hatch, M.D., Osmond, C.B. and Slatyer, R.O. (Eds.): Photosynthesis and Photorespiration, Wiley-Interscience, New York. pp. 288-293.

Bartholomew, B. (1973): Drought response in the gas exchange of *Dudleya farinosa* (Crassulaceae) grown under natural conditions. *Photosynthetica* 7: 114-120.

Bender, M.M., I. Rouhani, H.M. Vines & C.C. Black (1973): $13C/12C$ ratio changes in Crassulacean acid metabolism plants. *Plant Physiol.*, 52: 427-430.

Björkman, O. & C.B. Osmond (1974): Effect of oxygen on carbon dioxide fixation in *Kalanchoe daigremontiana*. *Carnegie Inst. Wash. Yearbook* 73: (in press).

Bradbeer, J.W., S.L. Ranson & M.L. Stiller (1958): Malate synthesis in Crassulacean leaves I. The distribution of $14C$ in malate of leaves exposed to $14CO_2$ in the dark. *Plant Physiol.*, 33: 66-70.

Brownwell, P.F. & C.J. Crossland (1974): Sodium requirement in relation to the photosynthetic options of a CAM plant. *Plant Physiol.*, 54: 416-417.

Dittrich, P., W.H. Campbell & C.C. Black (1973): Phosphoenolpyruvate carboxykinase in plants exhibiting Crassulacean acid metabolism. *Plant Physiol.*, 52: 351-361.

Kluge, M. (1969): Verandliche Markierungs muster bei $14CO_2$ Fütterung von *Bryophyllum tubiflorum*. I. Die $14CO_2$-Fixierung unter Beleichtung. *Planta (Berl.)*, 88: 113-129.

Kluge, M. (1975): This volume.

Kluge, M. & K. Fischer (1967): Über Zusammenhänge zwischen dem $CO_2$-Austausch und der Abgabe von Wasserdampf durch *Bryophyllum tubiflorum* Berg. *Planta (Berl.)*, 77: 212-223.

Kluge, M. & Osmond, C.B. (1972): Studies on phosphoenol pyruvate carboxylase and other enzymes of Crassulacean acid metabolism of *Bryophyllum tubiflorum* and *Sedum praealtum*. *Z Pflanzenphysiol.*, 66: 97-105.

Lerman, J.C., E. Deleens, A. Nato & A. Moyse (1974): Variation in the carbon isotope composition of a plant with Crassulacean acid metabolism. *Plant Physiol.*, 53: 581-584.

Lerman, J.C. & O. Queiroz (1974): Carbon fixation and isotope discrimination by a Crassulacean plant: dependence on photoperiod. *Science*, 183: 1207-1209.

Neales, T.F. (1973a): Effect of night temperature on the assimilation of carbon dioxide by mature pineapple plants, *Ananas comosus* (L.) Merr. *Aust. J. Biol. Sci.*, 26: 539-546.

Neales, T.F. (1973b): The effect of night temperature on $CO_2$ assimilation, transpiration, and water use efficiency in *Agave americana* L. *Aust. J. biol. Sci.*, 26: 705-714.

Neales, T.F., A.A. Patterson & V.J. Hartney (1968): Physiological adaptation drought in the carbon assimilation and waterloss of xerophytes. *Nature*, 219: 469-472.

Osmond, C.B., W.G. Allaway, B.G. Sutton, J.H. Troughton, O. Queiroz, U. Lüttge & K. Winter (1973): Carbon isotope discrimination in photosynthesis of CAM plants. *Nature*, 246: 41-42.

Osmond, C.B. & W.G. Allaway (1974): Pathways of $CO_2$ fixation in the CAM plant *Kalanchoe daigremontiana* I. Patterns of $^{14}CO_2$ fixation in the light. *Aust. J. Plant Physiol.*, (in press).

Queiroz, O. & C. Morel (1974): Photoperiodism and enzyme activity: towards a model for the control of circadian rhythms in the Crassulacean acid metabolism. *Plant Physiol.*, 53: 596-602.

Rouhani, I., H.M. Vines & C.C. Black (1973): Isolation of mesophyll cells from *Sedum telephium* leaves. *Plant Physiol.*, 51: 97-103.

Sutton, B.G. & C.B. Osmond (1972): Dark fixation of $CO_2$ by Crassulacean plants: evidence for a single carboxylation step. *Plant Physiol.*, 50: 360-365.

Szarek, S.R., H.B. Johnson & I.P. Ting (1973): Drought adpatation in *Opuntia basilaris*. Significance of recycling carbon trough Crassulacean acid metabolism. *Plant Physiol.*, 52: 539-541.

Szarek, S.R. & I.P. Ting (1974): Seasonal patterns of acid metabolism and gas exchange in *Opuntia basilaris*. *Plant Physiol.*, 54: 76-81.

Walter, H. & H. Leitz (1961): 'Klimadiagram Weltatlas'. Gustav Fischer Verlag, Jena.

Whelan, T., W.M. Sackett & C.R. Benedict (1973): Enzymatic fractionation of carbon isotopes by phosphoenolpyruvate carboxylase from $C_4$ plants. *Plant Physiol.*, 51: 1051-1054.

Winter, K. & D.J. von Willert (1972): NaCl- induzierter Crassulaceansäuerstoffwechsel bei *Mesembryanthemum crystallinum*. *Z. Pflanzenphysiol.*, 67: 166-170.

# HOW TO INTERPRET VARIATIONS IN THE CARBON ISOTOPE RATIO OF PLANTS: BIOLOGIC AND ENVIRONMENTAL EFFECTS*

J.C. LERMAN

*Centre des Faibles Radioactivités, Laboratoire Mixte CNRS-CEA, 91190 Gif-sur-Yvette, France*

*Abstract*

The relative abundance of the carbon isotopes in plants indicates the pathway by which the primary carbon is assimilated. Plants of the $C_3$ and $C_4$-types can be easily distinguished from each other by isotope analysis of tissue of any organ or of most metabolic fractions. Plants of the CAM type behave differently, the isotope composition of different metabolites and different organs may show large variations in the same plant specimen. Results obtained by different authors are compared and discussed on basis of isotope discrimination models attempting to assess how much of the observed variations in the $^{13}C/^{12}C$ ratio of CAM plants can be due to: 1. variations of environmental parameters as e.g. isotope composition of the atmosphere, and 2. variation in the carbon fixation pathway correlated with environmental parameters (e.g. temperature, illumination, water availability) and biologic parameters (e.g. age).

## Introduction

The relative abundance of the natural carbon isotope $^{13}C$ in plant tissues is a simple and reliable indicator of the pathway by which carbon has been fixed by a plant. Consequently, the isotope method is frequently chosen to determine the type of photosynthetic pathway which operates in a plant and especially to study changes in the carbon metabolism, such as those correlated with biological and environmental parameters. The large number of articles recently published on the subject reflects the impact of the isotope method on plant physiology.

One of the purposes of the present communication is to comment on some of the difficulties and pitfalls of the method. It easily happens when reading the results of isotope analyses that one overlooks some of the assumptions on which the isotope method is based. Procedures to avoid misleading conclusions will be worked out here on basis of examples taken from the recent literature.

Other purposes of this communication are to generate interdisciplinary feedback 1. to facilitate the development of the isotope method, e.g. to attain more quantitative evaluations of the metabolic flow of carbon in plants, and 2. to

---

*Abbreviations:* $C_3$-plant = plant of the Calvin-Benson metabolic type; $C_4$-plant = plant with the Hatch-Slack metabolism; CAM plant = plant with the crassulacean acid metabolism; D = isotope discrimination; $\delta$ = isotope composition; PEPC = phosphoenolpyruvate carboxylase; RuDPC = ribulose-1,5-diphosphate carboxylase.

*Notes:* Numerals within parentheses indicate technical notes found at the end of the text.

* Work supported in part by the Laboratoire de Géologie Dynamique, Université de Paris VI, and the Laboratoire de Physiologie Cellulaire Végétale, Université de Paris-Sud.

promote the extension of the method to other disciplines, e.g. paleoclimate and paleoecology, which might derive useful information from the environmental effects preserved in the isotope record of fossil organic matter.

## Isotope Effects

Before discussing how to use the isotope method for the study of biologic and environmental effects on carbon metabolism, we need to recall some of the basic concepts and assumptions that will be needed in the following sections.

*Isotope composition ($^{13}C/^{12}C$ ratio and $\delta$ value)*
The natural abundance of the stable isotope $^{13}C$ relative to the isotope $^{12}C$ in a given sample is expressed by the ratio of the number of atoms of $^{13}C$ to the number of atoms of $^{12}C$ in the sample. This $^{13}C/^{12}C$ ratio is about 1.1 percent. As the natural variations of this ratio span a few percent (of this 1.1 percent), natural carbon isotope compositions are usually expressed as deviations in *per mil* (‰) from the isotope ratio in a standard. These deviations, or relative differences, are symbolized by $\delta^{13}C$, or simply $\delta$, defined as follows:

$$\delta = \frac{R_{sample} - R_{standard}}{R_{standard}} \times 1000, \quad \text{or} \quad \delta = \left[ \frac{R_{sample}}{R_{standard}} - 1 \right] \times 1000,$$

where $R = {}^{13}C/{}^{12}C$ (1). The usual standard to which the measurements are referred to is the PDB (Pee Dee Belemnite): $CO_2$ obtained from the carbonate shell of a Cretaceous mollusc, *Belemnitella americana* from the Pee Dee Formation in South Carolina (Craig, 1957). As organic matter and atmospheric $CO_2$ are depleted in $^{13}C$ when compared with carbonates, all $\delta$ values we deal with are negative numbers. A less negative figure means richer in $^{13}C$, or 'heavier' (2).

*Isotope analyses*
The $\delta$ values are obtained by analyzing with special mass-spectrometric techniques the $CO_2$ produced by quantitative combustion of few milligrams of sample. The combustion and subsequent analysis in a double collector/double inlet mass-spectrometer are performed in a couple of hours, and the operation cost is relatievely cheap (tens of U.S. dollars per sample). The accuracy of measurements of $CO_2$ is about 0.02‰ but combustions of organic matter usually provide results less reproducible by one order of magnitude.

*Isotope discrimination (or fractionation)*
When different isotopic species of a compound diffuse or react, each species shows slightly different physicochemical properties. This means, concerning the carbon isotopes in which we are interested, that those molecules containing $^{13}C$ diffuse or react slower than molecules containing $^{12}C$ (molecules containing $^{14}C$ are even slower than those containing $^{13}C$). The ratio of the diffusion or reaction rates of two isotopic species is approximately proportional to the square root of their mass ratio. This difference between the properties of isotopic species produces differences between the $\delta$ values of a reaction product and the initial carbon

source. This phenomenon is called isotope discrimination (or fractionation). Its magnitude, hereafter called D, is approximately equal to the difference between the respective $\delta$ values of the source and product and it is also expressed in *per mil* (3). Thus, for land plants, D will be approximately the difference between $\delta$ values of the atmospheric $CO_2$ and the $\delta$ values of the plant tissues. It is worth to mention here that discrimination against the heavy isotope occurs when the reaction draws its carbon from a large source. This is due to the fact that the first molecules produced by the reaction will be effectively depleted in $^{13}C$ by the value D while the source, when of finite size, will result enriched in $^{13}C$. Thus, the following molecules produced by the reaction will be relatively richer in $^{13}C$ than the first molecules, because the source becomes progressively richer in $^{13}C$. Thus in this case the $\delta$ value of the product will be higher (and D lower) than in the case when the product is the result of a reaction which uses carbon from an infinite source. In the particular case of total conversion of the carbon in the source, the product contains all the $^{13}C$ and $^{12}C$ atoms which were present in the source thus reproducing an identical $\delta$ value in the product and the source (D = 0).

*Carbon isotope discrimination by land plants*

The carbon fixed by land plants appears depleted in $^{13}C$ (and relatively more depleted in $^{14}C$) when compared with the isotope composition of the atmospheric $CO_2$. In 1960, Park & Epstein proposed a mechanism to explain the photosynthetic isotope discrimination. In their model the largest discrimination step appears to be at the level of the enzymatic fixation of $CO_2$. Their model explains a very wide range of isotope compositions in plant tissues (Epstein, 1969). However, examination of the distribution of $\delta$ values for several hundreds of $C_3$ and $C_4$-plants (Fig. 1a) shows that each one of these photosynthetic types discriminates against $^{13}C$ in a characteristic way: $\delta$ values in $C_3$-plants differ from the $\delta$ value of atmospheric $CO_2$ by about 20%, whereas for $C_4$-plants the difference between the $\delta$ value in the plants and the atmosphere is about 4‰. These values, considered as known within an interval $\pm 2$‰, are indicated in Fig. 1b. They are shifted to the side of each histogram with somewhat higher $\delta$ value (i.e. less negative). The reason for this shift is that such values are more likely to represent the carbon fixed in a 'free atmosphere', i.e. without admixture of $CO_2$ from biologic or industrial activity. To some extent the histograms represent $\delta$ values of plants not grown in such a free atmosphere, especially many of those which in the $C_3$ histogram appear largely depleted (J. Raynal & Lerman, in prep.). $C_3$ and $C_4$ plants from different families do not differ significantly in isotope composition (4). This fact, among others, suggests 1. that discrimination by the other two processes considered by the model of Park & Epstein (diffusion and transport of dissolved $CO_2$ within the cell and the plant) is small, and 2. that the drastic difference between the $\delta$ values of the two photosynthetic types of plants, $C_3$ and $C_4$, is mainly due to a difference in the isotope discrimination produced by the primary carbon reduction. In each of these plant types the primary fixation of carbon is catalized by a different carboxylating enzyme: RuDPC for the $C_3$-plants and PEPC for the $C_4$-plants. Investigation of the isotope discriminations produced *in vitro* has qualitatively confirmed that the enzymatic discrimination against $^{13}C$ is different for the $C_3$ and the $C_4$ reactions. The results obtained *in vitro* cannot be compared quantitatively with the observations *in vivo*

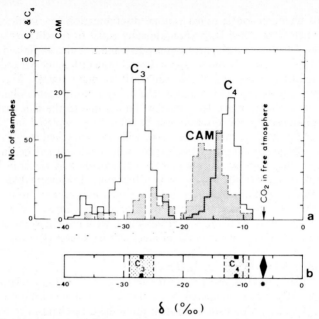

*Fig. 1.* Carbon isotope composition of land plants grown in natural environments. (a) The histograms show the distribution of $\delta$ values in $C_3, C_4$ and CAM plants (from a review by Lerman & Troughton, in preparation). (b) Assumed $\delta$ ranges for $C_3$ and $C_4$-plants grown in a free atmosphere ($\delta = 7‰$). A copy of this scale might be useful as measuring ruler in the subsequent discussions. (*) The diamond indicates the $\delta$ value of the free atmosphere. It will be used as reference point to normalize results under atmospheres with $\delta$ values deviating from -7‰.

as discussed elsewhere (Deleens et al., 1974). Hence, in the following discussions we will use for the $C_3$ and $C_4$ carboxylation reactions the values for D deduced from the respective types of plants (Fig. 1).

The carbon metabolism after synthesis of the first carbohydrate is not associated with large discrimination. The $\delta$ values of different plant metabolites (lipids excepted) in the $C_3$ and $C_4$-plants analyzed are not too different from each other (Park & Epstein, 1961; Whelan et al., 1970; Whelan, 1971). This was expected because 1. the molecular masses involved are large, and 2. the intermediate pools of carbon are small thus rendering the discrimination small.

Variations of $\delta$ values in $C_3$ and $C_4$-plants correlated to environmental and biological parameters are much smaller and less understood than in CAM plants. Although some knowledge on the metabolic and pure physicochemical effects has been acquired in the last 20 years (cf. review by Lerman, 1974), we are not yet able to separate the two types of effects and thus apply the method to study physiological variations in $C_3$ and $C_4$-plants.

## Variations of $\delta$

When studying variations we must keep in mind Medawar's principle which say that 'it is not informative to study variations unless we know beforehand the norm from which the variants depart' (Medawar, 1967). For our purpose, i.e. the study of plant metabolism by means of the natural carbon isotope composition, the accuracy about the norms of depart of two variants should not be neglected as it occurs frequently in the current literature.

First, the atmosphere in which the plants were grown: lack of information about the $\delta$ value of the carbon source is probably the result of the belief that the atmosphere $\delta$ is a constant (5).

Second, the part of the plant analyzed: different organs and different metabolites of a same plant do not have necessarily the same isotope composition as has been frequently assumed.

These two variants will be discussed in the following referring to recent literature on metabolic variations of CAM plants, because these were expected to vary their carbon metabolism following environmental changes (cf. review by Ting et al., 1972). These environmental changes would be reflected in the $\delta$ values of the plants (Lerman, 1972), as e.g. the variations observed in *Opuntia acanthocarpa* from different elevations in Arizona by A. Long (unpublished).

### Growth atmosphere

The $\delta$ value of the carbon source from where the plant draws its carbon is obviously the base value from where the isotope discrimination D is measured (3). The $\delta$ value of the atmosphere may vary by admixture of $CO_2$ from biological or industrial activity. Calculated $\delta$ values for such a 'mixed atmosphere' are given in Fig. 2. One sees that depletions of 10‰ in the $\delta$ value are easily possible. This must be taken into account when studying the meaning of the $\delta$ values of plants which are not grown in a free atmosphere.

Analyses of three species of *Sedum* (plotted in Fig. 3a) were cited as to span the whole range of $\delta$ values from $C_3$ to $C_4$ values (Bender et al., 1973; Black, 1973). This conclusion must be re-examined because the plants were not grown in the same atmosphere, thus making conclusions about the observed range difficult. *S. telephium* and *S. telephoides* were grown in a greenhouse in Georgia while *S. rubrotinctum* grew in a greenhouse in Wisconsin. From analyses of $C_3$ and $C_4$-plants grown in the Georgia greenhouse (Rouhani, 1972), I estimate that the atmosphere might have had a $\delta$ value of -10‰ if the plants were simultaneously grown and the night and day atmospheres had identical $\delta$ values. By sliding along the abscissae the ruler of Fig. 1b until the diamond faces the value -10‰ a $\delta$ scale normalized to the free atmosphere is constructed. In this new scale the $\delta$ value of *S. telephoides* becomes -19‰. The comparison of the $\delta$ value of *S. rubrotinm* to the other plants is, however, impossible as the $\delta$ value of the atmosphere in the Wisconsin greenhouse is unknown.

Similar comments apply to the data on Bromeliaceae, grown in a greenhouse in Venezuela, by Medina & Troughton (1974). Their results are plotted in Fig. 3b. As we know that one type of possible extreme values for CAM plants are $C_4$-like $\delta$ values (Fig. 1a), we may superpose the $C_4$ range of our ruler to the -14‰ mode of Fig. 3b. We find that the diamond faces a value around -9.5‰, reasonable for the atmosphere in a ventilated tropical greenhouse. Assuming this normalization, the 25‰ mode in the original scale becomes ca. -22.5‰ in the normalized scale.

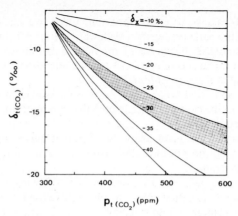

*Fig. 2.* Carbon isotope composition of a 'mixed atmosphere'. Carbon dioxide with isotope composition $\delta_x$ is added to an atmosphere with 300 ppm of $CO_2$ at -7‰ to obtain a total partial pressure $P_{t(CO2)}$ with a resultant isotope composition $\delta_{t(CO2)}$. The shaded zone indicates the $\delta$ values obtained when admixture is provided by fuel combustion or by biological activity such as respiration or decay of $C_3$- plants.

Consequently most of the values in the '$C_3$-histogram' of Fig. 3b appear to be intermediate between the characteristic $\delta$ values for $C_3$ and $C_4$-plants.

Information about the $\delta$ value of the carbon source is also lacking in the experiments by Osmond et al. (1973). It is apparent that a normalization of the $\delta$ scale is necessary because the conditions of the growth atmosphere have been in general different for each experiment. Their results are plotted in Fig. 3c. They represent increase of thermoperiod, shortening of photoperiod, increase of draught and increase of salinity. The variations of the $\delta$ values, indicated by the arrows, are in the direction expected from physiologic considerations. However, they seem rather small as to support the suggested good correlation between the $\delta$ values and the principal mode ($C_3$ or $C_4$-like) of $CO_2$ fixation (Osmond et al., 1973). In fact, data on the $\delta$ values of $C_4$-plants grown presumably in similar growth conditions (Osmond, this Symposium) would indicate that the atmosphere in which the plants were grown might have been largely depleted in [13]C, i.e. with $\delta$ values *per mil* more negative than those of the free atmosphere.

Similar comments can be made on the data published by Bender et al., (1973). A few of their examples have been plotted in Fig. 3d, which shows the variation in the $\delta$ values of the plants when the environment is modified by providing continuous temperature (and illumination). In this case, indirect evidence on the $\delta$ value of the growth atmosphere is given by information on the $CO_2$ concentration in the heated greenhouse, which attained up to 500 ppm by admixture of $CO_2$ presumably produced by the combustion of fuel. According to Fig. 2, this 500 ppm mean a $\delta$ value for the resulting atmosphere of -14 to -16‰. Hence the solid points of Fig. 3d might appear artificially more negative up to 7 to 9‰ indicating larger variations than real if the open points correspond to growth in an unheated or less heated greenhouse.

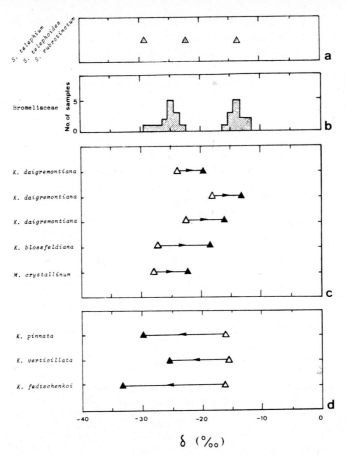

*Fig. 3.* Biologic and environmental variations in the carbon isotope composition of total tissue or leaf material of CAM plants. The grey points indicate plants grown in uncontrolled conditions. The open points indicate plants grown in controlled environments and the solid points the results after varying the growth conditions of clonal plants. The δ values of the respective atmospheres are unknown. (a) Three species of *Sedum* (data from Rouhani, 1972, p. 79; Black, 1973, p. 280; Bender et al., 1973). (b) 27 species of Bromeliaceae (data from Medina & Troughton, 1974) (c) *Kalanchoe* and *Mesembryanthemum* plants (data from Osmond et al., 1973). (d) *Kalanchoe* plants (data from Bender et al., 1973, Tables 2 and 3).

*Part of the plant*

The other variant that we should care of when studying variations of δ values is of biologic origin. The almost constancy of δ values in $C_3$ and $C_4$ plants which are independent of the plant organ and chemical fraction does not seem to apply to some CAM plants we analyzed (Deleens et al., 1974; Lerman & Queiroz, 1974; Lerman et al., 1974).

In experiments on two Crassulaceae plants we could show significant differences in isotope composition between the insoluble (mainly carbohydrates) and the water extracts (mainly malate) of the same leaf. Figs. 4a and 4b show the results obtained on *Bryophyllum* grown in a greenhouse under natural conditions and on *Kalanchoe* grown in controlled conditions. These results are already normalized to the free atmosphere, thus the ruler of Fig. 1b should be placed with its diamond in coincidence with the diamonds in Fig. 4.

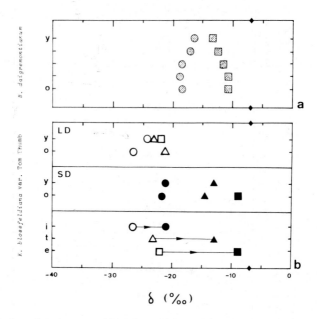

*Fig. 4.* Biologic and environmental variations in the carbon isotope composition of different fractions of CAM plants. The shading of the points is explained in Fig. 3. The fractions are symbolized as follows: □ = e = extract; $\triangle$ = t = total tissue; ○ = i = insoluble. All the results are normalized to an atmosphere with a $\delta$ = -7‰, indicated by a diamond in the abscissae. (a) Analyses of all leaves from oldest (o) to youngest (y) in *Bryophyllum* (data from Deleens et al., 1974; Lerman et al., 1974). (b) Analyses of young (y) leaves and old (o) leaves of *Kalanchoe* grown in long days (LD) and short days (SD); the variations in the $\delta$ value of the three plant fractions are indicated by arrows in the summary graph (bottom section of figure) (data from Lerman & Queiroz, 1974).

Fig. 4a shows the $\delta$ value varying with a biologic parameter: the leaf age (Deleens et al., 1974, Lerman et al., 1974). There is an apparent paradox because the leaves which fix their carbon by the CAM pathway as e.g. the oldest leaves (confirmed by their malate with a $\delta$ value indicative of carbon fixation by PEPC) have their insoluble shifted in the direction of $C_3$ values. This has been explained by the CAM model proposed in Fig. 5.

Similar results have been observed in *Kalanchoe*, correlated with a change in photoperiod (Lerman & Queiroz, 1974). This work clearly showed the importance of the choice of fraction analyzed when investigating metabolic variations. In particular we pointed out the loss of information when analyzing total tissue and the problem of the memory effect when analyzing insoluble fractions (see also Osmond et al., 1973).

# Models and Interpretation

We tried to explain the behaviour of CAM-plants by a model given in Fig. 5. As CAM plants function with the same biochemical reactions of $C_3$ and $C_4$-plants,

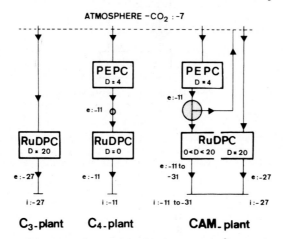

*Fig. 5.* Tentative models proposed to explain the characteristic $\delta$ values of $C_3$, $C_4$ and CAM plants. The figures indicate $\delta$ values of two raw fractions: (e) extracts and (i) insoluble. They are expressed in per mil (‰) within ca. $\pm 2$‰ (Adapted from Deleens et al., 1974; Lerman & Queiroz, 1974; Lerman et al., 1974).

and appear to be able to behave in particular cases as each one of those types, we discuss our CAM model by comparing it with models for $C_3$ and $C_4$-plants. Each box represents the value D apparent in each enzymatic reaction, including the corresponding diffusion and transport steps. The shaded circles represent the carbon pools (malate) and the arrows indicate the flow of carbon. The D values for the $C_3$ and $C_4$ reactions were taken from the results *in vivo* (Fig. 1b) with the assumption that the total isotope fractionation observed in a $C_4$-plant is only due to the PEPC step: the RuDPC step would be quantitative and thus not discriminating, as described in the introduction of this communication. The model for the CAM plant is a combination of the same boxes of the $C_3$ and $C_4$ models with important differences: the presence of a large carbon pool and the possibility of carbon loss to the atmosphere.

We attempted to obtain some quantitative information on 1. the relative proportion of carbon flowing by each of the possible pathways in a CAM-plant and 2. the magnitude of the D value of the RuDPC step which is in series with the PEPC step in the CAM pathway. The figures shown in Fig. 6 represent carbon fixation by the youngest leaf of the *Bryophyllum* experiment (Fig. 4a) (Deleens et al., 1974). The attempt is very rough because the loss of $CO_2$ to the atmosphere has not yet been taken into account. But we believe that this type of estimation coupled with other information, e.g. from gas exchange experiments, would give a more rigorous and quantitative picture.

Meanwhile, I try to offer here a tentative graphic aid for the diagnosis of the photosynthetic operation of a plant in which one or several fractions have been

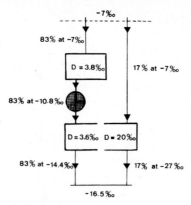

*Fig. 6.* Rough estimation of the flow of carbon in the youngest leaf of the *Bryophyllum* in Fig. 4a, from the atmosphere to the insoluble fraction (cf. CAM model in Fig. 5). The figures represent relative amount of carbon in percent at a certain $\delta$ value in per mil. (Adapted from Deleens et al., 1974).

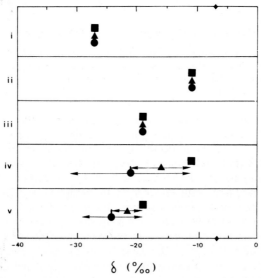

*Fig. 7.* Graphical aid to determine the type of photosynthetic pathway of a plant, on basis of $\delta$ values of the three described fractions: ■ = extract; ▲ = total tissue; ● = insoluble. Five different examples described in the text have been constructed from the models of Fig. 5. By comparing experimental results, e.g. Figs. 3 and 4, with these figures i to v, the relative proportions of different metabolic pathways can be deduced. The scale has been normalized to an atmosphere of $\delta = -7‰$, indicated by the diamonds. All $\delta$ values are assumed within a range of $\pm 2‰$.

analyzed. The different examples ( i to v ) illustrate the following cases (cf Fig. 7):

    i   Either a $C_3$-plant, or a CAM-plant operating exclusively by the $C_3$ pathway i.e. assimilating its primary carbon only via RuDPC.

    ii Either a $C_4$- plant, or a CAM- plant operating exclusively by the CAM pathway (i.e. assimilating all its primary carbon by the PEPC in night) and with total re-assimilation of the pool or endogenous carbon in daytime via RuDPC (= super-CAM plant, cf. Neales, this Symposium).

iii Plant assimilating its primary carbon by equal parts via RuDPC and PEPC, i.e. 50% via the $C_3$ pathway and 50% either via the $C_4$ pathway or via a CAM pathway without loss of endogenous carbon.

iv Plant assimilating all its primary carbon by the CAM pathway with partial loss of endogenous $CO_2$. The arrows indicate the possible range of $\delta$ values for the insoluble fraction according to the relative size of the pool (or extent of pool carbon loss). This insoluble carbon might have a paradoxical $C_3$ value even when all the primary carbon is fixed via the PEPC reaction. This is due to the fact that the second discriminating step (RuDPC) might discriminate by any value within the limits indicated (Fig. 5), depending on the relative size of the pool or carbon loss. The $\delta$ value of total tissue obviously depends on the proportion of extracts and insoluble. The arrows indicate the possible $\delta$ range. The arrow representing the $\delta$ range of total tissue behaves as attached to the insoluble point, thus when the insoluble point moves, it carries along the total tissue range.

v Plant assimilating its primary carbon by equal parts via the $C_3$ pathway and via a CAM pathway with loss of some of the endogenous carbon, i.e. with a relatively large pool of malate. The meaning and behaviour of the arrows is the same as explained in the previous example (iv).

## Conclusions

From the results commented in the previous section it appears that the isotope method is useful to study metabolic variations in CAM-plants Variations in the $\delta$ value in plant constituents correlated to biologic parameters as e.g. age, and environmental parameters as e.g. photoperiod, thermoperiod, draught and salinity have been observed and are larger than the pure physicochemical variations. It seems that a better picture will emerge 1. when one eliminates the contribution of external variations of the $\delta$ value (like growth atmosphere) and 2. when purer chemical fractions than total tissue will be analyzed. Until now our CAM-plant model based on the behaviour of two crassulacean plants has proved useful to predict some of the observed variations of $\delta$ values in CAM-plants It is to be expected that the isotope technique interacting with other plant physiologic approaches will help to provide a more accurate understanding of the environmental and biological effects on plants.

## NOTES

(1) Some authors quote erroneously R in the formula as ratio of mass 45 to mass 44 (Tregunna et al., 1970; Smith & Epstein, 1971). The mass spectrometric measurements are done on $CO_2$ in which the relative contents are mass 45 is compared to mass 44, but the $\delta$ values are calculated on basis of $^{13}C/^{12}C$ ratios, obtained after correction for contribution of $^{17}O$ to the mass 45. Corrections for contribution of most abundant isotopes $^{12}C$ and $^{16}O$ indicated by Medina & Troughton (1974) must be the result of a misprint as also their formula for $\delta$ was mistakenly printed.

$\delta^{13}C/^{12}C$ (Kanai & Black, 1972: Black, 1973) where the meaning is, instead, $\delta^{13}C$;
$\delta^{13}C$ ‰(Bender et al., 1973) where the meaning is only ‰;
$^{13}C/^{12}C$ of the PDB sample set at zero (Bender et al., 1973), when this ratio is about 1.1. percent.

(2) Higher $\delta$ values are less negative. A frequent mistake is to forget the algebraic sign.

(3) The correct relationship is $D\ (‰) = \dfrac{\delta_{source} - \delta_{product}}{1 - \delta_{source}} \times 10^{-3}$

Confusions in the literature range from simply semantic, e.g. calling the isotope composition (or $\delta$) a discrimination (Tregunna et al., 1970; Hatch et al., 1972; Osmond et al., 1973) to conceptual e.g. interpreting the $\delta$ value as a discrimination factor (Raven, 1970).

(4) Error in drafting Fig. 1 of Smith & Epstein (1971) makes appear a difference in $\delta$ values between mono- and dicotyledons of $C_4$-type. The original results do not show that difference (Table 1). However, the misleading graph has been reproduced several times since then (Rouhani, 1972; Bender et al., 1973).

(5) Although Smith & Epstein (1971) and Whelan (1971) seem to indicate that only urban air has more negative values than the free atmosphere, atmospheric $CO_2$ does change with geography and topography (Keeling, 1961a, 1961b).

# REFERENCES

Bender, M.M., I. Rouhani, H.M. Vines & C.C. Black, Jr. (1973): $^{13}C/^{12}C$ ratio changes in crassulacean acid metabolism plants. *Plant Physiol.,* 52: 437-430.

Black, C.C., Jr. (1973): Photosynthetic carbon fixation in relation to net $CO_2$ uptake. *Ann. Rev. Plant Physiol.,* 24: 253-286.

Craig, H. (1957). Isotopic standards for carbon and oxygen and correction factors for mass-spectrometric analysis of carbon dioxide. *Geochim. Cosmochim. Acta,* 12: 133-149.

Deleens,E., J.C. Lerman, A. Nato & A. Moyse (1975): Carbon isotope discrimination by the carboxylating reactions in $C_3, C_4$ and CAM plants. Proc. 3rd Internat. Congr. Photosynthesis, Rehovot 2-6 Sept. 1974, (Elsevier Publ. Co., Amsterdam).

Epstein, S. (1969). Distribution of carbon isotopes and their biochemical and geochemical significance. In: Forster, R.E., Edsall, J.T., Otis, A.B., & Roughton, F.J.W. (Eds.): Proc. Sympos. on $CO_2$: Chemical, Biochemical, and Physiological Aspects, Haverford, Pennsylvania, U.S.A., August 1968 (NASA SP-188) p. 5-14.

Hatch, M.D., C.B. Osmond, J.H. Troughton & O. Björkman, (1972). Physiological and biochemical characteristics of $C_3$ and $C_4$ *Atriplex* species and hybrids in relation to the evolution of the $C_4$ pathway. *Carnegie Inst. Year Book* 71, p. 135-141 (Carnegie Inst., Washington, D.C.).

Kanai, R., & C.C. Black, Jr. (1972). Biochemical basis for $CO_2$ assimilation in $C_4$-plants. In: Black, C.C. (Ed.): Net Carbon Dioxide Assimilation in Higher Plants, Sympos. South Sect. Amer. Soc. Plant Physiol., pp. 75-93 (Cotton Inc., Raleigh, N.C.).

Keeling, C. (1961b): The concentration and isotopic abundances of carbon dioxide in rural and marine air. *Geochim. Cosmochim. Acta,* 24: 277-298.
Lab. di Geol. Nucl., Pisa).

Keeling, C. (1961b). The concentration and isotopic abundances of carbon dioxide in rural and marine air. *Geochim. Cosmochim. Acta,* 24: 277-298.

Lerman, J.C. (1972) Carbon-14 dating: Origin and correction of isotope fractionation errors in terrestrial living matter. In: Rafter, T.A. & T. Grant-Taylor (Eds.), Proc. 8th Internat. Conf. on Radiocarbon Dating, p. H 16-H 28 (Royal Soc. New Zealand, Wellington).

Lerman, J.C. (1974): Isotope „paleothermometers" on continental matter: Assessment. In: Les Methodes Quantitatives d'Etude des Variations du Climat au Cours du Pleistocene, Coll. Internat. CNRS, No. 219: 163-181.

Lerman, J.C. & O. Queiroz (1974). Carbon fixation and isotope discrimination by a crassulacean plant: Dependence on the photoperiod. *Science,* 183: 1207-1209.

Lerman, J.C., E. Deleens, A. Nato & A, Moyse (1974). Variation in the carbon isotope composition of a plant with crassulacean acid metabolism. *Plant Physiol.,* 53: 581-584.

Medawar, P.B. (1967): The Art of the Soluble (Methuen, London) p. 109, cited by Tinbergen, N. (1974): Ethology and stress diseases. *Science.* 185: 20-27.

Medina, E., & J.H. Troughton, (1974): Dark $CO_2$ fixation and the carbon isotope ratio in Bromeliaceae. *Plant Sci. Lett.,* 2: 357-362.

Osmond, C.B., W.G. Allaway, B.G. Sutton, J.H. Troughton, O. Queiroz, U. Lüttge & K. Winter (1973): Carbon isotope discrimination in photosynthesis of CAM plants. *Nature,* 246: 41-42.

Park, R & S. Epstein, (1960): Carbon isotope fractionation during photosynthesis. *Geochim. Cosmochim. Acta,* 21: 110-126.

Park R. & S. Epstein (1961): Metabolic fractionation of $C^{13}$ and $C^{12}$ in plants. *Plant Physiol.,* 36: 133-138.

Raven, J.A.(1970). Exogenous inorganic carbon sources in plant photosynthesis. *Biol. Rev..,* 45: 167-221.

Rouhani, I. (1972): Pathways of carbon metabolism in spongy mesophyll cells, isolated from *Sedum telephium* leaves, and their relationship to crassulacean acid metabolism plants. Ph. D. Diss. University of Georgia, Athens, Georgia. (University Microfilms, Ann Arbor, Michigan. No. 72-34, 140: 169 pp.).

Smith, B.N. & S. Epstein (1971): Two categories of $^{13}C/^{12}C$ ratios for higher plants. *Plant Physiol.,* 47: 380-384.

Ting, I.P., H.B. Johnson & S.R. Szarek (1972): Net $CO_2$ fixation in crassulacean acid metabolism plants. In: Black, C.C. (Ed.): Net Carbon Dioxide Assimilation in Higher Plants, Sympos. South. Sect. Amer. Soc. Plant Physiol., p. 26-53 (Cotton Inc., Raleigh, N.C.).

Tregunna, E.B., B.N. Smith, J.A. Berry & W.J.S. Downton, (1969): Some methods for studying the photosynthetic taxonomy of the angiosperms. *Can. J. Bot.,* 48: 1209-1214

Whelan, T., W.M. Sackett & C.R. Benedict (1970): Carbon isotope discrimination in a plant possessing the $C_4$ dicarboxylic acid pathway. *Biochem. Biophys. Res. Commun.,* 41: 1205-1210.

Whelan, T., III (1971): Stable carbon isotope fractionation in photosynthetic carbon metabolism. Ph.D. Diss.Texas A&M University, College Station, Texas (University Microfilms, Ann Arbor, Michigan, No. 72-13, 258: 93 pp.).

# CONTROL OF GLYCOLYSIS IN SUCCULENT PLANTS AT NIGHT

B.G. SUTTON[1]

*Department of Environmental Biology, Research School of Biological Sciences, Australian National University, Canberra, A.C.T. Australia*

## Abstract

The major flow of carbon in succulent plants at night, from storage carbohydrate to malic acid, takes place via glycolysis. The storage carbohydrate has been identified as glucan, some of which conforms to an operational definition of starch. Free sugar pools contribute little to malic acid production and were shown to lie outside the mainstream of carbon flow. Further experiments suggested that environmental factors which limit malic acid accumulation at night do so by limiting production rather than by stimulating consumption.

The enzymes of glycolysis were isolated from leaves of succulent plants and the activities measured were sufficient to accomodate measured rates of acid accumulation. Phosphofructokinase from the succulent plant was shown to be 100-fold less sensitive to inhibition by PEP than the same enzyme from a non-succulent plant. Together with other known regulatory systems in succulent plants, this feature of phosphofructokinase seems particularly important in allowing continued synthesis of carboxylation substrate and the accumulation of high concentrations of malic acid. The properties of the other important regulatory enzyme of glycolysis in succulent leaves, phosphorylase, were consistent with the path of carbon flow outlined above.

## Introduction

The most distinguished feature of CAM is the extensive fixation of $CO_2$ at night into organic acids, mainly malic acid (Ranson & Thomas, 1960). Largely as a result of the rigorous work of Pucher & Vickery 20-30 years ago, who demonstrated an extremely significant relationship between the diurnal fluctuations of starch and malic acid (see, for example, Pucher et al., 1949), starch has been identified as the reserve carbohydrate which is precursor of the PEP used in $CO_2$ fixation at night. This report attempts to outline the path followed by carbon from reserve carbohydrate to malic acid at night and to suggest probable regulatory features of this pathway.

[1]Present Address: Department of Biology, University of California, Riverside, California, 92502, U.S.A.

*Acknowledgement:* I wish to thank Dr. C.B. Osmond for stimulating and helpful discussion during the course of this work. I was supported by a Commonwealth Post-graduate Research Award.

*Abbreviations:* BSA, bovine serum albumin; CAM, Crassulacean acid metabolism; F-6-P, fructose-6-phosphate; MES, 2-(N-morpholino)-ethanesulphonic acid; PEP, phosphoenolpyruvate; PGA, 3-phosphoglycerate; PVP, polyvinylpyrrolidone; RudP, ribulose-1,5-diosphate; TCA, tricarboxylic acid cycle.

## Materials and Methods

Experimental plants, *Bryophyllum tubiflorum* Harv. and *Kalanchoe daigremontiana* Hamet et Perrier were grown as described previously (Sutton & Osmond, 1972). Starch was assayed by the method of Pucher et al. (1948) and all other tissue components were extracted and assayed as described previously (Sutton & Osmond, 1972). All carbohydrates were estimated as glucose equivalents by the anthrone method of Fairbarn (1953). Carbohydrates were isolated from the perchloric acid extract used in the starch assay by chromatography on Dowex-1 formate (200-400 mesh) resin, hydrolysed with 1.5 N HCl at 100 C for 2 hours, (Pirt & Whelan, 1951) and chromatographed in the upper phase of n-butanol: ethanol: water (5:1:4, v:v) (Hirst & Jones, 1949). Procedures used in the exposure of tissue to $^{14}CO_2$, separation of products and degradation of malic acid-$^{14}$C have been described elsewhere (Sutton & Osmond, 1972). Glycolytic enzymes with the exception of phosphofructokinase, were extracted from *K. daigremontiana* by grinding in 100 mM bicine, pH 8.0 containing 30 mM 2-mercaptoethanol and 1% PVP and desalted on Sephadex G-25, equilibrated in 100 mM bicine pH 8.0 containing 30 mM 2-mercaptoethanol. The same procedures were used for phosphofructokinase, but 1% BSA was added to the extraction buffer and the gel buffer was 100 mM bicine pH 8.0, containing 5 mM $(NH_4)_2SO_4$. For more detailed studies phosphorylase was prepared by desalting the protein which precipitated from the crude homogenate between 35% and 65% ammonium sulphate saturation on a Sephadex G-25 gel in the standard buffer and was assayed by the procedure of Gilboe et al (1972). Phosphofructokinase was prepared from the crude homogenate by isolating, from the protein which precipitated between 25% and 45% ammonium sulphate saturation, a second fraction which precipitated in 40% ammonium sulphate. This protein was desalted in the absence of 2-mercaptoethanol.

## Results and Discussion

A number of points can be made concerning the probable structure of the path followed by carbon from reserve carbohydrate to malic acid at night. Firstly, there is a growing body of evidence which indicates that PEP used in malic acid synthesis at night is not produced by a preliminary carboxylation reaction (Bradbeer et al., 1958). According to this earlier scheme, carbon from reserve carbohydrate was diverted through the oxidative pentose phosphate pathway and RudP was formed from intermediates of this pathway. The RudP was consumed in a carboxylation reaction which produced 2 molecules of PGA which were converted to PEP. However, the experiments of Sutton & Osmond (1972), in which malic acid-$^{14}$C isolated from succulents after short term (5 sec) dark $^{14}CO_2$ fixation was found to be predominantly ( > 95%) C-4 labelled (compared with the 66% C-4 labelling predicted by the earlier scheme) indicate that only one carboxylation reaction, catalysed by PEP carboxylase, is involved in malic acid synthesis at night in succulent plants. This view is supported by Lerman & Queiroz (1974) who found the carbon isotope discrimination ratio of a CAM plant indicated night-time $CO_2$ uptake via PEP carboxylase (Whelan et al., 1973)

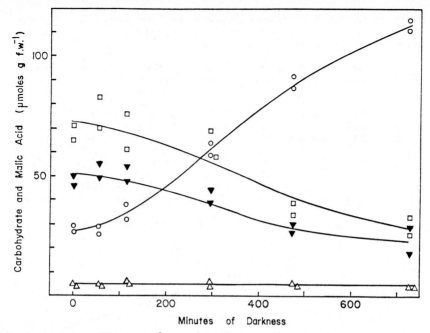

*Fig. 1.* Total glucan (□), starch (♦), water soluble sugars (△) and malic acid (○) changes in *B. tubiflorum* in darkness.

Secondly, starch may not always be the sole donor of carbon for PEP production at night. Figure 1 shows that in *B. tubiflorum,* malic acid levels increased by 88 $\mu$moles g f.w.$^{-1}$ of tissue during a 10 hour night, but loss of carbon from starch was sufficient to account for only 58 $\mu$moles of acid. The soluble sugar pool cannot have acted as a carbon reservoir as the level of the pool did not change. However, the level of one other fraction, the carbohydrates soluble in perchloric acid, did change. This fraction contained no free sugars and was immobile when chromatographed. Starch was precipitated from this fraction by iodine. Glucose was the only product of hydrolysis. On the basis of these characteristics, this fraction was termed total glucan and starch was operationally defined as that portion on the glucan fraction which could be precipitated with iodine. Loss of carbon from glucan was suffient to account for the observed acid increase. Therefore, this fraction was considered the source of carbon for PEP production at night, in the remaining experiments.

This assumption was supported by examination of the utilization of labelled photosynthate during the dark. Carbohydrates were labelled by exposing phyllodes of *B. tubiflorum* to $^{14}CO_2$ for 2 hours immediately prior to darkness. Table 1 shows that malic acid was the major labelled product and always contained about 90% of the total $^{14}C$. During the 12 hours dark, the specific activity of malic acid declined (Fig. 2). The specific activity of glucan also declined sharply for the first 2 hours, but remained constant after that. The initial decline is analogous to the pattern of carbon utilization observed in partially labelled tobacco starch (Porter et al., 1959) and stems apparently from the complicated

*Table 1.* Distribution of $^{14}C$ in components of *B. tubiflorum* during 12 hours in the dark following 2 hours light $^{14}CO_2$ fixation. Values shown are % of total $^{14}C$ incorporated and are means of two determinations.

| Time (mins) | Amino Acids | Malic Acid | Citric Acid | Free Sugars | Other Soluble Compounds | Glucan | Residue |
|---|---|---|---|---|---|---|---|
| 0 | 0.6 | 88.4 | 1.0 | 6.5 | 2.0 | 0.8 | 0.7 |
| 10 | 0.45 | 86.5 | 1.1 | 8.6 | 2.0 | 0.95 | 0.45 |
| 30 | 0.7 | 80.2 | 2.1 | 11.8 | 2.95 | 1.45 | 0.75 |
| 120 | 0.45 | 89.1 | 1.95 | 5.8 | 1.75 | 0.5 | 0.5 |
| 240 | 0.4 | 91.0 | 3.0 | 2.95 | 1.4 | 0.5 | 0.75 |
| 480 | 0.25 | 90.6 | 5.2 | 1.95 | 1.05 | 0.45 | 0.6 |
| 720 | 0.2 | 90.8 | 5.85 | 1.45 | 0.7 | 0.35 | 0.65 |

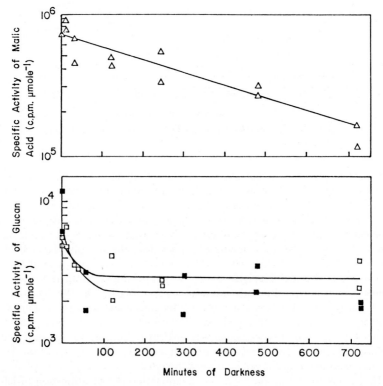

*Fig. 2.* Specific activities of malic acid ($\triangle$) and glucan ($\square$) during 12 hours darkness following 2 hours $^{14}CO_2$ photosynthesis by *B. tubiflorum*. Results of a duplicate determination of glucan specific activity are shown ($\blacksquare$).

process of reserve carbohydrate deposition on the starch grain, which is a combination of layering of material on the outside of the grain and insertion of new carbohydrate inside the existing grain structure. If the initial decline is neglected and calculations are based on events in the final 10 hours, it can be shown that if PEP for malic acid production had been derived solely from glucan of the observed specific activity, 2.9 x $10^3$ c.p.m. $\mu mole^{-1}$ then the final specific activity of malic acid should have been 2.03 x $10^5$ c.p.m. $\mu mole^{-1}$. This agrees reasonably well with the observed value of 1.7 x $10^5$ c.p.m. $\mu mole^{-1}$.

Thirdly, carbon from glucan does not flow through the free sugar pool between glucan breakdown and glycolysis. Because glucose from glucan had a much lower $^{14}C$ content than glucose in the free sugar pool initially (2.9 x $10^3$ c.p.m. $\mu mole^{-1}$ compared with 2.75 x $10^5$ c.p.m. $\mu mole^{-1}$), the $^{14}C$ content of the free sugar pool would have been extensively diluted if all the carbon from glucan had flowed through it. The final activity, in this case, would have been 1.57 x $10^4$ c.p.m. g.f.w.$^{-1}$ compared with the observed value of 3.5 x $10^5$ c.p.m. g.f.w.$^{-1}$ (Fig. 3).

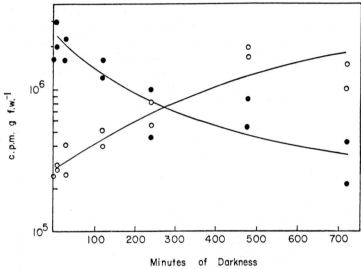

Fig. 3. Behaviour of $^{14}C$ levels in free sugars (●) and citric acid (○) in *B. tubiflorum* in darkness following a 2 hour light $^{14}CO_2$ fixation.

Since the final experimental value could have been achieved by a flow of a small fraction only (⟨15%) of the carbon lost from glucan through the free sugar pool, it was judged that the major flux of carbon from glucan at night was directly to glycolysis.

Forthly, the increase in the $^{14}C$ content of citric acid probably occured by exchange with malic acid through reactions of the TCA cycle. The 13.4 x $10^6$ c.p.m. g.f.w.$^{-1}$ increase (Fig. 3) was too great to have accrued from glycolytic intermediates. The only fraction with sufficient $^{14}C$ at the beginning of the dark to have donated this much label was the malic acid. As photosynthetically produced malic acid is symmetrically labelled (Gibbs, 1953), the data in Table 2 suggest that more than 90% of the $^{14}C$ would have been lost as $^{14}CO_2$ if carbon from malic acid had been transferred to citric acid by decarboxylation of the former to pyruvate and conversion of this acetyl CoA, prior to incorporation into the TCA cycle. A more likely mechanism of exchange of $^{14}C$ between malic and citric acids would involve reactions of the TCA cycle.

Finally, the effect of temperature on the utilization of reserve carbohydrate

*Table 2.* Percent ${}^{14}$C in the C-4 carboxyl group of malic acid at indicated times in the dark, following 2-hour light ${}^{14}CO_2$ fixation by *B. tubiflorum*. Means of two determinations.

| Time (mins) | % ${}^{14}$C in C-4 |
|---|---|
| 0 | 45.6 |
| 30 | 42.1 |
| 120 | 43.6 |
| 480 | 45.5 |

and malic acid accumulation at night provides an indicator of the degree of interaction between acid production and other cell processes in CAM plants at night. An important model in this area of CAM physiology has been that of Brandon (1967) who suggested that depressed malic acid gains under elevated night-time temperatures were due to a greater enhancement of consumption than production. The data presented in Fig. 4 however, which show the response of the

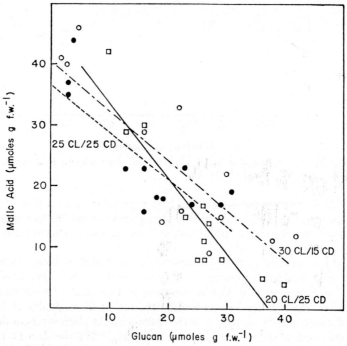

*Fig. 4.* Fitted linear regressions of malic acid on total glucan for samples of *K. daigremontiana* harvested in the dark from the indicated temperature regimes.

glucan-malic acid relationship in *K. daigremontania* with increasing night temperatures, relative to the day regimes, do not agree with this model. To do so, the curves would have to reflect a decreasing malic acid gain for similar glucan consumption, with increasing temperature and would therefore tend to the line y = a. The data, instead, suggest decreased glucan utilization and malic acid gains at higher night temperatures. Vickery (1954) showed that starch consumption in *B. calycimum* was greatest when malic acid accumulation was greatest and Kluge (1969) found that both malic acid gain and starch loss were depressed at high night temperatures or when exogenous $CO_2$ was not supplied. Thus, in CAM plants at night, it appears that the major flux of carbon is from reserve carbohydrate through glycolysis to malic acid. Carbon is, to a large extent, retained in this sequence, so that if incorporation into acid is limited, mobilization from glucan is also limited.

In this study, the problem of regulation of this pathway was approached by attempting to identify differences between CAM and non-CAM plants which allow the former to accumulate relatively massive quantities of malic acid at night. The first possibility considered was that insufficient carbon to support acid production might be lost from reserve carbohydrate at night in non-CAM plants. The data in

*Table 3.* Starch losses and acid gains for several species during a 12-hour night.

| Plant | Predominant CO2 Fixation Mode | Starch Loss ($\mu$ moles glucose) g f.w.$^{-1}$) | Malic Acid Gain ($\mu$ moles g f.w.$^{-1}$) |
|---|---|---|---|
| *Atriplex hastata* | $C_3$ | 50.66 | −6.01 |
| *A. spongiosa* | $C_4$ | 50.36 | −10.21 |
| *B. pinnatum* | CAM | 96.54 | +126.43 |
| *K. daigremontiana* | CAM | 33.50 | +104.50 |

Table 3 show that comparable losses of a major reserve carbohydrate occur in both types of plant at night. The fate of this carbon in the CAM plants was malic acid production, but it is probable that in the non-CAM plants, the carbon was lost from glycolysis, to free sugars.

A second possibility was that the capacity of the glycolytic enzymes in non-CAM plants was insufficient to allow adequate supply of PEP for acidification. The maximal rates for glycolytic enzymes measured under optimal conditions, from both CAM and non-CAM plants, are shown in Table 4. Rates for amylase and pyruvate kinase are not shown here as the former enzyme could not be demonstrated in a CAM and $C_3$ plant and the activity of the latter in the CAM plant was only 1.7% of the activity of PEP carboxylase. For most of the enzymes, the activities were similar in both the CAM and non-CAM plants. PEP carboxylase activity was low in the $C_3$ plant and the activities of phosphorylase and phosphofructokinase tended to be lower in the non-CAM plants. However, while the low activity of PEP carboxylase might limit acid production in the $C_3$ plant, this qualification does not apply to the $C_4$ plant. The data in Table 4, then, suggest that the accumulation of malic acid in CAM plants at night is probably a property of the kinetic properties of the glycolytic enzymes in these plants.

Two enzymes which were considered to be possible regulators of glycolysis in CAM and $C_4$ plants, phosphorylase and phosphofructokinase, were examined and their properties compared. The effects of some important metabolites on the

Table 4. Activities of enzymes associated with glycolysis and malic acid synthesis in darkened leaves of several plants. Activities are given in $\mu$moles min$^{-1}$ mgChl$^{-1}$. In most instances, means of two or three determinations are given, but 95% confidence limits are shown where four or more determinations were made.

| Plant | Atriplex hastata | A. spongiosa | B. pinnatum | K. daigremontiana |
|---|---|---|---|---|
| ENZYMES COMMON TO GLYCOLYSIS AND THE CALVIN CYCLE | | | | |
| Aldolase | 1.61 | 3.86 | 3.08 | 1.71 |
| Triose phosphate isomerase | 312.40 | 413.35 | 536.23 | 352.90 |
| Phosphoglycerate kinase | 91.12 | 132.54 | 91.65 | 37.08 |
| ENZYMES COMMON TO GLYCOLYSIS AND CARBOHYDRATE SYNTHESIS | | | | |
| Phosphorylase | 0.68 | 1.40 | 4.81 | 1.84 ± 0.04 |
| Phosphoglucomutase | 9.24 | 18.87 | 15.60 | 11.48 |
| Phosphoglucoisomerase | 3.44 | 10.33 | 8.38 | 5.74 |
| ENZYMES UNIQUE TO GLYCOLYSIS | | | | |
| Phosphofructokinase | 0. 32 ± 0.16 | 0.30 ± 0.14 | 1.33 | 0.80 ± 0.25 |
| NAD-glyceraldehyde phosphate dehydrogenase | 12.50 | 10.68 ± 1.03 | 39.76 | 11.89 ± 3.8 |
| Phosphoglyceromutase-Enolase | 0.94 | 2.06 ± 0.39 | 6.08 | 1.95 ± 1.9 |
| ACIDIFICATION | | | | |
| PEP carboxylase | 0.71 | 10.59 ± 4.60 | 9.74 | 4.64 ± 0.62 |
| OTHER ENZYMES | | | | |
| Glucokinase | 0.26 | 0.54 | 0.68 | 0.33 |
| Glucose-6-P dehydrogenase | 0.36 | 0.63 | 0.34 | 0.42 |

activity of phosphorylase from *K. daigremontiana,* assayed in 100 μl of 25 mM MES-citrate buffer, pH 6.0, containing 0.5 mg of dispersed dialysed soluble starch are shown in Figure 5. $P_i$ (1 mM) inhibited competitively, PEP (2 mM) inhibited

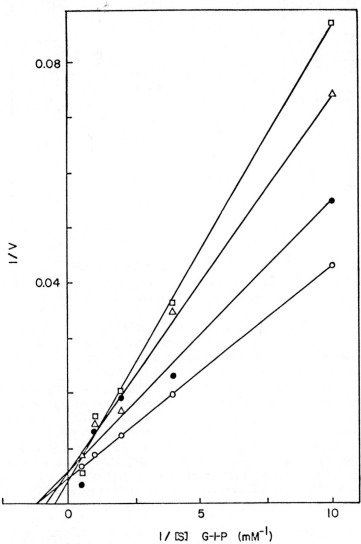

*Fig. 5.* Double reciptocal plots of the effect of 2 mM PEP (●), 4 mM glucose (△) and 1mM $P_i$ (□) on the rate of phosphorylase from *K. daigremontiana* (○).

non-competitively and glucose (4 mM) demonstrated mixed type inhibition. The phosphorylase from *Atriplex spongiosa* was inhibited less by $P_i$ and not at all by glucose or PEP. The important difference between the kinetic properties of phosphofructokinase from both plants was the relative sensitivity of the enzymes

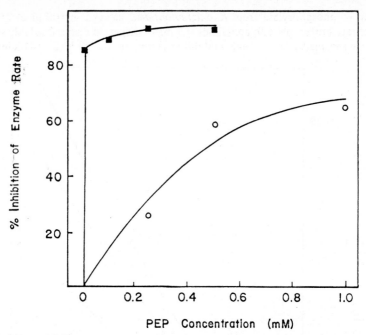

*Fig. 6.* Effect of PEP on the rate of phosphofructokinase of *K. daigremontiana* (● 0.5 mM F-6-P in assay) and *A. spongiosa* (■, 0.25 mM F-6-P in assay).

to inhibition by PEP. The data in Figure 6 show that when the enzyme from both plants was assayed at $K_m$ levels of F-6-P and otherwise under optimal conditions, phosphofructokinase from *A. spongiosa* was about 100 times more sensitive to inhibition by PEP than the enzyme from *K. daigremontiana*.

The reactions through which carbon flows from glucan to malic acid in CAM plants at night were outlined earlier. It is a reasonable assumption that any malic acid production at night in a $C_4$ plant would proceed through the same reactions. The activity of phosphofructokinase can be controlled by cytoplasmic malic acid concentration and the primary event in this would be the inhibition of PEP carboxylase by malic acid, with a subsequent rise in the level of the PEP pool, which would, in turn, inhibit phosphofructokinase activity. As a consequence of the greater sensitivity of this enzyme from a $C_4$ plant to inhibition by PEP, it is likely that this inhibition would only occur at much higher cytoplasmic levels of malic acid in a CAM plant than a $C_4$ plant. It is probable that carbon loss from starch in a $C_4$ plant is by transfer of sugar from hexose phosphates to the free sugar pool. Two controls exist to prevent this loss in a CAM plant. The first is the inhibition of phosphorylase by PEP and the second is the inhibition of this enzyme by glucose.

## REFERENCES

Brandon, P.C. (1967): Temperature features of enzymes affecting Crassulacean acid meta bolism. *Plant Physiol.,* 42: 977-984.

Fairbarn, H.J. (1953): A modified anthrone reagent. *Chem. Ind. (Lond.),* p.86

Gibbs, M. (1953): Effect of light intensity on the distribution of [14] C in sunflower leaf. metabolites during photosynthesis. *Arch. Biochem. Biophys.,* 45: 156-160.

Gilboe, D.P., K.L. Larson & F.Q. Nuttall (1972): Radioactive method for the assay of glycogen phosphorylases. *Anal Biochem.,* 47: 20-27.

Hirst, E.L. & F.K.N. Jones (1949): The application of partition chromatography to the separation of sugars and their derivatives. *Faraday Society, Discussions,* 7: 268-274.

Kluge, M. (1969): Zur Analyse des $CO_2$ - Austausches von *Bryophyllum.* II Hemmung des nächtlichen Stärkeabbaus in $CO_2$ verarmter Atmosphäre. *Planta (Berl.),* 86: 142-150.

Lerman, J.C. & O. Queiroz (1974): Carbon fixation and isotope discrimination by a Crassulacean plant: dependence on the photoperiod. *Science,* 183: 1207-1209.

Pirt, S.J. & W.J. Whelan (1951): The determination of starch by acid hydrolysis. *J. Sci. Food Agr.,* 2: 224-228.

Porter, H.K., R.V. Martin & F.F. Bird (1959): Synthesis and dissolution of starch labelled with [14]C in tobacco leaf tissue. *J. Exp. Bot.,* 10: 264-276.

Pucher, G.W., C.S. Leavenworth & H.B. Vickery (1948): Determination of starch in plant tissues. *Anal. Chem.,* 20: 850-853.

Pucher, G.W., H.B. Vickery, M.D. Abrahams & C.S. Leavenworth (1949): Studies in the metabolism of Crassulacean plants: diurnal variation of organic acids and starch in excised leaves of *Bryophyllum calycinum. Plant Physiol.,* 24: 610-620.

Ranson, S.L. & M. Thomas (1960): Crassulacean acid metabolism. *Annu. Rev. Plant Physiol.,* 11: 81-110.

Sutton, B.G. & C.B. Osmond (1972): Dark fixation of $CO_2$ by Crassulacean plants: evidence for a single carboxylation step. *Plant Physiol.,* 50: 360-365.

Vickery, H.B. (1954): The effect of temperature on the behaviour of malic acid and starch in leaves of *Bryophyllum calycinum* cultured in darkness. *Plant Physiol.,* 29: 385-392.

Whelan, J., W.M. Sackett & C.R. Benedict, C.R. (1973): Enzymatic fractionation of carbon isotopes by phosphoenolpyruvate carboxylase from $C_4$ plants. *Plant Physiol.,* 51: 1051-1054.

# EFFECT OF PHOTOPERIOD ON THE $CO_2$ AND $O_2$ EXCHANGES IN LEAVES OF *BRYOPHYLLUM DAIGREMONTIANUM* (Berger).

R. MARCELLE

*Laboratory of Plant Physiology, Research Station of Gorsem, B-3800 Sint-Truiden, Belgium*

*Abstract.*

Attached leaves from *Bryophyllum daigremontianum* Berger were used to study the effect of photoperiod on the $CO_2$ and $O_2$ exchanges. It was found that in LD and SD a net evolution of oxygen occurred in the light with a maximum rate during the period in which no $CO_2$ exchange could be measured in young leaves. The effect of photoperiod on $CO_2$ exchanges were more quantitative than qualitative: the $CO_2$ balance per 24 hr was larger in LD than in SD; the ratios 'night balance/day balance' were also modified by photoperiod and by the age of leaves.

The effects of $CO_2$-free air application on the pattern of gas exchange in the light were also investigated. It was found that $CO_2$-suppression could inhibit or not the oxygen emission according to the moment of suppression. It was concluded that the photosynthetic activity measured by oxygen emission depended only on either the atmospheric $CO_2$ or the endogeneous $CO_2$ pool supplied by the decarboxylation of stored malate. In LD but not in SD, the light dependent decarboxylation of malate presented a lag phase; in both daylengths, an other lag phase seemed to occur between the beginning of decarboxylation and the onset of the oxygen emission.

## Introduction

It is well known that Crassulacean Acid Metabolism (CAM) is characterized by a nightly acidification of the leaves due to the accumulation of malate. In the light, deacidification occurs through decarboxylation of stored malate and the liberated $CO_2$ is photosynthesized and converted into carbohydrates (for reviews, see Beevers et al., 1966 and Ranson, 1965).

The $CO_2$ exchange pattern of Crassulaceae is also well documented, especially in *Bryophyllum* (Neurnbergk, 1965; Kluge & Fischer, 1967; Marcelle, 1970). Adult leaves of this species assimilate large quantities of $CO_2$ during the night in LD as well as in SD contrary to the leaves of *Kalanchoe* where the dark assimilation of $CO_2$ is mainly induced by SD (Gregory et al., 1954).

Little is known about the effect of photoperiod on the $O_2$ exchange pattern of Crassulaceae. The results of Brunnhöfer et al., (1968) concerned detached leaves measured in SD but taken from plants first cultivated in LD. In our opinion, the $CO_2$ exchange pattern reported by Brunnhöfer typically concerned plants transferred from LD to SD (see Marcelle, 1970) and perhaps also plants for which the beginning of the light phase has been shifted. The importance of circadian rhytms for the regulation of CAM metabolism has been underlined by many authors (see

*Acknowledgement:* Thanks are due to the I.W.O.N.L. for financial support.

*Abbreviations:* LD: long days; SD: short days.

Queiroz, 1974). However, it is obvious that the photosynthetic utilization of stored malate resulted in a net evolution of oxygen as also shown by Denius & Homann (1972). This last work was made on leaf slices from *Aloe arborescens* Mill. to lessen, or eliminate, problems of stomata and diffusion of gases.

The first part of the present investigation was conducted to ascertain the effects of photoperiod on the daily fluctuation of oxygen and carbon dioxide in leaves of *Bryophyllum daigremontianum*. The second part is devoted to the effect of $CO_2$-free air on the pattern of gas exchanges in the light. It was expected that the suppression of $CO_2$ in the air surrounding the leaves could differently influence the oxygen fluctuation according to the moment of $CO_2$ suppression.

## Material and Methods

Clonal plants of *Bryophyllum daigremontianum* (Berger) were cultivated in greenhouse as described earlier (Marcelle, 1970). The plants obtained from epiphyllous buds were always maintained either in LD (from 5 a.m. to 9 p.m.) or in SD (from 9 a.m. to 5 p.m.). Temperature of the greenhouse was not controlled except for 15°C minimum during the night. For measurements of gas exchanges, the plants were transferred to a growth chamber: 22°C day temperature, 17°C night temperature, 50% relative humidity, $25.10^3$ ergs $sec^{-1}$ $cm^{-2}$ light intensity at the level of leaves no. 2[a] (fluorescent tubes 'Phytor').

$CO_2$ exchanges were measured in an open system with an IRGA, 'Uras 1' (Marcelle, 1969; full details of the installation on request). Measurements of oxygen exchanges were performed with a paramagnetic oxygen analyzer 'Oxygor' placed right after the IRGA. The rate of air flow inside the cuvettes, each enclosing a single leaf attached to the plant, was adjusted to 40 liters per hour. In the light, the temperature inside the cuvettes raised to 28° ± 1°C (measured by Ysi thermistors and telethermometer) but was maintained at 17° ± 0.5°C in the dark. Chlorophyll content of leaf disks was determined according to Mackinney (1941).

## Results and Discussion

The effects of photoperiod on the $CO_2$ and $O_2$ exchange patterns of leaves no 2 are shown in fig. 1A and B. In both daylengths, the diurnal fixation of atmospheric $CO_2$ was restricted to the beginning and the second half of the light phase; between these two fixation periods no $CO_2$ exchange could be measured in leaves of such an age. Identical fluctuation of the $CO_2$ exchange pattern was found by Kluge (1968b) in 12 hr photoperiod. The dark $CO_2$ fixation was very important in both daylengths but its rate was different just before the lights were turned on. When the light phase was beginning, the possibility of dark $CO_2$ fixation was exhausted in SD but not at all in LD.

As far as the $O_2$ exchanges were concerned, fig. 1A and B show that oxygen was evolved in the light in both SD and LD, with a maximum rate during the

[a] Numbering of leaves from the top, see Marcelle, 1970.

350

period of $CO_2$ compensation. In darkness oxygen was clearly taken up but the sensitivity of our analyzer did not allow us to exactly evaluate the fixation rate and its possible fluctuations (we have arbitrarily represented a steady fixation rate on fig. 1.).

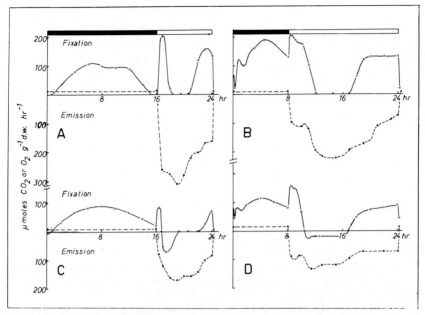

*Fig. 1.* Effect of photoperiod (SD: A.C; LD: B.D.) on the $CO_2$ and $O_2$ exchanges of young (A.B) and old (C.D) leaves of *Bryophyllum daigremontianum* ○———○ : $CO_2$ exchanges; ○- - - -○ : $O_2$ exchanges). Mg chlorophyll $g^{-1}$d.w.: A: 11.56; C: 7.82; B: 7.81; D: 4.02.

Keeping in mind that the transpiration fluctuations parallel the consumption of atmospheric $CO_2$ (Kluge & Fischer, 1967, confirmed by our own unpublished results), we have to admit that stomata are probably closed or almost closed in light during the period of $CO_2$ compensation (see also Nishida, 1963). A net evolution of oxygen is however measured during the same period. We can conclude that, in our conditions the leaf stomatal resistance do not seem to play an important part in oxygen diffusion; it is also the case for $CO_2$ (Kluge & Fischer, 1967).

It is known that the $CO_2$ exchanges patterns in Crassulaceae are largely controlled not only by the environmental conditions of temperature (Kluge, 1968b; Brunnhöfer et al., 1968) intensity of light (Kluge, 1968b) or water stress (Kluge & Fischer, 1967) but also by the biological conditions of plants. As an illustration, fig. 1C and D show the effect of leaf age on $CO_2$ and $O_2$ daily fluctuations. Except for quantitative differences, young and old leaves presented the same pattern of exchanges but in the light, $CO_2$ was evolved by the old leaves.

The $CO_2$ balances per 24 hr so as the ratios 'night balance/day balance' were also influenced by the photoperiod applied to the plants (Table 1). The quantity of $CO_2$ fixed in 24 hr was always larger in LD than in SD while the ratios 'night/day' indicated that $CO_2$ was mainly fixed during the night in SD plants but in the light in LD plants. It was also obvious that the $CO_2$ balance per 24 hr decreased in old leaves; the dark $CO_2$, fixation became predominant mainly because the $CO_2$ evolution observed in old leaves during the light phase decreased the day balance.

*Table 1.* Effect of photoperiod on the $CO_2$ balance per 24 hr and the ratio 'night balance (N)/ day balance (D)'

| Photoperiod | Age of leaves | $CO_2$ balances per 24 hr ($\mu$ moles $CO_2$) | | Ratios N/D |
|---|---|---|---|---|
| | | /g f.w. | /g d.w. | |
| 8 hr | no 2 | 58.9 | 1537 | 2.75 |
| | no 5 | 35.6 | 1078 | 16.2 |
| 16 hr | no 2 | 111.0 | 2008 | 0.80 |
| | no 5 | 85.4 | 1503 | 1.12 |

The photosynthetic $CO_2/O_2$ quotient found to average 1.26 in air with 21% $O_2$ (Fock et al., 1972) was essentially variable in *Bryophyllum* leaves. In young leaves no 2, the $CO_2/O_2$ ratio at the first $CO_2$ fixation peak in the morning was close to 1.00 in SD but approached 2.00 in LD; just before the night, values close to 1.00 and 1.50 were respectively found in SD and LD.

To investigate the effect of $CO_2$-free air on the gas exchanges, the light phase was divided into three periods in both daylengths: the first and the third ones corresponding to the fixation of atmospheric $CO_2$ in normal air, the second being the period of $CO_2$ compensation or $CO_2$ evolution according to the age of leaves.

The most striking effects of $CO_2$ suppression were observed during the first period. In SD (fig. 2A), $CO_2$ was evolved with a maximum peak about 1 1/2 hr after the lights were turned on. The position and intensity of this peak did not depend on the moment of $CO_2$-free air application contrary to the $CO_2$ evolution rate at the end of the treatment. From many experiments, it was also obvious that the rate of $CO_2$ evolution at the end of the treatment by $CO_2$-free air was always larger in young than in old leaves.

The effect of $CO_2$ suppression on the oxygen evolution is shown on fig. 2 B. At the three experimental moments of $CO_2$ suppression, an inhibition of oxygen evolution was observed; this inhibition was total when $CO_2$ was suppressed before the lights were turned on but only partial when $CO_2$ was later suppressed. At the end of the experiment, the mean rate of $O_2$ evolution was the same in all three treatments.

The same effects of $CO_2$-free air were approximately observed in LD (fig. 3A and B) but due to the expansion of the first period of the light phase, some features appeared more clearly in LD than in SD. The suppression of $CO_2$ just before the lights were turned on resulted in a weak $CO_2$ emission which increased after 1 1/2 hr to reach a maximum peak 2 hr later (fig. 3A). The evolution of oxygen, however, became measurable only after about 3 hr (fig. 3B). These result

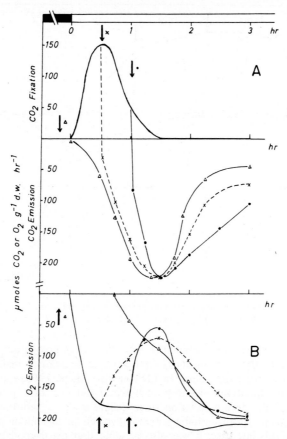

*Fig. 2.* Effect of $CO_2$-free air application during the first period of the light phase on the $CO_2$ (A) and $O_2$ (B) exchanges in SD leaves. The results obtained with the same leaf but on four successive days have been put together on the figure. On the first day, $CO_2$ and $O_2$ exchanges in normal air were recorded (continuous lines with no symbols); on the second ($\triangle$), third ($\times$) and fourth ($\bullet$) days, $CO_2$ was daily suppressed at different moments (represented by the arrows). The rather good reproducibleness of the results allowed us to consider that the traces obtained in normal air remained the same as long as $CO_2$-free air was not applied. In fig. 2A, the lines between the traces in normal and $CO_2$-free air ($\times$- - - -$\times$; $\bullet$ —— $\bullet$) were drawn to make the figure easier but they really represented a period of instability of the IRGA due to the $CO_2$ suppression.

were confirmed by suppressing $CO_2$ later in the first period of the light phase (fig. 3A and B).

If we assume that the decarboxylation of stored malate is responsible for the increasing $CO_2$ evolution seen in $CO_2$-free air, our results could suggest the following conclusions:

1. the decarboxylation of stored malate presented a lag phase in LD but not in SD; this lag phase might be due to the fact that the pool of malate to be decarboxylated was not saturated by the dark $CO_2$ fixation as it was in SD; this pool should be filled up by the photosynthetic activity during about the first two hours of the

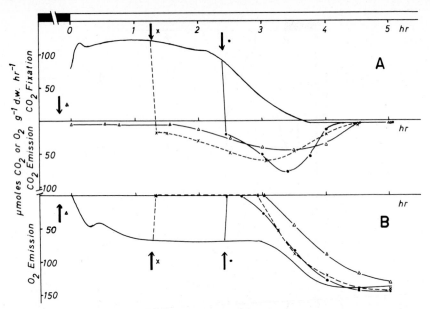

*Fig. 3.* Effect of $CO_2$-free air application during the first period of the light phase on the $CO_2$ (A) and $O_2$ (B) exchanges in LD leaves no. 2. Details as in fig. 2.

light phase before the beginning of the decarboxylation. Preliminary measurements of the total malate content show an increase in malate during the first two hours of the light phase in LD (see also Kluge, 1968) but a decrease in SD. The filling up of the malate pool however did not seem to be an absolute prerequisite to the decarboxylation since malate is also decarboxylated in $CO_2$-free air when the photosynthetic activity involved in this filling up was totally inhibited. In this last case, it was also possible that decarboxylation happened only by entrainment of the circadian rhythm.

2. A lag phase also occured between the beginning of the decarboxylation and the utilization of the liberated $CO_2$ by photosynthesis in both daylengths; this lag phase was best seen in $CO_2$-free air.

The effects described on fig. 2 and 3 clearly prove the total dependence of photosynthesis on atmospheric $CO_2$ during the first 30-45 minutes of the light phase in SD and during the first 90-180 minutes in long days. Later on photosynthesis depends progressively on the mobilization of the stored $CO_2$ pool.

The suppression of $CO_2$ during the second period of the light phase (period of $CO_2$ compensation or evolution) resulted in a weak emission of $CO_2$ (about 20 $\mu$ moles $g^{-1}$ d.w. $hr^{-1}$) in both daylengths but the oxygen evolution did not apparently change at least during the maximum duration of the tested $CO_2$ suppression (1 hr). During this second period, photosynthesis only occurred at the expense of the stored $CO_2$ pool, the $CO_2$ compensation point ($\Gamma$) being then the $CO_2$ concentration of the normal air surrounding the plant at least in young leaves presenting no fixation or emission of $CO_2$ during this second period of the light phase.

When $CO_2$ was suppressed during the third period of the light phase (period o atmospheric $CO_2$ fixation), a weak $CO_2$ emission (20 to 30 $\mu$ moles g$^{-1}$ d.w. hr$^{-1}$) was also observed in both daylengths but the oxygen evolution was entirely inhibited. As well as in the beginning of the light phase, photosynthesis in the third period only depended on the atmospheric $CO_2$, the internal $CO_2$ pool being exhausted or no longer available.

At the end of this paper, one question is arising: are the measurements of oxygen exchanges able to bring forward something more than the measurement of $CO_2$ exchanges to explain what happens during the light phase in CAM plants? We think that the answer is affirmative for some reasons:

1. keeping in mind, as suggested by Kluge (1968a), that the compartmentation of $CO_2$ metabolism in the cells allows $CO_2$ production and consumption to occur independently of one another (see the lag phase in oxygen emission when $CO_2$ was suppressed during the first period of the light phase), it is not possible to separate the two phenomenons if only $CO_2$ exchanges are measured;

2. if we assume that oxygen emission is only linked with the $CO_2$ fixation by RudP carboxylase (the light $CO_2$ fixation by PEP carboxylase being then exactly what it is in dark, it is to say a fixation without oxygen emission), the measurements of oxygen and $CO_2$ exchanges should allow us to know what are, in total $CO_2$ fixation, the part of photosynthetic fixation and the part of non autotrophic fixation as defined by Ting (1971). For the same reason, it seems to us that it is easier to study the oxygen sensitivity of photosynthesis in CAM plants by measuring $O_2$ and $CO_2$ exchanges.

# REFERENCES

Beevers, H., M.L. Stiller & V.S. Butt (1966): Metabolism of the organic acids. In Steward, F.C. (Ed.): Plant Physiology, A Treatise, Vol. IV B. Pp. 119-262. Academic Press, New-York.

Brünnhöfer, H., H. Schaub & K. Egle (1968): Der Verlauf des $CO_2$- und $O_2$-Gaswechsels bei *Bryophyllum daigremontianum* in Abhängigkeit von der Temperatur. *Z. Pflanzenphysiol.*, 59: 285-292.

Denius, Jr., H.R. & P.H. Homann (1972): The relation between photosynthesis, respiration, and Crassulacean Acid Metabolism in leaf slices of *Aloe arborescens* Mill. *Plant Physiol.*, 49: 873-880.

Fock, H., W. Hilgenberg & K. Egle (1972): Kohlendioxid- und Sauerstoff-Gaswechsel belichteter Blätter und die $CO_2/O_2$- Quotienten bei normalen und niedrigen $O_2$-Partialdrucken. *Planta (Berl.)*, 106: 355-361.

Gregory, F.G., I. Spear & K.V. Thimann (1954): The interrelation between $CO_2$ metabolism and photoperiodism in Kalanchoe. *Plant Physiol.*, 29: 220-229.

Kluge, M. (1968): Untersuchungen über den Gaswechsel von *Bryophyllum* wahrend der Lichperiode. I. Zum Problem der $CO_2$-Abgabe. *Planta (Berl.)*, 80: 255-263.

Kluge, M. (1968): Untersuchungen über den Gaswechsel von *Bryophyllum* während der Lichtperiode. II. Beziehungen zwischen dem Malatgehalt des Blattgewebes und der $CO_2$-Aufnahme. *Planta (Berl.)*, 80: 359-377.

Kluge, M. & K. Fischer (1967): Uber Zusammenhänge zwischen dem $CO_2$-Austausch und der Abgabe von Wasserdampf durch *Bryophyllum daigremontianum* Berg. *Planta (Berl.)*, 77: 212-223.

Mackinney, G. (1941): Absorption of light by chlorophyll. *J. biol. Chem.*, 140: 315-322.

Marcelle, R. (1969): Recherches sur la croissance, le bourgeonnement épiphylle et le développement de *Bryophyllum tubiflorum* (Harv.) et *Bryophyllum daigremontianum*

(Berger); comparaison des effets de la gibbérelline et de la photopériode. Mémoire de Doctorat, Université de l'Etat, Liège, 227 pp.

Marcelle, R. (1970): Sur l'absence d'une relation simple entre le type d'échange du $CO_2$ et l'induction de la floraison chez *'Bryophyllum daigremontianum'*. In Bernier, G. (Ed.): Cellular and molecular aspects of floral induction. Pp. 243-251. Longman Group Ltd, London.

Nishida, K. (1963): Studies on stomatal movement of Crassulacean plants in relation to the acid metabolism. *Physiol. Plantarum,* 16: 281-298.

Nuernbergk, E.L. (1955): Uber den zeitlichen Verlauf der Photosynthese bei Gewächspflanzen. *Gartenbauwiss.,* 19: 391-398.

Queiroz, O. (1974): Circadian rhythms and metabolic patterns. *Annu. Rev. Plant Physiol.,* 25: 115-134.

Ranson, S.L. (1965): The plant acids. In Bonner, J. & Varner, J.E. (Ed.): Plant Biochemistry. Pp. 493-525. Academic Press, New-York.

Ting, I.P. (1971): Nonautotrophic $CO_2$ fixation and Crassulacean acid metabolism. In Hatch, M.D., Osmond, C.B. and Slatyer, R.O. (Eds.): Photosynthesis and Photorespiration. Pp. 169-185. Wiley-Interscience, New-York.

# RHYTHMICAL CHARACTERISTICS AT DIFFERENT LEVELS OF CAM REGULATION: PHYSIOLOGICAL AND ADAPTIVE SIGNIFICANCE

O. QUEIROZ

*Laboratoire du Phytotron, Centre National de la Recherche Scientifique, 91190 Gif-sur-Yvette, France*

## Abstract

Overt circadian rhythms typical of CAM operations in *Kalanchoe blossfeldiana* (rhythms in $CO_2$ exchanges and in malate content) are shown to be in good correlation to rhythms of enzyme capacity (viz. PEP carboxylase) as a function of photoperiodic conditions. The photoperiod-enzyme relationships differ in long days and in short days showing that a central timing mechanism is involved: in physiological long days (i.e. long days or short days + red light flash interruption of the night) the peak of PEP carboxylase capacity is fixed in time of the day; in short days the rhythm of enzyme capacity shifts according to complex transients and shows only relative coordination to the day-night rhythm. The hypothesis that sensitivity of the plant to light-dark signals changes progressively under short days but not under long days is presented and discussed in terms of adaptive significance. Labelling experiments suggest the operation of a regulatory mechanism combining enzyme circadian rhythmicity and feed-back by malate. Experimental and theoretical implications of these findings are discussed.

## Introduction

*Responses to light signals and metabolic coordination*
The aptitude of timing metabolic events to the appropriate part of each day-night cycle is a fundamental adaptive aspect of the Crassulacean Acid Metabolism (CAM).

It is generally assumed that metabolic characteristics result from evolutionary adaptation to climate conditions. And since most environmental parameters are periodical (daily and annual cycles of light, temperature, etc.), the 24 hr cycling in a physiological process is indicative of its evolutionary origin, locking together the biological rhythm and the day-night rhythm (Borthwick & Hendricks, 1960). Therefore adaptation necessarily implies the operation of mechanisms able to synchronize metabolic functions to environmental rhythms.

Light certainly is the most universal biological synchronizer, and obviously is the most determinative in plant metabolic behaviour. But it must be noted that most studies on light effects neglect the necessary distinction between two fundamental classes of response: in some cases light acts through simple photoactivation and hence metabolic response can be obtained at any time of the 24 hr cycle, i.e. will be entrained by any light-dark schedule; in other cases, on the contrary, physiological functions appear to be able to synchronize only to a limited range of light-dark periodicities, i.e. will 'accept' or 'refuse' to respond, or will respond in a different way, according to the time of the 24 hr cycle at which the light stimulus is given (limited entrainment range). The fundamental physiological difference between these two classes of responses to light is that in the first case metabolic functions are controlled simply by light activation of key-reactions, whereas in the second case the implication of an endogenous time-measuring mechanism is inescapable.

This latter class of responses was shown to comprehend a very great number of periodic events in higher organisms. It has never been observed in prokaryotes. In plants and in animals it appears to be ubiquitously present throughout all levels of organization, from the infracellular level (for example rhythms in enzyme population, in shape and number of organelles, etc.) up to the morphogenetic and behavioral levels (for example rhythms in flower induction, in motility). In Table 1 are listed some of the better known circadian rhythms in CAM; since they have

Table 1. Earliest references for some of the most studied circadian rhythms in CAM plants

| Rhythm | References |
|---|---|
| Stomatal opening | Schwabe, 1952 |
| Respiration rates | Schmitz, 1951 |
| $CO_2$ output | Wilkins, 1959 |
| $CO_2$ fixation | Queiroz, 1970 |
| $CO_2$ compensation point | Jones & Mansfield, 1972 |
| Organic acid content | Becker, 1953 |
| Enzyme capacity | |
| - phosphatases | Ehrenberg, 1954 |
| - CAM enzymes | Mukerji, 1968; |
| | Queiroz, 1968; Morel, |
| | Celati & Queiroz, 1972 |
| Flowering induction | Carr, 1952; Schwabe, |
| | 1955; Bünning & |
| | Engelmann, 1960; |
| | Engelmann, 1960 |
| Petal movements | Bünsow, 1953 |

been discovered all these rhythms received extended attention and were the object of a large number of papers during the recent years. It is a fact that even in systems in which key-enzymes are photoactivable it appears that adaptation has evolved, and superimposed to the simple photoactivation control, a much more complex solution, the so-called endogenous clock or timesense property. For example, photosynthesis, enzyme activity and RNA synthesis in chloroplasts of unicells exhibit typical endogenous clock properties (Vanden Driessche, 1966, 1970; Vanden Driessche & Bonotto, 1969; Walther & Edmunds, 1973).

As discussed elsewhere (Pittendrigh, 1965, 1973; Queiroz, 1974) a selective advantage of this mechanism probably lies in its coordinating ability and homeostatic value. The property of limited entrainment range i.e. the ability of 'accepting' or 'refusing to follow' external signals according to the time of day or the periodicity of the signal, is obviously significative of an homeostatic aptitude to maintain the necessary metabolic coordination.

*Physiological timing and environmental periodisms in CAM*

When studying CAM regulation it is convenient to make a clear distinction between rhythms in 'terminal' events (malate production, malate utilization, connected $CO_2$ exchanges) and rhythms in the underlying machinery, i.e. enzyme capacity. The former afford an evaluation of the actual activity of the pathway, the latter give information on its potentialities at different moments under different conditions. This distinction between actual activity and potentiality has proved to be a very useful tool for studying physiological timing in CAM and other materials, particularly in *Chlorella* and animal organs and tissues.

It is known that photoperiodism has a regulatory role in the physiology of Crassulaceae. This effect has been first shown in flower induction and in organic acid variation. Roughly speaking, two main groups of Crassulaceae received extended attention from the point of view of response to photoperiodism: 'short day plants' (for ex. *Kalanchoe blossfeldiana*) and 'long-short day plants' (for ex. *Bryophyllum daigremontianum*). The former are vegetative in long days (15 hr days is an usual schedule) and show low or very low CAM activity, particularly in young leaves; in short days CAM activity is drastically increased together with flowering. In the case of *Bryophyllum daigremontianum* and the other long-short day plants the effect of photoperiod is much more complex and less clearcut: the effects of long days and of short days are not opposite but complementary and long transients have been observed after a change of light schedule.

It must be noted that in a short-day schedule, interruption of the night even by a dim, low energetic white or red light during as short as a few minutes will destroy the short day effect and produce a long day effect. This action is the most effective when the light flash is given during the middle part of the night, but can then be reversed by a flash of far-red radiation. These classic data establish that, in the case of CAM control, 1) phytochrome is involved, 2) an endogenous rhythm of sensitivity to light is involved, and, 3) the direct cause of the photoperiodic effect is the timing of the light/dark cycle within the 24 hr cycle and not the actual duration of photosynthetic activity.

Light pulses, dark pulses and also temperature pulses can strongly modify the phase of CAM rhythms, i.e. the hour of the maximum activity. But the effect produced by a given pulse can be very different according to the moment of the 24 hr cycle at which it is applied (see Fig. 2).

## Results and Discussion

### $CO_2$ exchanges

It has been established in earlier work that $CO_2$ fixation during the first part of the night and $CO_2$ production during the first part of the day show properties typical of endogenous circadian rhythms (Wilkins, 1959; Queiroz, 1970). When plants are placed under continuous light or continuous darkness both rhythms keep operating, with progressive damping; period of the free-running rhythms is slightly different from 24 hr. Figure 1 shows that a flash of light given at about 9 p.m. in a continuous darkness experiment with *K. blossfeldiana* resets the rhythm. A systematic study by Wilkins (1960, 1965) of phase-shifts produced by light or

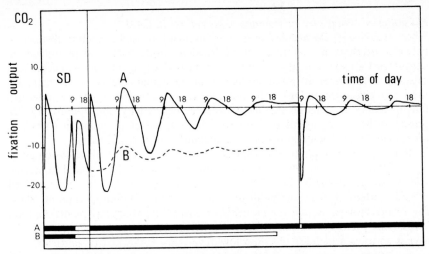

**Fig. 1.** Rhythms of $CO_2$ output and $CO_2$ fixation by leaves of *Kalanchoe blossfeldiana*. Short day conditions (SD) are changed into continuous darkness (curve A) or continuous light (curve B). Light schedules are shown by the bars in the bottom of the graph. In experiment A leaves were subjected to a short light treatment after damping of the rhythms. Ordinates: $CO_2$ measurements by infrared gaz analyzer expressed in % of $CO_2$ in the air leaving the plexiglas chambers containing the attached leaves.

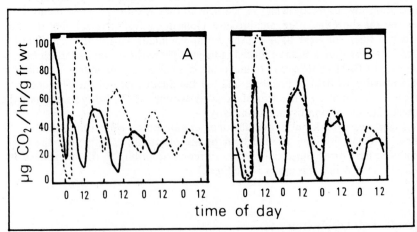

**Fig. 2.** The effect of short light treatments (3 or 5 hours) on the phase of a rhythm in darkness. A: treatment in the trough between two peaks. B: treatment at the crest of a peak. Broken line: control in continuous darkness. (modified from Wilkins, 1960).

temperature pulses on *Bryophyllum* established the endogenous origin of the free-running rhythms (Fig. 2).

*Day-night variations in malate content*
At the level of malate, the effect of photoperiodism is strictly consistent with $CO_2$ exchanges as shown in preceding work (Queiroz, 1965). Malat

content is low in the young leaves of long day grown *K. blossfeldiana* (about 0.10 mmoles/g dry wt) and increases exponentially after the transfer of the plants to short day conditions (up to 5 mmoles after 40-50 short days). This increase in malate content starts after a lag time of about 7 short days (under a thermoperiod of 27° C/17° C day/night). It can be noted incidentally that lag periods are frequent in photoperiodically induced processes. Experiments with different thermoperiods established that malate synthesis during the night depends on the temperature during the preceding day (Queiroz, 1966).

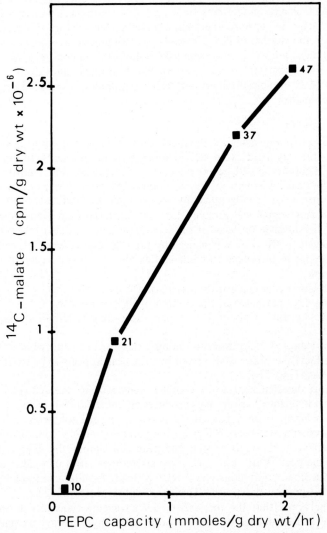

*Fig. 3.* Correlation between increase in PEP carboxylase capacity and increase in $^{14}C$-malate produced during $^{14}CO_2$-feeding experiments in short days. Data obtained at the beginning of light for the 10 th, 21st, 37th and 47th short days. (modified from Queiroz & Morel, 1974).

## Circadian rhythms in enzyme capacity

The action of photoperiodism affords a deeper insight on the enzymic mechanisms underlying the overt CAM variations. The observed exponential increase of malate production during the first 40 to 50 short days after lag time correlates with an exponential increase in PEP carboxylase (Fig. 3) as proved by labelling experiments (Queiroz & Morel, 1974). Increase in PEP carboxylase capacity goes with increasing amplitude of daily oscillation (Queiroz, 1968; Morel, Celati & Queiroz, 1972; Queiroz & Morel, 1974). The same authors have shown that all the other enzymes of the pathway (malate dehydrogenase, malic enzyme, aspartate and alanine aminotransferases) present rhythmical, non synchronized circadian changes in capacity under photoperiodic control (Fig. 4). Recent results confirm that the rhythm of PEP carboxylase is also present in young leaves grown under long days, with very low mean value (about 0.2 mmoles/hr/g dry wt) and amplitude (less than 0.05 mmoles) as compared to maximum values attained in short days (4-10 mmoles/hr/g dry wt, 1.2- 2.0 mmoles, resp., according to different experiments).

## Flexibility of CAM

It is obvious that physiological significance of metabolic rhythms lies on phase relation to the day-night cycle and to other periodic functions. In other terms, metabolic flexibility could be obtained in a very precise and sensitive way by shifts of the peak of activity of key reactions. As stated above, one of the most typical properties of endogenous rhythms is their well-defined 'phase-response curve': the same signal will produce different responses (and even contradictory responses, viz. advance or delay of the activity peak) according to the moment of the 24 hr cycle at which it is perceived by the organism. Thereafter, endogenous rhythms are apt to introduce coordinated flexibility in the operation of the metabolic network.

Brulfert, Guerrier & Queiroz (in preparation) have shown that:
- under long days the phase of the rhythm of PEP carboxylase is fixed, the peak taking place steadily at about 4 to 5 p.m., according to different experiments (Fig. 5A);
- under short days + night interruption by a flash of red light (which results in a long day effect) the phase is also fixed but the peak appears to be strictly locked to the red flash (Fig. 5B);
- under short days, in contrast, a complex behaviour is observed (Fig. 5C): the change from long days to short days produces an immediate advancing shift of the peak, which jumps from 5 p.m. to 12 noon; during the following short days, a series of advancing transients brings up the phase to about 9 a.m.; then follows a series of delaying transients shifting the peak throughout the day; at the 14th short day the peak is at 8 p.m., when begins another series of weaker advancing transients. A slow, continuous phase delay is finally established after about 20-25 short days, as previously reported (Queiroz, Celati & Morel, 1972).

It can be noted that the first series of 7 advancing transients during which mean value and amplitude of the PEP carboxylase rhythm stays unchanged correspond to the 7 short days lag observed for malate variation.

Alternating series of advancing and delaying shifts under the action of a 24 hr repeated new light schedule could be explained by a periodically changing sen-

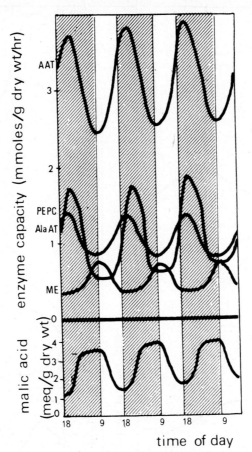

*Fig. 4.* Typical pattern of variations in malic acid content and in enzyme capacity for PEP carboxylase (PEPC), aspartate aminotransferase (AAT), alanine aminotransferase (Ala AT) and malic enzyme (ME) during 3 consecutive short days. Plants were previously given 32 short days.

sitivity of the plant to photoperiod. In other words, the 'phase response curve' would be steady under long days and progressively modified by short days.

*The hypothesis of feedback control*

It is known that feedback control necessarily produces oscillatory outputs. Hence the hypothesis that the observed rhythms in PEP carboxylase could be the result of negative feedback by malate operating as end-product has been proposed in an early paper (Queiroz, 1968) since malate was shown to inhibit the enzyme extracted from *Kalanchoe*. This hypothesis seems at first sight to be consistent with the typical day-night kinetics of malate variation and of PEP carboxylase variations in CAM. But more recent and more complete results show that the rather simple feedback hypothesis cannot account for

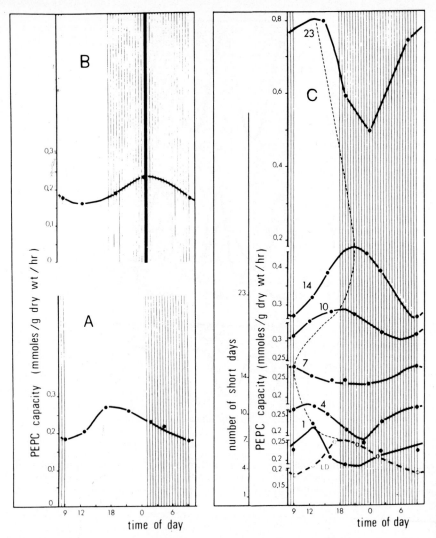

*Fig. 5.* Rhythms of PEP carboxylase capacity. A: under long days; B: under short days + night interruption by a flash of red light (black bar); C after a change from long days (broken curve) to short days (data for the 1st, 4th, 7th, 10th, 14th and 23rd short days). In A and B the phase of the rhythm is stable. In C the phase shows complex transient shifts, pointed out by the dotted line.

- the rhythms of CAM enzymes other than PEP carboxylase
- the phase shifts as described in previous papers and in the preceding section of the present paper
- the 24 hr periodicity of the rhythms.

The existence of a 24 hr periodicity is a strong objection against the hypothesis that feedback would be the direct cause of CAM rhythms. Malic enzyme is present in the tissue and obviously the relaxation time for a malate process cannot be expected to account by itself for a 24 hr period. Moreover, the 24 hr period is a constant component of CAM: it is found in all CAM plants and in leaves of all ages, which obviously present different levels of enzyme activity, different malate contents, different storage capacity of vacuols; if the rhythm was the result of feedback effects arising from the combination of these factors, different periodicities should be expected according to intrinsic differences in species and in leaves; this is never the case. A 24 hr periodicity could be achieved by the effect of a 'lights-on' signal on one of the steps of the process (malic enzyme activity, changes in vacuole permeability and in local malate concentrations) but at our knowledge such effect has not been demonstrated as yet. Anyway a light effect would hardly explain the observed phase behaviour of the different CAM enzymes.

A survey of the published discussion on this problem shows some degree of confusion between measurements of enzyme activity and the assumed feedback effects *in vivo*. An experiment was designed to distinguish between variations in these two components of the system: entering of label into malate during short pulse experiments with $^{14}CO_2$ was assumed to provide an evaluation of changes in actual PEP carboxylase activity at different moments of the day-night cycle and at different number of short days as compared to simultaneous measurements of enzyme capacity.

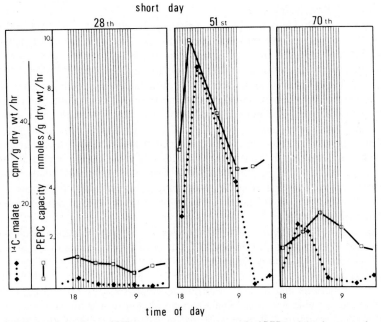

*Fig. 6.* Compared rhythms of PEP carboxylase capacity and of PEP carboxylase actual activity in vivo, evaluated by the production of labelled malate in short-pulse $^{14}CO_2$-feeding experiments. Data for the 20th, 51st and 70th short days. See text.

Data in Fig. 6 show that daily variations in enzyme capacity and in activity of $^{14}C$ incorporation into malate appear to be in phase during the 28th and 51st cycles; in contrast, at the 70th short day, the peak of actual activity precedes the peak of capacity (by about 6 hours) showing a modified relationship between the two variables. It must be noted that these data confirm earlier results which showed that PEP carboxylase capacity decreases after the 50th short day and also the amount of accumulated malate in the leaves stays very high. The ratio between malate content and maximum PEP carboxylase capacity is much higher at the 70th short day than at the 50th short day.

These results suggest a modification of the initial feedback hypothesis. At the beginning of the short day treatment PEP carboxylase capacity and consequently enzyme activity are low: in ,this situation the rhythm of $CO_2$ incorporation into malate would be directed by the rhythm of enzyme capacity. Cumulative effect of short days and the resulting increasingly high enzyme activity produce increasing malate content and the vacuolar storage capacity probably becomes limiting as suggested by Kluge & Reininger (1973). Negative feedback would then be expected to operate as soon as the addition of newly produced malate to the bulk of previously accumulated malate exceeds storage capacity. It results that the peak of *in vivo* activity would no more be synchronized with the rhythm of enzyme capacity measured in the extracts.

In summary, when feedback by malate operates it would be superposed as a modulator on a permanent enzymic oscillator.

## Conclusions: adaptive significance

The results presented in this paper emphasize the flexibility of CAM and have both practical and theoretical implications. From the practical point of view they show that any clearcut distinction between 'short-term events' and 'long term events' in CAM can be misleading. In CAM plants large amplitude rhythmicity cuts across all levels of biological complexity. Fundamental metabolic parameters present very large variations according to the time of day, the time of year, the degree of ageing and the preceding story of the plant. The conflicting data observed in $^{14}C$ distribution between $C_1$ and $C_4$ positions in malate, as discussed elsewhere in this volume, are good example of the disturbing effect that experimentally non-controlled 'long-term' variables can have on the results of 'short term' experiments. The possibility of long term phase transients of the peak of enzyme activity resulting from light or temperature signals or changes in the light-dark cycle cannot be neglected *a priori* without serious risk of misinterpretation of the results.

From the theoretical point of view, considering endogenous timing mechanisms is inescapable in modern physiology as far as adaptive and homeostasic processes are concerned. Endogenous timing appears to be a generalized property of biological systems allowing precise and flexible definition of phase relations between biological functions and external rhythms as emphasized by Pittendrigh (1965). The 'limited entrainment range' characteristic obviously is a fundamental tool in metabolic coordination following a change in external conditions. If, as suggested by the results of Civen, Ulrich, Trimmer & Brown (1967) on liver tissue, inducible enzymes are enzymes displaying circadian rhythmicity, then rhythmic ability could be a necessary implication of key metabolic functions.

The results show that the photoperiod-enzyme relationships are intrinsically different according to the photoperiod: the peak of enzyme capacity is strictly phase-coupled to the long day photoperiod but shows uncoupling, or only relative coordination, to the short day periodicity. This difference could be meaningful in the physiology of a short day plant.

# REFERENCES

Becker, T. (1953): Wuchstoff- und Säuerschwankungen bei *Kalanchoe blossfeldiana*, in verschiedenen Licht-Dunkelwechseln. *Planta (Berl.)*, 43: 1-24.

Borthwick, H.A. & S.B. Hendricks (1960): Photoperiodism in plants. *Science*, 133: 1223-1228.

Brulfert, J., D. Guerrier & O. Queiroz (in preparation)

Bünning, E. & W. Engelmann (1960): Endogen-tagesperiodische Schwankungen der photoperiodischen Hellrot-Empfindlichkeit bei *Kalanchoe blossfeldiana*. *Naturwiss.*, 47: 332.

Bünsow, R. (1953): Endogene Tagesrhythmik und Photoperiodismus bei *Kalanchoe blossfeldiana*. *Planta (Berl.)*, 42: 220-252.

Carr, D.J. (1952): A critical experiment of Bünning's theory of photoperiodism. *Z. f. Naturf.*, 7b: 570-571.

Civen, M., R. Ulrich, B.M. Trimmer & C.B. Brown (1967): Circadian rhythms of liver enzymes and their relationship to enzyme induction. *Science*, 157: 1563-1564.

Ehrenberg, M. (1954): Einfluss verschiedenen Licht-Dunkelwechsels auf die Rythmik des Phosphatasenaktivität in den Blättern von *Kalanchoe blossfeldiana*. *Planta (Berl.)*, 43: 528-536.

Engelmann, W. (1960): Endogene Rhythmik und photoperiodische Blühinduktion bei *Kalanchoe*. *Planta (Berl.)*, 55: 496-511.

Jones, M.B. & T.A. Mansfield (1972): A circadian rhythm in the level of carbon dioxide compensation in *Bryophyllum fedtschenskoi* with zero values during the transient. *Planta (Berl.)*, 103: 134-146.

Kluge, M. & Reinnger, B. (1973): Untersuchungen über den Efflux von Malat aus den Vacuolen der assimilierenden Zellen von *Bryophyllum* und mögliche Einflüsse dieses Vorganges auf den CAM. *Planta (Berl.).*, 113:333-343.

Morel, C., C. Celati & O. Queiroz (1972): Sur le métabolisme acide des Crassulacées. V. Adaptation de la capacité enzymatique aux changements de photopériode (métabolisme de l'acide malique et transaminations). *Physiol. vég.*, 10: 743-763.

Mukerji, S.K. (1968): Four hour variations in the activity of malate dehydrogenase 'decarboxylating' and phosphoenolpyruvate carboxylase in the cactus (*Nopalea dejecta*) plant bud. *Ind. J. Biochem.*, 5: 62-64.

Pittendrigh, C.S. (1965): On the mechanism of the entrainment of a circadian rhythm by light cycles. In Aschoff, J. (Ed.): Circadian clocks. Pp. 276-297. North Holland, Amsterdam.

Pittendrigh, C.S. (1973): Circadian oscillations in cells and the circadian organisation of multicellular systems. In Schmitt, F.O. (Ed.): The neurosciences: Third Study Program.MIT Press.

Queiroz, O (1965): Sur le metabolisme acide des Crassulacées. I. Action à long terme de la température de nuit sur la synthèse d'acide malique par *Kalanchoe blossfeldiana* „Tom Thumb" placée en jours courts. *Physiol. vég.*, 3: 203-213.

Queiroz, O. (1966): Sur le métabolisme acide des Crassulacées. II.: Action à long terme de la température de jour sur les variations de teneur en acide malique en jours courts. *Physiol. vég.*, 4: 323-339.

Queiroz, O. (1968): Sur le métabolisme acide des Crassulacées. III. Variations d'activité enzymatique sous l'action du photopériodisme et du thermopériodisme. *Physiol. vég.*, 6: 117-136.

Queiroz, O. (1970): Sur le métabolisme acide des Crassulacées. IV. Réflexions sur les phénomènes oscillatoires au niveau enzymatique et sur la compartimentation métabolique sous l'action du photopériodisme. *Physiol. vég.*, 8: 75-110.

Queiroz, O (1974): Circadian rhythms and metabolic patterns. *Annu. Rev. Plant Physiol.*, 24: 115-134.

Queiroz, O., C. Celati & C. Morel (1972): Sur le métabolisme acide des Crassulacées. VI.

Photopériodisme et rythmes circadiens d'activité enzymatique. Systèmes régulateurs. *Physiol. vég.,* 10: 765-781.

Queiroz, O. & C. Morel (1974): Photoperiodism and enzyme activity. Towards a model for the control of circadian metabolic rhythms in the Crassulacean acid metabolism. *Plant Physiol.,* 53: 596-602.

Schmitz, J. (1951): Über Beziehungen Zwischen Blütenbildung in Verschiedenen Licht–Dunkelkombinationen und Atmungsrythmik bei Wechselnden photoperiodischen Bedingungen. *Planta (Berl.),* 39: 271-308.

Schwabe, W.W. (1952): Effects of photoperiodic treatment on stomatal movements. *Nature,* 169: 1053-1055.

Schwabe, W.W. (1955): Photoperiodic cycles of lengths differing from 24 hours in relation to endogenous rhythms. *Physiol. Plantarum,* 8: 263-278.

Vanden Driessche, T., (1966): The role of the nucleus in the circadian rhythm of *Acetabularia mediterranea. Biochim. Biophys. Acta.,* 126: 456-470.

Vanden Driessche, T. (1970): Circadian variation in ATP content in the chloroplasts of *Acetabularia medeterranea. Biochim. Biophys. Acta.,* 205: 526-528.

Vanden Driessche, T. & S. Bonotto (1969): The circadian rhythm in RNA synthesis in *Acetabularia mediterranea. Biochim. Biophys. Acta.,* 179: 58-66.

Walther, W.G. & L.N. Edmunds, Jr. (1973): Studies on the control of the rhythm of photosynthetic capacity in synchronized cultures of *Euglena gracilis* (Z.). *Plant Physiol.,* 51: 250-258.

Wilkins, M.B. (1959): An endogenous rhythm in the rate of $CO_2$ output of *Bryophyllum.* I. Some preliminary experiments. *J. expt. Bot.,* 10: 337-390.

Wilkins, M.B. (1960): An endogenous rhythm in the rate of $CO_2$ output of *Bryophyllum.* II. The effects of light and darkness on the phase and period of the rhythm. *J. exp. Bot.,* 11: 269-288.

Wilkins, M.B. (1965): The influence of temperature and temperature changes on biological clocks. In Aschoff J. (Ed): Circadian clocks, Pp 146-163. North Holland, Amsterdam.

# A PERSONAL ASSESSMENT OF THE STATE OF KNOWLEDGE OF CRASSULACEAN ACID METABOLISM (CAM)

J.W. BRADBEER

*Department of Plant Sciences, University of London King's College.*

To a former worker with Crassulacean plants both the resurgence of interest in CAM and the quality of the work presented at this symposium have provided much personal satisfaction. Although the main impetus to this renewed interest has probably resulted from discoveries about C-4 plants there seem to be two other important factors involved. One of these is the application of infra-red gas analysis to the study of the $CO_2$ uptake and output of plants. The other factor is the development of what might be described as 'ecophysiology', of which the leading practitioner would seem to be Dr. O. Björkman, who has made an important contribution to the symposium. This subject involves the extension of precise laboratory and field investigations to a wide range of natural plant populations, beyond the usual concentration of the interests of plant physiologists and bio-chemists on cultivated species.

The organisers of this symposium should be congratulated for bringing together such substantial CAM contributions. As a postgraduate student in this area just 20 years ago I felt that CAM was generally regarded as something of a scientific backwater. My supervisor, Meirion Thomas, had deduced from the literature that the synthesis of malic acid in acidifying darkened CAM plants involved the fixation of large quantities of $CO_2$. These deductions were firmly established by experimental work by H. Beevers in Thomas's laboratory. Thomas took the somewhat unusual course of publishing their results for the first time in the third edition of his textbook (Thomas, 1947) and it wasn't until sometime later that the full data were published (Thomas & Beevers, 1949) alongside a comprehensive analysis of the literature relating to the gaseous exchanges of CAM plants (Thomas, 1949). This first publication in a text book was in fact appropriate as Thomas had made his deductions while he was working on the manuscript of his third edition. For a number of years Thomas's department became the main centre of work on CAM under the joint direction of Professors Thomas and Ranson who summarized the main findings up to 1959 in a review (Ranson & Thomas, 1960). In this present symposium volume Bradbeer, Cockburn and Ranson have included some previously unpublished data to establish firmly that when CAM plants synthesize malic acid in the dark in the presence of $^{14}CO_2$ both carboxyl carbons of the malic acid become labelled with approximately twice as much label in carbon atom 4 as in carbon atom 1. The Newcastle tradition in this field is maintained by important new work described in the paper by Cockburn.

In the present volume the contributions of Cockburn, Kluge, Osmond and Sutton all contain substantial pieces of evidence which are inconsistent with the hypothesis that dark acid fixation in CAM plants proceeds by a double car-boxylation pathway. However it was accepted that the 2:1 distribution of label between the 4 and 1 carboxyls of malate synthesized in the dark by the $^{14}CO_2$ fixation of CAM plants was a genuine phenomenon and not an artifact of the degradation as had been suggested previously (Sutton & Osmond, 1972). The consistency with which the 2:1 distribution of label was found had led Bradbeer et al. (1958) to propose the double carboxylation pathway as the simplest possible explanation of the data. The main piece of inconsistent evidence is the finding by Cockburn that no more than a negligible number of malate molecules, which were synthesized in the presence of $^{13}CO_2$, contained more than one atom of $^{13}C$. As Cockburn's experimental procedure was of necessity somewhat complicated, he will have to provide the most rigorous checks of his methods together with copies of the mass-spectra so that his findings can be conclusively established. The remaining evidence is less direct, for example both Cockburn and Osmond found many cases where the distribution of label fixed from $^{14}CO_2$ diverged appreciably from the 2:1 ratio. Osmond also found it difficult to explain his pulse-chase data on the basis of a double carboxylation. Kluge found that exogenous phosphoenolpyruvate and and 3-phosphoglycerate gave a substantial promotion of malate synthesis without altering the 2:1 distribution in malate of the simultaneously fixed $^{14}CO_2$. The weight of this evidence suggests strongly, but does not conclusively establish, that the double carboxylation hypothesis is incorrect.

We do not yet have a firmly formulated hypothesis to replace the double carboxylation hypothesis however. There is very little information so far about the enzymology and the properties of the organelles of CAM plants. At the present time any attempt to make a precise description of the pathways of metabolism of CAM plants and to define the subcellular location of each reaction has to depend to a very large extent on assumed analogies with non-CAM plants. In the discussions at this meeting Dr. P.C. Brandon provided some insight into the difficulties of working with organelle preparations from CAM plants. The technical difficulties arising from the high organic acid content of CAM plants together with the presence of tannins etc. has proved a severe deterrent to progress. It is to be hoped that further experimental investigations on the isolated enzymes, chloroplasts, mitochondria and peroxisomes of CAM plants will follow in the future.

The papers, in this volume, by Neales, Marcelle and Queiroz also point to a considerable difficulty in the investigation of the mechanism of CAM by *in vivo* labelling experiments. Is is evident from the data that steady state metabolism is not to be expected in CAM leaves and as a consequence labelling experiments will tend to be extremely difficult to analyse.

In the papers by Ting and Neales very valuable information about CAM plants has come from the careful physiological study of the plants in relation to their natural environment. Dr. Ting has shown that under water stress the CAM plant is able to virtually eliminate gaseous exchange with the environment while its internal metabolism is continued. The photoperiodic complexities of CAM plants have been described, if not accounted for, in the papers by Marcelle and Queiroz. Finally in the paper by Lerman and the discussion contribution by Dr. Troughton the findings and potentialities of the application of mass spectrometry in the precise study of the discrimination by CAM plants against naturally occurring $^{13}CO_2$ has been covered. It seems possible that the experimental use of this technique may provide a new line of evidence relating to the mechanism of CAM

# REFERENCES

Bradbeer, J.W., S.L. Ranson & Mary Stiller (1958): Malate synthesis in Crassulacean leaves. I. The distribution of $C^{14}$ in malate of leaves exposed to $C^{14}O_2$ in the dark. *Plant Physiol.,* 33: 66-70.

Ranson, S.L. & M. Thomas (1960): Crassulacean acid metabolism. *Ann Rev. Plant Physiol.,* 11: 81-110.

Sutton, B.G. & C.B. Osmond (1972): Dark fixation of $CO_2$ by Crassulacean plants. Evidence for a single carboxylation step. *Plant Physiol.,* 50: 360-365.

Thomas, M. (1947): Plant physiology, 3rd. edition. J. & A. Churchil Ltd. - Publishers, London.

Thomas, M. (1949): Physiological studies on acid metabolism in green plants. I. $CO_2$ fixation and $CO_2$ liberation in Crassulacean acid metabolism. *New Phytol.,* 48: 390-420.

Thomas, M. & H. Beevers (1949): Physiological studies on acid metabolism in green plants. II. Evidence of $CO_2$ fixation in *Bryophyllum* and the study of diurnal variation of acidity in this genus. *New Phytol.,* 48: 421-447.

# LIGHT LEVEL AND THE MEAN SPEED OF TRANSLOCATION IN *ZEA MAYS* LEAVES

John H. Troughton

*Physics and Engineering Laboratory,
Department of Scientific and Industrial Research, Private Bag, Lower Hutt, New
Zealand.*

## Abstract

The effect of light on the mean speed of translocation of carbon compounds in *Zea mays* leaves was investigated using the short-lived isotope carbon-11. This isotope was generated using a 3.0 MeV Van der Graaf accelerator and converted to $^{11}CO_2$ before it was photosynthesised by a restricted, 2 cm wide, segment of a leaf attached to a plant. The plant was in a controlled environment chamber at $28 \pm 1°C$ and 80% relative humidity. Carbon-11 transported along the leaf in the phloem was detected externally at several positions by NaI scintillation counters attached to photomultipliers. The pulses from the photomultipliers were passed through single channel analysers, amplified and counted, and the results corrected by computer for the half-life and background radiation.

The results established that the shape of the pulse was primarily determined by the loading phenomena in the region fed isotope. A mean speed of translocation determined from the time delay in arrival of the mid-point of the front of the pulse at successive counters was proportional to the light level, although a component accumulated in the light influenced the speeds in the dark. The maximum speed recorded was 11.0 cm min$^{-1}$.

## Introduction

The subject of the long distance transport of carbon compounds in plants has prompted much speculation and debate on all aspects of the topic: anatomy, mechanism, kinetics and composition of the fluid in the phloem. The inability to quantitatively describe this process has hindered progress in understanding plant productivity because translocation is an essential link in the carbon pathway between photosynthesis and carbon utilization in cell division, cell wall deposition, new leaf production, root growth and the production of economic yield. To some extent progress in solving these problems has been slow because of the unavailability of appropriate techniques to investigate the kinetic aspects of the process.

Most kinetic studies of carbon transport have utilised the isotope carbon-14 but alternative carbon isotopes are available and the work reported here was initiated to evaluate the short-lived isotope carbon-11. There are several potential advantages of this isotope for carbon translocation studies.

1. Use of a carbon isotope ensures labelling of the components in the phloem which are of most interest, that is sucrose. Comparison in this laboratory of the

The author acknowledges the assistance and co-operation during this project of Mr R. More and the Accelerator Section of the Institute of Nuclear Sciences, D.S.I.R. and Mrs K. Card and Mr B.G. Currie of the Physics and Enginering Laboratory, D.S.I.R.

C-11 isotope with other short-lived isotopes, such as technetium-99m and [24]Na which are also phloem mobile have re-emphasised the value of directly labelling the carbon compounds.

2. The half-life of carbon-11 is 20.4 minutes and even with the activities that can be produced on small accelerators (3 MeV) the isotope can be monitored in plants for about three hours. The advantage of the short half-life is that numerous measurements can be made on the same leaf tissue even within the same day. After about three hours a leaf is isotopically 'clean' again and able to be used in further experiments.

3. Carbon-11 decays by emission of 0.96 MeV positrons which on annihilation yield two 511-KeV gamma rays. The ease and efficiency of detection of these protons allows *in vivo* measurements of isotopes to be made in plants. This removes the requirement that a plant has to be killed before the level of radioactivity in the tissue can be measured.

These advantages of carbon-11 allow the development of a system for almost continuous monitoring of translocation which is compatible with the continuous monitoring which can be carried out on water flow, transpiration and photosynthesis.

There are some disadvantages of using carbon-11 and these include the necessity to be close to an accelerator or cyclotron. Furthermore, the short life of the isotope prevents long-term carbon translocation or distribution problems to be investigated. It is anticipated, however, that both the stable, radio-active and short-lived carbon isotopes will be used in conjunction with each other in order to obtain the maximum information about translocation. A further problem which arose in using this isotope was the need for on-line computing facilities for both data acquisition, correction of the data for half-life and background radiation and for analysis of the data. In appreciating and overcoming these problems it is now possible to make routine measurements of the speed of translocation on different plants or the same leaf under a variety of environment conditions.

## Material and Methods

Several reactions are available to produce carbon-11 but in this laboratory the reaction $^{10}B$ (d,n)$^{11}C$ with a 3.0 Van der Graaf accelerator has been shown to be the most useful for this application (Troughton & More 1974). A deuteron beam with an energy of 2.75 MeV and a current of 30 $\mu$A was used to irradiate a solid $^{10}B$ enriched (90% $^{10}B$) disc while oxygen was used both as a reactant and as a sweep gas, resulting in the production of $^{11}CO$. This system provided continuous production of the isotope and could be used for up to 10 hours of irradiation time before there was any damage to the target. The form in which the isotope was most useful was as carbon dioxide and used in batches, i.e. for pulse feeding. A system was therefore designed which included a differential pumping system to recover the isotope from the accelerator and a furnace run at 600°C and containing CuO wire to convert $^{11}CO$ to $^{11}CO_2$. Contaminants that may have been present in the gas stream were removed from the $CO_2$ gas either by a dry ice-alcohol trap (for water) or a liquid oxygen trap which selectively removes $^{11}CO_2$. At the completion of an irradiation period the liquid oxygen trap was closed and warmed to room temperature and the $^{11}CO_2$ was recovered by a syringe in a form ready to be fed to a segment of the leaf of a plant.

All experiments were conducted with *Zea mays* (*cv morden 88*) which was grown from seed for about four weeks in a controlled environment of $28°C \pm 1°$, 80% relative humidity, 12 hour photoperiod and irradiance of $250Wm^{-2}$ (400-700nm) from HPLR mercury vapour lamps. The experiments were carried out on whole plants kept in controlled-environment cabinets under the same conditions. Mature leaves were selected for the experiments and an area of leaf about $15cm^2$ but in a 2cm wide strip about two thirds the way along the leaf from the base towards the tip was fed the radioactive carbon dioxide. The small leaf chamber in this position had a water jacket to control air temperature close to that of the whole plant chamber. Leaf and air temperatures were continuously monitored using copper-constantan thermocouples. The plants were kept well watered and at normal $CO_2$ concentration throughout all experiments.

The isotope feeding procedure was that of a pulse labelling whereby the plants were exposed to $^{11}CO_2$ for periods between two and five minutes depending on the light conditions during feeding. At the end of this period the small leaf chamber was flushed using normal air. Flushing was usually complete within one minute.

The isotope was detected using NaI scintillation counters. The counters were positioned at 10 cm intervals both downstream and upstream from the feeding chamber. The portion of the isotope moving towards the tip of the leaf was small and therefore only one counter was used upstream from the feeding chamber. This allowed three or four counters to be placed at 10 cm intervals from the fed portion of the leaf towards the leaf base. Each counter was surrounded by lead and collimated to allow a 2 cm wide strip of leaf immediately above each counter to be viewed by the counters. The output from all detectors was amplified, passed through single channel analysers and counted in scalers before being recorded on punched tape. All counters were measured simultaneously with counts accumulated over 20 secs, which allowed two measurements per counter per minute. The counts were corrected for background and radioactive decay and only transformed data is presented here.

## Results

A characteristic output from four counters downstream from the feeding chamber in a pulse label experiment is given in Fig. 1. The counts in this example are normalised to allow direct comparison between counters. Absolute counts can be obtained which would allow measurement of the loss of tracer with distance along the leaf. The shape of the relationship between the number of counts and time for any counter was of a pulse type. However, there was a dramatic broadening of the pulse feed to the leaf and that same pulse appearing at any of the counters. The half width of the pulse fed to the leaf was about two minutes, whereas it is evident from Fig. 1 that the half width of the pulse can be as much as 60 minutes. The half width of the pulse was shown in these experiments to be highly variable and to be between 20 minutes and approximately 120 minutes. There were periods when the half width did not change with distance along the leaf, which

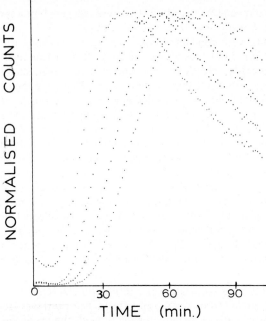

NORMALISED COUNTS

TIME (min.)

*Fig. 1.* The output from four counters given as a function of time following feeding the $^{11}CO_2$ to a 2 cm wide maize leaf segment. All counts are corrected for the half-life of the isotope and for background radiation, and are normalised to facilitate comparisons between counters. The counters were at 10 cm intervals downstream from the fed region and in the order shown by the time-displacement of the pulses.

would provide confirmation that the site determining pulse shape was primarily located in the fed area (Troughton, unpublished results).

The primary characteristic of carbon transport which was investigated in these experiments was the speed. This was obtained from the pulses by determining the delay in arrival of the midpoint of the front of the pulse at successive counters. This represents a mean speed and may differ substantially from speeds measured from the arrival of the first isotope at a counter or at the peak of the pulse and does not represent the mean speed for all the particles within the pulse. There was also evidence that the mean speed varied with distance along the leaf and therefore it is also necessary to qualify the mean speed reported with respect to the distance over which it is measured. In most examples in this study the mean speed was determined over a 20 cm distance, with the first counter 10 cm from the fed area.

Using the definition of a mean speed given above, the effect of light and dark, and in particular light level, on the mean speed of translocation in maize was investigated. It had previously been shown that there were transients in the mean speed of translocation during transfer of a plant from the dark into the light (Moorby et al 1974) and it was therefore necessary to investigate the time required for a leaf to reach a new equilibrium speed after a change of light level. The characteristic response pattern for two light levels is given in Fig. 2. In order to check the extent to which the leaf was standardized prior to the new treatment the speed of translocation was also measured in the dark. In all experiments reported here the speed of translocation in the dark was approximately 1 cm min⁻¹. Furthermore, to ensure that the plant had come to equilibrium at each light level the speed of translocation was taken when three consecutive measurements of the speed were equal. (Fig. 2). In the high light experiment it can be seen that there was a transient speed between that in darkness and after equilibrium is reached in full light but in both high and low light three consecutive measurements of the speed were constant.

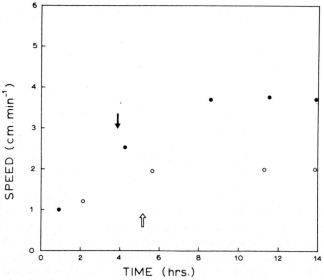

*Fig. 2.* The time-dependence of the development of equilibrium mean speeds of translocation in maize under constant environment conditions. The plants were initially in the dark and the time of transfer to the two different light regimes is shown by the arrows. The open circles and open arrow are for the experiment one (light level 10 Wm⁻²). and closed circles and closed arrow for a second experiment (light level 230 Wm⁻²).

As shown in Fig. 2 there was an effect of light level on the speed of translocation in maize. This effect is further established in the results presented in Fig. 3 where there is an approximately linear relationship between the mean speed of translocation and light level over the range from 0 - 240 Wm⁻². This result is similar to that previously reported for ice (Troughton et al 1974a) and therefore the dependence of speed on light level occurs in both C₃ and C4 type monocotyledons. The effect of light on the speed of translocation shown in Fig. 3 can be used to estimate the speed of translocation in the dark by extrapolation to zero light. In that example the speed in the dark was 1 cm min⁻¹. This is approximately the value shown in Fig. 2 which was obtained after a normal dark period of 12 hours.

To further investigate effects of long-term light and dark on the speed of translocation, measurements of the speed of translocation were made in the dark after a plant had been exposed to a long period (46 hours) of continuous light (100 Wm⁻²). It is not possible using these experimental methods to obtain measu-

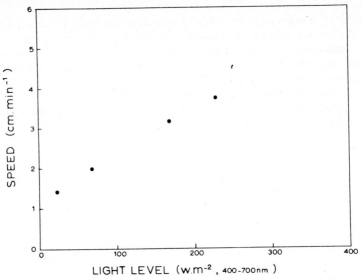

*Fig. 3.* The relationship between the light level at the maize leaf surface and the mean speed of translocation measured over a 20 cm length of leaf. Each reading is the mean value of three consecutive measurements which at each light level were constant to within 0.1 cm min$^{-1}$.

rements of translocation in the dark without exposing the plant to light during feeding of the isotope. In these experiments the plant was left in the light for a minimum period of time which was usually 7 minutes; three minutes before the isotope was fed to allow stomatal opening, followed up by four minutes during the feeding process. As soon as the feeding period concluded the lights were switched off. This period of light was small relative to the intervals between feedings and would amount to only about 4% of the total time in the dark. Nevertheless there may have been some effects of this light period on the recorded mean speeds although the treatment was consistent for all measurements and will allow the measurements in the dark to be compared. These results shown in Fig. 4 establish that there is an immediate and long term effect of darkness on the mean speed of translocation. In this example the pretreatment light period was 46 hours and the time taken to recover to the speed of translocation expected in the dark under a normal diurnal cycle was about 20 hours, with a half time of response of about five hours.

The previous measurements were made on a single leaf of a plant but the whole plant was either in the light or in the dark. To further investigate the site at which the effect of light on speed was operating, the same type of measurements were made on a single leaf. During these experiments the whole plant was kept in the light and the shade treatments were applied to one leaf; shading was obtained by using tinfoil. Leaf temperatures were continuously measured to ensure that there was no effect of the treatment on leaf temperature. The results shown in Fig. 5 indicate that the site of the stimulation of translocation by light is in the leaf on which the speed is being measured. In this example the leaf was exposed to light for six hours before the whole leaf was placed in darkness. In the light the mean speed of translocation came to a constant value but on shading there was in

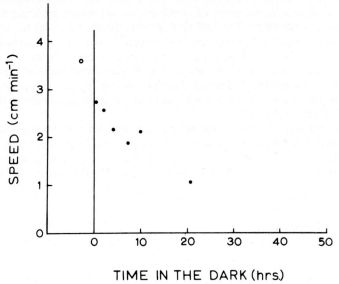

Fig. 4. Light-dark transients in the mean speed of translocation in maize. The open circle is the speed after 45 hours in continuous light and the closed circles the speed in the dark. Approximately 7 minutes of light were given during feeding of the $^{11}CO_2$ in the dark. In this experiment the whole plant was darkened.

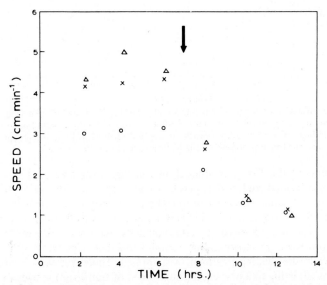

Fig. 5. Light-dark transients in the mean speed of translocation at three positions 15 cm (o), 25 cm (x), 35 cm (△) downstream from the fed region. The plant was previously kept under a 12 hour light regime and given 6 hours of high light before the dark period (indicated by the arrow). Only the leaf on which the measurements were made was darkened during this experiment.

mediate reduction in the speed of translocation and within six hours the speed had fallen to 1 cm min$^{-1}$. The half time for this change was about one hour. The difference in response between the two experiments is almost certainly to be related to the duration of the light pretreatment, 46 hours in the first example and six hours in the other. The results suggest that a compound or compounds were being accumulated during the light period and the size of the pool determines the period for which these substances significantly contribute to the speed of translocation in the dark.

These two sets of results indicate that the light level directly and indirectly influences the speed of translocation in maize leaves. To further investigate the site of the effect of light on the speed of translocation, experiments were designed to investigate the effects of localised shading within the leaf on the speed of

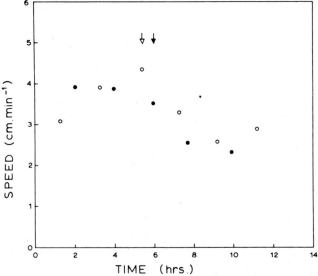

*Fig. 6.* The influence of localised shading of 20 cm length of leaf on the mean speed of translocation in maize. Results for two experiments are given (open and closed circles) and the arrows indicate the time of application of shading to the leaf kept at high light (approx. 250 Wm$^{-2}$). The shading was applied to the leaf directly above the counters.

translocation. In the first treatment shown in Fig. 6 the area directly above the counters was shaded and two experiments are reported. The speed of translocation prior to shading was allowed to come to equilibrium before a 20 cm length of leaf over the counters was shaded and the effect of this treatment on the speed of translocation was followed for approximately six hours. The results for the same leaf on two days shown in Fig. 5 established that there was a 30% reduction in speed when the shading treatment was applied. There was some evidence that the leaf was adjusting to a new equilibrium speed of translocation over this period. In comparison with the shading of the whole leaf this amounted to a significantly lower effect on speed.

To further investigate the effects of localised shading on the speed of translocation, a 10 cm segment of the leaf between two counters was shaded. A measurement of speed was obtained over 10 cm length of leaf upstream and downstream from the shaded portion as well as in the tissue which was shaded. The results are

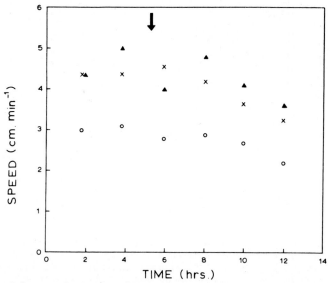

*Fig. 7.* The influence of a localised shading treatment (10 cm length of leaf) on the mean speed of translocation in a maize leaf in three 10 cm segments; immediately upstream (o), immediately downstream (△) and in the shaded region (x). The leaf was kept in high light (approx. 250 Wm$^{-2}$) for the twelve hour experiment and the arrow indicates the application of the shade.

given in Fig. 7 for the three positions and for 5 hours before and 6 hours after shading. There was some evidence for an overall reduction in the speed of translocation in all positions amounting to about 18%. This was significantly less than that induced by shading a 20 cm long leaf segment. There was no clear evidence from the results that the effect of shading was greater on any one of the three segments although there may have been a transient reduction of 20% on the segment immediately downstream from the shaded region.

The comparison of the whole leaf, 20 cm length or 10 cm length of shading suggested that the area of leaf shaded was likely to have a proportional effect on the speed of translocation. As a proposition for testing we assumed that the effect of shading on the speed of translocation was correlated with the proportion of the area shaded relative to the total area of the leaf upstream from the counters. In a further series of experiments the area shaded was kept at a 10 cm length of the leaf but the area involved was approximately 10% of the leaf area upstream from the counters. The mean speed of translocation on three consecutive days is given in Fig. 8. One significant feature was that on the three consecutive days the mean speed was the same for the leaf kept under the same light regime and within four hours of being exposed to light. This indicates the repeatability of the translocation system and of the carbon-11 technique used in this manner. The proportion of the leaf shaded in all cases was approximately 10% and the effect on the speed of translocation where either 10 cm upstream from the feeding area or

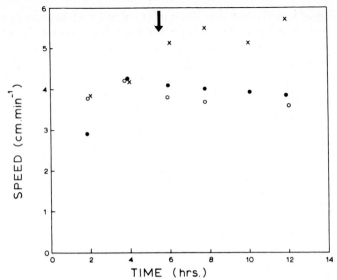

*Fig. 8.* The influence of shading treatments on the mean speed of translocation along a 20 cm length of maize leaf. All results were measured on the same leaf, but on three consecutive days. The plant was kept under constant environmental conditions, wth a daylength of 14 hours. The mean speed of translocation was the same at the 4 hour measurement on all days. On Day 1 (o) a 10 cm length of leaf upstream from the leaf chamber was shaded, on Day 2 (△) 10 cm over the counters was shaded and on Day 3 (x) the 10 cm between the feeding chamber and the first counter downstream was shaded. The arrow indicates the time the shading treatment was applied.

10 cm between the first two counters were shaded was small and resulted in a reduction in speed of less than 10%. In these results, however, there was a dramatic effect of shading on the speed of translocation when the shading was between the feeding chamber and the first counters. As shown in Fig. 8 the speed increased by about 25% and remained high for the period following shading. Further analysis of the variation in speed between the two counting positions indicated (Fig. 9) that there was some reduction in speed in the second counting position but in the 10 cm length immediately below the shaded region the speed increased from 3.5 cm min[-1] to about 6.5 cm min [-1]. In fact the speed estimated for one measurement was approximately 11.0 cm min[-1]. Because the distance between the counters was 10 cm it is not possible to exactly resolve the speed for that run. A high speed also occurred in one other experiment in which the estimated speed under normal conditions for a leaf shaded in the same manner reached 10 cm min[-1].

## Discussion

The mean speed of carbon translocation measured using carbon-11 as in this paper illustrates the value of this technique as a routine method of estimating a mean speed of translocation in plants. During the course of translocation experiments in

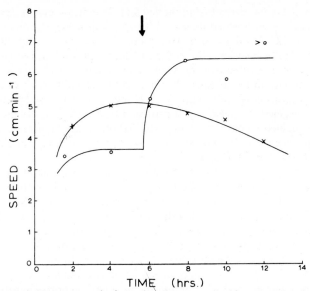

*Fig. 9.* Analysis of the mean speed of translocation over two 10 cm lengths of maize subject to the conditions shown in Figure 8 for Day 3. The arrow indicates the onset of the shading treatment. The open circles indicate the segment adjacent to the shaded area and the crosses the 10 cm segment immediately downstream from the open circle segment. The measurement indicated as $>$ was 11.0 cm min$^{-1}$ but the resolution would be $\pm 2$ cm min$^{-1}$ at that speed and with this arrangement of the equipment.

our laboratory, in excess of 250 measurements of speed have now been made which confirm our expectations that the carbon-11 technique will be useful in analysing some aspects of the speed of translocation and its response to environmental variables and genetic differences between plants. The technique has now been used on maize, tomatoes, soya beans, rice and wheat. The repeatability of the technique (and of the translocation process) is illustrated by the results in Fig. 2 where the speed of translocation was consistent for over six hours at two light levels and also in Fig. 8 where a mean speed of translocation on the same leaf was the same on three consecutive days when the environmental conditions were kept constant. For this latter leaf approximately 50 measurements of speed were able to be made on the same leaf. These results therefore support the use of short lived isotopes and gamma counting techniques in translocation investigations. Apart from the most easily derived parameter, the mean speed of translocation, it is also possible to define the shape of the pulse and to determine the effect on the shape of this pulse of the environment, genetic factors and position within the leaf.

A mean speed of translocation has been measured for many plants using carbon-14 techniques (Crafts & Crisp 1971). The range of values determined during these experiments with maize and using the carbon-11 technique has generally resulted in higher speeds than had been previously determined. There is no reason to suspect that there will be any difference in a speed calculated using a carbon-11 versus a carbon-14 technique other than that due to the way in which the results are analysed. It is difficult to determine a pulse of carbon using carbon-14 and therefore a speed is often determined from the arrival of the first radioactivity at a point distant from the fed leaf. This measurement clearly differs from the technique of analysis used in this paper which uses the midpoint of the front of the pulse. Another factor which may be involved in the difference in results of the carbon-11 versus the carbon-14 techniques is that the speeds in our experiments were measured over relatively short distances and effects of distance on speed were able to be obtained. In the case of the carbon-14 work the mean speed of translocation is often reported over longer distances and would include an effect of distance on speed. Investigations into the effect of distance on speed in maize has yielded inconsistent results, in that on occasions increase in speed of up to 75% between two 10 cm segments have occurred although in other experiments reductions in speed of up to 30% have been recorded.

One of the major reasons for the variation in speed between maize and many other species that have been studied may, however, be the relationship to the photosynthetic pathway. Maize is a $C_4$ type plant and may have consistently higher speeds than $C_3$ type plants, although at least one $C_3$ type plant, rice, has also been shown to have a relatively high mean speed of 3.5 cm per minute compared with an approximate mean speed in maize of 4.5 cm per minute (Troughton et al 1974b). These values, however, refer to moderately high light intensities and as the light levels in other studies have often not been reported it is not possible to comment to what extent the previously reported genetic variation may be related to the light intensity during the period of translocation. The light prehistory of the plant may also be involved. For example we have previously shown that keeping a plant in the dark for long periods prior to the measurement of the mean speed of translocation severely depressed the translocation for at least 10 hours after transfer to relatively high light levels (Moorby et al 1974). An evaluation of the genetic component in the speed of translocation will only be possible in further experiments under closely controlled conditions.

These experiments established that light affects the mean speed of translocation in maize leaves. The results confirm the relationship established for rice that the speed of translocation was proportional to the light level (Troughton et al 1974a). The presence of a significant mean speed of translocation in the dark, approximately 25% that at high light levels, is evidence that a pre-history component is contributing to the speed in the leaf. Pretreatment of the leaf for 6 or 46 hours in the light and following the speed in the subsequent dark periods led to the hypothosis that the mean speed of translocation in the dark is a function of the duration of the previous light period. These same light-dark transient measurements established that there were at least two time constants for the reduction in the mean speed of translocation on transfer from light to dark.

These results coupled with localised shading treatments on the leaf in which the speed was being measured confirmed that the mean speed of translocation was a function of the light level, duration of light and area of the leaf exposed to light. The latter component has been difficult to characterize because of conflicting results. If the speed were directly proportional to the area of leaf subtended then the mean speed would be expected to increase with distance down the leaf. Although some measurements confirm this relationship many results suggest the mean speed declines with distance and it is necessary to postulate that an anatomical or turgor mechanism is also involved. The localised shading treatments were designed to maintain a constant geometry but alter the proportion of the leaf in the light. The results from these experiments suggested that the speed was proportional to the area in the light, although one result showed a substantial increase in speed on shading.

Localised shading of the leaf immediately over the gamma counters established that the influence of light level on speed was not due to a direct effect of light in this region. Thus, if ATP generated by light reactions was involved in determining the speed, then the source of ATP must be outside the shaded region. It seems more likely that the light effects on translocation observed in this paper are primarily mediated through the influence of light on the levels of sucrose or other carbon compounds formed by the highly light dependent photosynthetic process.

# REFERENCES

Crafts, A.S. & Crisp, C.E. (1971): Phloem Transport in Plants. W.J. Freeman and Co., San Francisco

Moorby, J., J.H. Troughton & B.G. Currie (1974): Investigations of carbon transport in plants II The effects of light and darkness and sink activity on translocation. *J. exp. Bot.,* 25: 937-944.

More, R.D. & Troughton, J.H. (1973): Production of $^{11}CO_2$ for use in plant translocation studies. *Photosynthetica,* 7 : 271-274.

Troughton, J.H. Chang, F.H. & Currie, B.G. (1974a): Estimates of a mean speed of translocation in leaves of *Oryza sativa. Plant Sci. Letters*, 3: 49-54.

Troughton, J.H., Moorby, J., & Currie, B.G. (1974b): Investigations of carbon transport in plants I The use of carbon-11 to estimate various parameters of the translocation process. *J. exp. Bot.,* 25: 684-694.

# ENVIRONMENTAL EFFECTS ON THE MEMBRANE ASSOCIATED ELECTRON TRANSPORT REACTIONS OF PHOTOSYNTHESIS

John H. Troughton[1,2], D.C. Fork[2] & F.H. Chang[3]

[1]*Physics and Engineering Laboratory, Department of Scientific & Industrial Research, Private Bag, Lower Hutt, New Zealand,* [2] *Department of Plant Biology, Carnegie Institution of Washington, Stanford, California 94305,* [3]*Botany Department, Victoria University of Wellington, P.O.Box 190, Wellington, New Zealand*

## Abstract

Short-term effects of the environment on the photochemical events in the thylakoid membrane were monitored in intact leaves of $C_3$, $C_4$, and CAM pathway plants by measuring the light-induced absorbance changes at 510 nm and 420 nm. Significant light adaption responses were observed which established the existence of a higher quantum efficiency state in the dark than in the light adapted state. Photochemical reactions were also observed at 740 nm in dark adapted leaves which suggested that photosystem I was operating efficiently under those conditions. The measurements at 420 nm established that light adaption resulted in a reorganization in the rates of some reactions in the electron transport chain, leading to an inhibition of the rate of cytochrome f oxidation and enhancement of cytochrome f reduction.

Long-term effects of a range of temperatures on plant production of several $C_3$ and $C_4$ plants established that adverse temperatures were accompanied by a reduction in chlorophyll a and b and galactolipid contents of the leaves. Electron microscope examination revealed substantial disorganization of the chloroplast membranes under these conditions which would result in seriously impaired electron transport reactions.

## Introduction

The metabolic processes and structures associated with the growth and survival of plants continually vary in response to fluctuations in the natural environment. Plant adjustment to environment variation occurs at the community, single plant, cellular and molecular level and with a time constant which can vary from years to nanoseconds. Photosynthesis is particularly responsive to environmental conditions, as shown by numerous experiments on the influence of light, water stress, humidity, nutrients and temperature on $CO_2$ exchange. Many anatomical and physiological features of plants are involved in determining the rate of $CO_2$ exchange by a leaf but in this paper emphasis is given to the membrane associated electron transport reactions.

Electron transport reactions occur in the thylakoid membrane of the choroplast. This association with a membrane suggests the process will depend both on the integrity and cooperation of the components of the electron transport chain and on the structure and chemical composition of the matrix, ie the membrane. Basic information about membrane structure and function in relation to the environment, particularly temperature, pH and ionic affects, suggests that changes in the membrane may be expected to alter the photochemical reactions. In this

One of us, F.H.C., appreciated the use of the Climate Laboratory facilities of the Department of Scientific and Industrial Research, Palmerston North.

387

paper we report investigations into the short and long term effects of environmental variables on the reactions associated with photosynthetic membranes. More detailed descriptions of the techniques and analysis of the results and references to other workers are provided elsewhere (Fork & Troughton 1974a and b, Troughton & Fork 1974a and b, Chang 1975).

## Material and Methods

Measurements of the short-term influence of the environment were made by following the kinetics of the 510 and 420 nm light induced absorbance changes in intact leaves of higher plants and in the green algae *Ulva*. The 510 nm change is produced on functioning of both light reactions of photosynthesis and one hypothesis suggests the change is caused by a light induced electrical field across the thylakoid membrane due to the transfer of electrons. The 420 nm change has been established as being due to cytochrome f oxidation and therefore can be used to monitor some of the activities associated with both photosystems 1 and 2.

The rapid kinetics of the light induced absorbance changes were measured in intact tissue using a monochromatic measuring beam of sufficiently low intensity to prevent significant photochemical effects. The leaves of intact plants were kept in controlled environment chambers close to the surface of the photocathode. The output from the photocathode was amplified and recorded on a high speed Brush recorder or passed to a signal averager and onto an x-y recorder.

The measuring beam had a half bandwidth of 2 nm for all measurements. The exposure time for actinic light was 3-5 s. Broadband red or far-red wavelengths was obtained by filtering the light from a 650 W quartz-iodine lamp (type DWY) through 37 mm of water and either Schott glass filter RG2 (3 mm) or RG8 (3 mm). The red light had wavelengths extending from about 620-750 mm and an intensity of about $10.8 \times 10^5$ ergs cm$^{-2}$ s$^{-1}$. The far-red light had a wavelength band extending from about 685 mm to 750 mm and an intensity of about $8 \times 10^5$ ergs cm$^{-2}$ s$^{-1}$. In measurements reporting specific wavelength, narrow band interference filters having a half bandwidth of 5 nm were used.

In preliminary experiments, it was shown that the rates of change in magnitude of the 510 nm and 420 nm light induced absorbance changes could be substantially altered by short-term light pretreatment of the sample. This effect was termed light adaption. Light adaption in this paper refers to plants exposed to at least 2 minutes of high intensity red light, whereas dark adapted plants were left in the dark for a minimum of 20 minutes before measurements.

Tomato, bean, *Ginkgo biloba*, Kiwi berry (*Actinida chinensis*) and *Kalanchoe* plants were grown in soil in the lath or greenhouse. *Sorghum* was grown in Hoagland's solution in a growth chamber. *Disporum* was collected locally from the floor of a redwood forest. The marine green alga *Ulva expansa* was collected from the ocean at Half Moon Bay, California and maintained in the laboratory in seawater at 12° C under low intensity fluorescent light (15 nE cm$^{-2}$ s$^{-1}$). The *Ulva* was kept in 5 ml of seawater during measurements while all other leaves were kept in normal air.

Several species of plants were chosen for long term experiments on the effect of temperature on plant growth. The selection was based on the geographical origin of the plants and on their photosynthetic pathway. The $C_3$ group included the temperate species *Lycopersicum esculentum* cv. Moneymaker and *Triticum aestivum* cv. Arawa and the sub-tropical rice *Oryza sativa* cv Hungarian No. 1 and a tropical rice *Oryza sativa* cv. IR8. The $C_4$ plants were the tropical species *Gomphrena globosa* and the subtropical *Zea mays*. All species were grown from seed for 21 days in a glasshouse before transfer to controlled environment rooms at constant day/night temperatures of 15, 22, 28 and 35°C. Other conditions in the growth chambers included a $CO_2$ concentration of 330-360 ppm., air speed of 0.3 to 0.5 ms$^{-1}$, a vapour pressure deficit during the light period of -10 mb and a light level of 160 Wm$^{-2}$. The plants were watered with Hoagland's solution during the experiments.

The first harvest was after a seven day equilibration period in the controlled conditions and for most species four harvests at weekly intervals were made for growth analysis. The growth rates were determined from the change in dry weight over each weekly period.

The chorophyll, galactolipid and water contents of the leaves were obtained, although at fewer intervals than the growth rate measurements. The chorophyll a and b determinations were made as previously reported (Chang & Troughton 1972). The galactopid content was determined using the techniques of Roughan & Batt (1968). The techniques used for electron microscopy followed those described in Troughton & Card (1974) and for scanning electron microscopy Troughton & Donaldson (1972).

## Results

1. Short-term Influence of the Environment on Electron Transport Reactions.
A characteristic light minus dark difference spectrum for *Ulva* is shown in Fig. 1. The features of this spectrum relevant to this paper are the two peaks, one positive at 510 nm and the other negative at 420. Subsequent measurements reported in this paper refer to absorbance changes occurring at either of these two wavelengths.

The characteristic light adaption effect is shown in Fig. 2 for *Sorghum vulgare*. At both high intensity and low intensity actinic light was a significant effect of the degree of light adaption on the magnitude, and partly shape, of the light induced absorbance change. This same effect was shown to occur in all the species tested, which included a gymnosperm (*Ginkgo biloba*), $C_3$ species (bean, Kiwi berry and tomato) $C_4$ species (*Sorghum vulgare*), CAM species (*Kalanchoe sp.*) and a deep shade plant *Disporum smithii*. The mean depression of the latter versus he first light exposure was 82% for high actinic light and about 300% for low ctinic light (Troughton & Fork 1974a). There was no evidence of any distortion f the difference spectra about 518 nm between dark and light adapted leaves.

The magnitude of the light adaption effect on the 510 nm absorbance change a leaves was shown to be dependent on both the degree of light and dark laption and on the intensity of the actinic light. A most significant feature of the 'fect was the influence of adaption on the light intensity response curve, given r *Ulva* in Fig. 3, which indicated major differences in the efficiency of light to oduce the 510 nm change in the two adaption states. In the dark state about

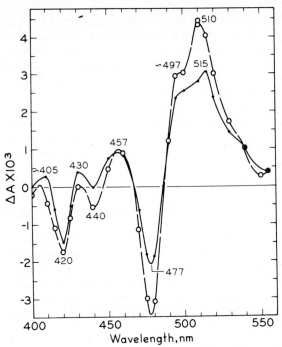

*Fig. 1.* Light-minus-dark difference spectra in light-adapted *Ulva*. Value at 2-3 sec (o) and 6 sec (●) of high intensity red actinic light.

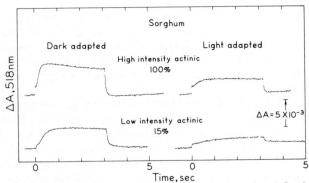

*Fig. 2.* The 518 nm absorbance change in dark and light adapted leaves of *Sorghum vulgare* at two intensities of red actinic light.

1.5% of full sunlight was required to achieve half the maximum absorbance change whereas in the light adapted state 12.5% was required. This suggested there were different quantum efficiency states for these processes depending on the degree of light adaption.

The action spectra shown in Figs. 4 and 5 for *Ulva* show significant difference between the dark and light adapted states. In the dark adapted state (Fig. 4), the action spectra was featureless, irrespective of the actinic light intensity, but in the

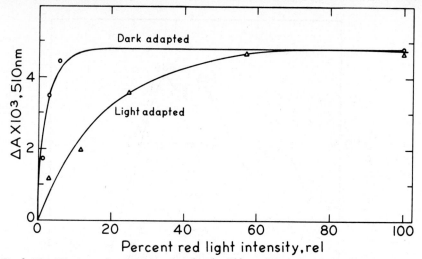

*Fig. 3.* The light intensity response curve for the 510 nm light induced absorbance change in dark and light adapted *ulva*.

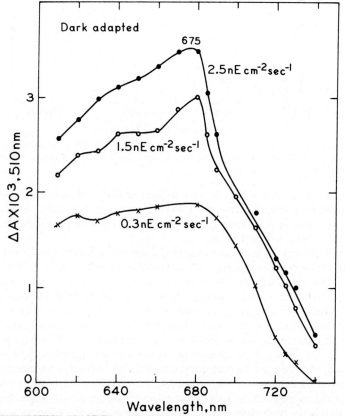

*Fig. 4.* The action spectra for the 510 nm absorbance change measured in dark adapted *Ulva*; the spectra is shown for three levels of actinic light.

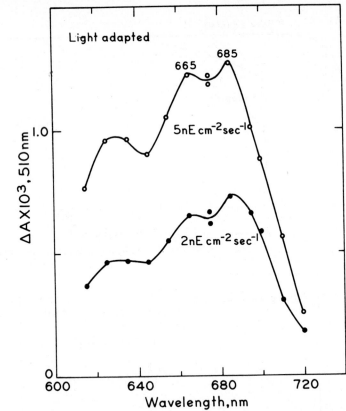

*Fig. 5.* The action spectra for the 510 nm absorbance change measured in light adapted *Ulva* at two light levels. A comparison of Figures 4 and 5 illustrates the difference between the dark and light adapted states.

light adapted state peaks (Fig. 5) were evident of 630, 665 and 685 nm. Furthermore, in the dark adapted state there was evidence of a significant photo-induced reaction at wavelength as long as 740 nm, which is normally considered outside the photosynthetic response range.

The 510 nm light induced absorbance change was also shown to be sensitive to other environmental parameters. $CO_2$ concentration had a relatively small effect on either the magnitude or the shape of the 510 nm light induced absorbance change as shown for tomato in Fig. 6. One interpretation of this response is that oxygen can also act as an electron acceptor, taking over this role in the absence of carbon dioxide. Removal of oxygen however had a most pronounced effect on the 510 nm absorbance change as shown in Fig. 7 for tomato. This effect was independent of the presence or absence of carbon dioxide and the oxygen response was characterized by the lack of development of the slower light induced absorbance change. This response was independent of the photosynthetic pathway as it also occurred in sorghum. In view of the effect of oxygen on the photorespiratory process in leaves, other than on the carboxylation events, the effect reported here may well be significant in linking the light reactions with the role of the glycolate

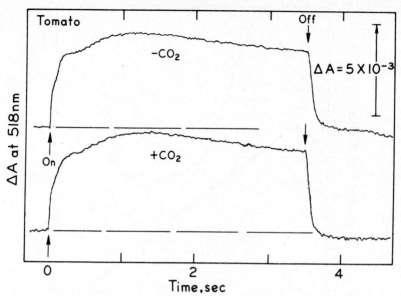

*Fig. 6.* The influence of $CO_2$ on the 518 nm light induced absorbance change in tomato leaves. High intensity red light was used as actinic and the leaf was dark adapted.

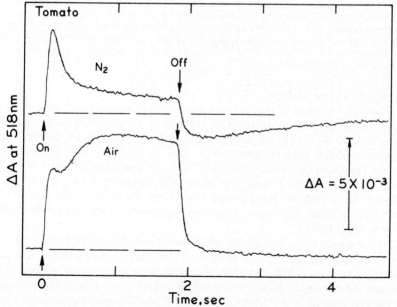

*Fig. 7.* A comparison of the 518 nm absorbance change in air and nitrogen in a tomato leaf. High intensity red light was used as actinic. A similar response was also observed in the $C_4$ plant *Sorghum vulgare*.

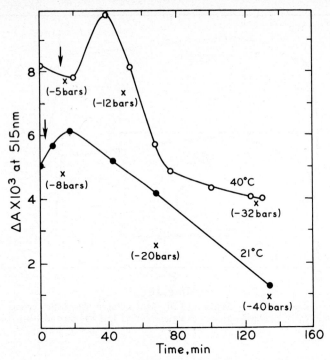

*Fig. 8.* The influence of water stress in tomato leaves on the photochemical reactions induced by high intensity red actinic light and monitored by the 515 nm absorbance change. The experiments were run at 21°C (●) and 40°C (o). The difference in response between temperatures was due to differences between leaves. The relative leaf water content in the same leaf was measured at several times (x) and related to the water potential term by calibration using a Scholander pressure bomb.

pathway in leaf metabolism.

The water relations of the leaf were also shown to have dramatic effects on the light induced membrane reactions at 515 nm in tomato leaves at two temperatures, 21 and ·40° C (Fig. 8). At both temperatures there was a small increase in magnitude of this light induced change after the onset of water stress in a leaf severed from the plant, but subsequently there was a significant decline in the 510 nm change. To test the reversability of the effect a leaf of one plant was severely stressed (to at least -40 bars) but on re-watering the plant rapidly recovered to yield light induced absorbance changes of approximately the same magnitude as prior to water stress.

The evidence from the 510 nm light induced absorbance change indicated that the membrane was particularly sensitive to alteration of the environmental parameters which were tested. In view of the significance in magnitude of some of the changes the investigations were extended to include more direct analysis of some of the components in the electron transport chain. The spectrophotometric technique makes possible the analysis of the kinitics of one electron transport component, cytochrome f, in intact tissue of *Ulva*. From the light induced absorbance change at 420 nm it was possible to estimate the rate of cytochrome f oxidation

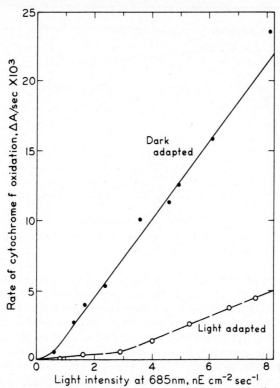

Fig. 9. The rate of cytochrome f oxidation as a function of light intensity for *Ulva* in the dark and light adapted states.

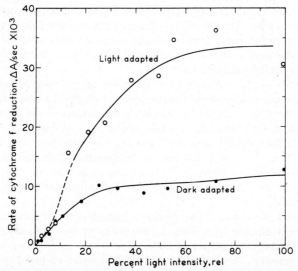

Fig. 10. The rate of cytochrome f reduction as a funtion of light intensity for *Ulva* in the dark and light adapted states.

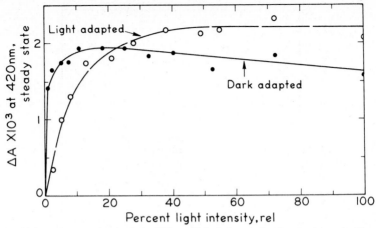

Fig. 11. The steady state level of the 420 nm light induced absorbance change in *Ulva* in the light and dark adapted states.

the rate of cytochrome f reduction and the extent of oxidation of the cytochrome f pool during a 3 second light exposure. The results for these three components are shown in Figs. 9, 10 and 11 where a comparison is also made of these responses between light and dark adapted states. There was a reduction in the rate of cytochrome f oxidation in the light compared with the dark adapted state at all light intensities but this was due to a displacement of the curve and not apparantly due to a change in quantum efficiency (Fork & Troughton 1974b). The effect was irrespective of the wavelength of light used. The response of cytochrome f reduction to light intensity was complex (Fig. 10) but the results are interpreted as indicating that at low light the size of the oxidized cytochrome pool limits the rate of reduction whereas at high light the rate is greater in a light compared with a dark adapted sample. The size of the oxidized cytochrome f pool is a result of both the rate of cytochrome f oxidation and reduction and, as shown in Fig. 11, the cytochrome f pool saturated at significantly lower light intensities in the dark compared with the light adapted state,

Measurements of the influence of temperature on the rates of cytochrome f oxidation and reduction illustrate the extent to which the electron transport may contribute to the well established temperature dependence of the photosynthetic reactions (Fork & T 1974b). The rate of cytochrome f oxidation was unaffected by temperature from 4-50°C but at 52°C was completely inhibited. The rate of cytochrome f reduction as a function of temperature followed an Arrhenius response up to 26°C but above 30°C there was a rapid decline in the rate presumably as the membranes became more fluid and influenced the organisation of the photoreactions.

Measurements were also made of the influence of both ionic and osmotic effects on the cytochrome f reactions. A comparison of the light adaption rates or the rate of cytochrome f reduction with $2 \times 10^5$ ergs cm$^{-2}$ s$^{-1}$ red actinic light with *Ulva* in seawater, distilled water and distilled water +5 mM $MgCl_2$ reduced the rate compared with the dark control by 67%, 35%, and 66% respectively. The rate of cytochrome f reduction however was approximately the same in both the dark and light adapted states in distilled water, but on addition of magnesium chloride the normal increase due to light adaption was observed. This result established that cations or anions were required to increase the rate of reduction of cytochrome f on light adaption.

Increasing the osmotic concentration of the solution using NaCl surrounding the *Ulva* tissue had relatively small effects on the rate of cytochrome f reduction until the osmotic pressure was less than -50 bars. The rate of cytochrome f oxidation however was more sensitive to NaCl and part of the light adaption induced decrease was overcome by decreasing the osmotic pressure to -35 bars. Further reductions in osmotic pressure resulted in a decrease in both the light and dark adapted values.

## 2. Long-term Environmental Effects

The measurement of the growth rates of the species under the four temperature regimes established that there were significant differences between the species in response to the thermal regime. In general the tropical and sub-tropical species had higher temperature optimum than those of the temperate species. Both maize and Gomphrena had a temperature optimum of $28°C$ while the tropical rice species IR8 had a broad optimum between 22 and $28°C$. The sub-tropical rice species Hungarian No. 1 and the temperate species tomato had a temperature optimum of $22°C$ while wheat had an optimum of $15°C$, which is to be expected for a temperate species. The division between $C_3$ and $C_4$ type plants was biased in this selection so that the two $C_4$ species were both from tropical or sub-tropical origins and had high temperature optimum for growth.

A comparison was also made of the relative growth rates of the different species at their optimum temperature. The highest relative growth rate was that of *Gomphrena* which was 23.5% at $28°C$. The relative growth rates of the plants declined in the order *Gomphrena*, tomato, maize, wheat, rice (Hungarian No. 1) and rice (IR8) with the relative growth rates expressed with respect to *Gomphrena* as 1.0, being 0.87, 0.72, 0.68, 0.60 and 0.55 for the other species respectively. There was no close correlation in these results between growth rate and photosynthetic pathway as the growth rate of tomato ($C_3$ plant) was higher than that of maize ($C_4$ plant).

Most species showed a pronounced optimum temperature for maximum growth rate measured on a dry weight basis. The characteristic response of two species to temperature are shown in Fig. 12A. Tomato, a temperate species, shows an optimum temperature of $22°C$ and a pronounced drop in growth rate over all temperatures up to $35°C$. In contrast, *Gomphrena* exhibited a maximum growth rate at $28°C$ with a pronounced reduction in growth rate at temperatures below this value. Further investigations were therefore undertaken to elucidate the reason for the reduced growth rate of tomato at high temperatures and Gomphrena at low temperatures.

The aspect of the investigations reported in this paper concern the influence of long term environmental effects, in this case temperature, on the membrane associated with electron transport. Measurements were made of both chlorophyll *a* and *b* contents but the responses were relatively similar for both components and only the a component is reported here. As shown in Fig. 12B the chlorophyll *a* content of the leaves expressed as a proportion of the dry weight was reduced at

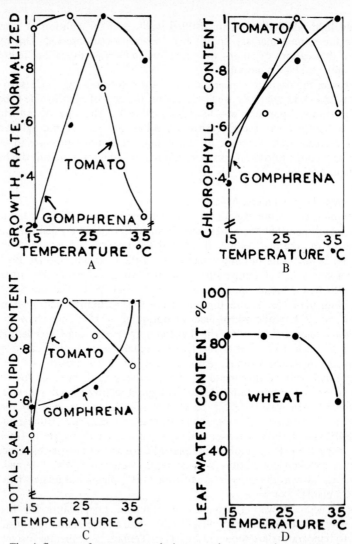

*Fig. 12.* The influence of temperature during growth on several parameters in tomato, *Gomphrena* and wheat. A. Normalised growth rate as a function of temperature. B. Normalised chlorophyll a content as a function of temperature. C. Normalised total galactolipid content as a function of temperature. D. Leaf water content in wheat leaves as a function of temperature.

low temperatures although it was also substantially reduced in tomato at the highest temperature of 35°C. At 35°C the chlorophyll a content of *Gomphrena* was the maximum. The reduced chlorophyll *a* content at high temperatures in tomato may be associated with the reduced growth rate but it would not explain the high growth rate with low chlorophyll a content at low temperature.

The total galactolipid content expressed on a dry weight basis was greatest for both species about their approximate temperature optimum. In tomato there was a decline in galactolipid content with increasing temperatures above 22°C but there was also a low level at 15°C where the relative growth rate of tomato was high. Similarly in *Gomphrena* there was little change in galactolipid content between 15, 22 and 28°C even though there was a pronounced change in growth

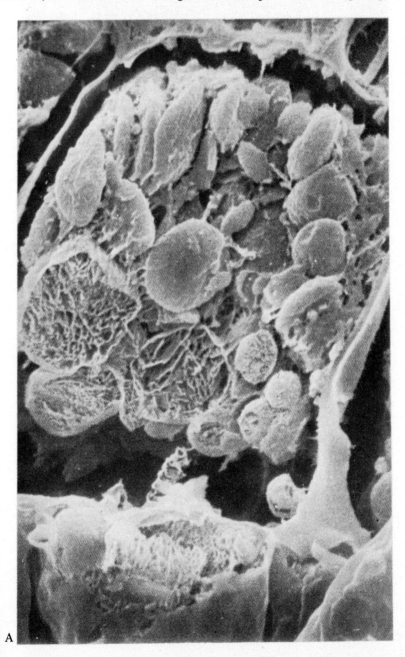

A

rate over this temperature range (Fig. 12C).

In most experiments where plants are grown at a range of temperatures in-

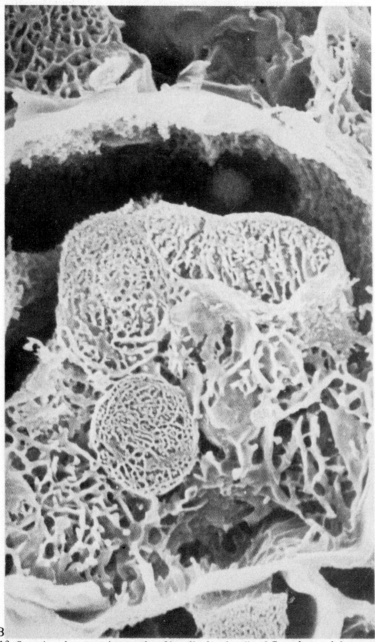

B

*Fig. 13.* Scanning electron micrographs of bundle-sheath cells of *Gomphrena globosa* grown at two temperatures. A. Growth temperature 28°C: B. Growth temperature 15°C.

400

cluding a high temperature, any detrimental effects of high temperature on growth or a component of the plant are normally directly attributed to a temperature effect. However, at high temperature some effects may be due to water stress induced by high evaporative demands or due to breakdown of membranes. In the case of wheat in these experiments it was evident that the plants grown at 35°C were significantly depressed in their growth but were also exhibiting signs of water stress. Measurements of the water content of the leaf expressed as a percent of the dry weight are shown in Fig. 12D and illustrate that there were significant reductions in water content at the highest temperatures. Thus, as well as temperature effects occurring on the thylakoid membrane it may be that the temperature was influencing fluidity of other cell membranes which are responsible for maintaining the integrity of the water system in the leaves.

Direct observations using the scanning electron microscope and electron microscope were made of the membrane structures of the chloroplast, particularly in plants under extreme environment conditions. An indication of the low temperature effect on chloroplast numbers and distribution in the cells is given in Fig. 13A and B for the bundle sheath cell of *Gomphrena*. At low temperature the number of chloroplasts in the bundle sheath cells were dramatically reduced compared with those in cells grown at normal temperatures and there was also some

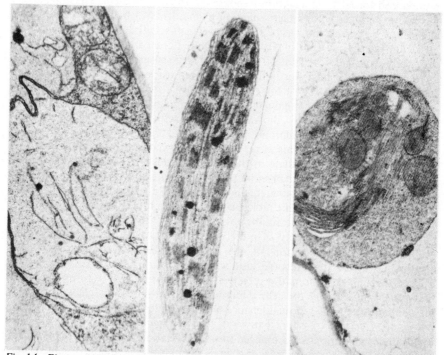

*Fig. 14.* Electron micrographs of the ultrastructure of chloroplasts illustrating the adverse effects of both high and low temperature. A. Tomato. Mesophyll cell chloroplasts from plant grown at 35°C. B. Normal appearance of a chloroplast from the mesophyll cell of *Gomphrena* plants grown at 28°C. C. Chloroplast from bundle-sheath cell of a *Gomphrena* grown at 15°C.

evidence of degeneration of the internal membrane structure. A similar reduction in chloroplast number was noted in tomato leaves grown at high temperature. Ultrastructural investigations of the thylakoid membranes in chloroplasts of tomato grown at 35°C and in *Gomphrena* grown at 15°C showed a marked degeneration of most aspects of chloroplast structure when compared with plants of either species grown at the optimum temperature. This is shown in Fig. 14 (A, B, C). These results were common for plants grown outside their normal temperature range and was independent of whether the choroplasts were in the mesophyll or bundle sheath cells in the $C_4$ plants. At temperatures intermediate between the optimum and the extreme there were also significant changes in ultrastructure but in many cases were due to pronounced accumulation of starch, particularly in the bundle sheath cells of *Gomphrena*.

## Discussion

The spectrophotometric measurements reported in this paper indicate the rapid reactions which occur within the thylakoid membrane on exposure to light and also establish that these reponses are sensitive to many environmental effects. As well as environmental effects on the instantaneous light reactions there are also pronounced longer term adaption responses to the environment. The notable one in these experiments was a light adaption phenomena. Light induced absorbance changes at both 510 nm and 420 nm wavelengths indicated a higher quantum efficiency of the reaction in tissue dark adapted compared with the same tissue previously exposed to at least 10 minutes of high intensity red light. The highest quantum efficiency state was in the dark adapted tissue which led to the proposal that the rate limiting step of the electron transport chain shifted during exposure to light and allowed all components of the chain to come into a new equilibrium. This adaption measured by the cytochrome f responses was evident as an inhibition of the rate of cytochrome f oxidation and an enhancement in the rate of cytochrome f reduction. This latter reaction may have occurred as a result of a build-up of some intermediate in the electron transport chain between photosystems 1 and 2. The mechanism responsible for the light adaption process was not elucidated in these experiments but it seems likely that it is associated with structural reorganisation at the molecular level in the thylakoid membranes.

The significance of these results to plant production has yet to be evaluated but some evidence is worthy of more extensive experimental tests. It has been proposed (Fork & Troughton 1974b) that light absorbed at the longer wavelengths and in the dark adapted state with the high quantum efficiency may be utilised through a cyclic pathway activated by photosystem 1 to produce ATP that can be utilized by the cell. Under adverse light levels this source of energy may be important for cell maintenance and to supplement energy derived via the non-cyclic electron transport pathway. This mechanism would be particularly useful under conditions of prolonged low $CO_2$ fixation as occurs in shade plants.

In natural plant communities the large fluctuations in the environment may lead to prolonged periods of adverse climatic conditions. The extent to which plants may adapt to these conditions is limited and the influence of adverse temperature conditions was examined in the second part of this paper. The collective response reported of adverse temperatures reducing chlorophyll $a$ and $b$ contents, galactolipid contents and membrane integrity as illustrated by electron micrographs, suggests that the potential for electron transport processes in these plants would be severely inhibited. The inability of different species to tolerate these conditions will not only reside in the electron transport reactions but also in the carboxylation reactions of photosynthesis and the innumerable processes responsible for the growth of the plants. The results presented here, for example, suggest that at low temperature the restriction is not associated with membrane composition even though some degeneration of the chloroplasts was evident.

Further evidence of the role of the membrane matrix in regulating the photochemical behaviour of the thylakoid membranes is· vividly demonstrated in a recent study on *Anacystis* and some higher plant chloroplasts (Murata et al 1975). There was a high correlation between the kinetics of the temperature dependence of membrane spin label responses, chlorophyll $a$ fluorescence, oxidation-reduction reactions of $P_{700}$, $O_2$ evolution and Pigment State 1 and State 2 shift. These responses were directly linked to differences in the state of the membrane; either the liquid crystalline or the mixed liquid crystalline - solid state.

## REFERENCES

Chang, F.H. (1975): Environmental control of carbon metabolism in plants. Ph.D. Thesis. Victoria University of Wellington, Wellington, New Zealand.

Chang, F.H. & Troughton, J.H. (1972): Chlorophyll a/b ratios in $C_3$ - and $C_4$ - plants. *Photosynthetica*, 6: 57-65.

Fork, D.C. & Troughton, J.H. (1974a): Quantum efficiency states of photosynthesis; II. Light adaption and the light-induced 510 nm absorbance change in *Ulva*. Carnegie Institution of Washington Year Book 74 (In Press).

Fork D.C. & Troughton J.H. (1974b): Quantum efficiency states of photosynthesis: IV. Transients and the rate of cytochrome f oxidation with far-red and red actinic light in *Ulva* in the light adapted state. Carnegie Institution of Washington Year Book 74 (In Press).

Murata, N., Troughton J.H. & Fork, D.C. (1975): The relationship between the transition of the physical phase of membrane lipids and photosynthetic parameters in *Anacystis nidulans* and lettuce and spinach chloroplasts. *J.Biol. Chem.*, (submitted).

Roughan, P.G, & Batt, R.D. (1968): Quantitive analysis of sulfolipid (Sulfoquinovosyl diglyceride) and galactolipids (Monogalactosyl and diglycerides) in plant tissues. *Anal. Bioch.*, 22: 74-88.

Troughton, J.H. & Card, K.A. (1974): Leaf anatomy of *Atriplex buchananii*. *N.Z.J. Bot.*, 12: 167-177.

Troughton, J.H. & Donaldson, L.A. (1972): Probing plant structure. A.H. and A.W. Reed, Wellington, New Zealand.

Troughton J.H. & Fork, D.C. (1974a): Quantum efficiency states of photosynthesis: I 510 nm light induced absorbance changes in higher plant leaves. Carnegie Institution of Washington Year Book 74 (In Press).

Troughton, J.H. & Fork, D.C. (1974b): Quantum efficiency states of photosynthesis: III Light adaption of the cytochrome f reactions in *Ulva*. Carnegie Institution of Washington Year Book 74 (In Press).

# AUTHOR INDEX

# SUBJECT INDEX